T0337883

Self-Commutating Converters for High Power Applications

Self-Commutating Converters for High Power Applications

Jos Arrillaga
University of Canterbury, Christchurch, New Zealand

Yonghe H. Liu
Inner Mongolia University of Technology, China

Neville R. Watson
University of Canterbury, Christchurch, New Zealand

Nicholas J. Murray
Mighty River Power Limited, Auckland, New Zealand

A John Wiley and Sons, Ltd., Publication

This edition first published 2009

Registered office
John Wiley & Sons Ltd, The Atrium, Southern Gate, Chichester, West Sussex, PO19 8SQ, United Kingdom

For details of our global editorial offices, for customer services and for information about how to apply for permission to reuse the copyright material in this book please see our website at www.wiley.com.

Library of Congress Cataloging-in-Publication Data

Self-commutating converters for high power applications / J. Arrillaga ... [et al.].
p. cm.
Includes bibliographical references.
ISBN 978-0-470-74682-0 (cloth)
1. Commutation (Electricity) 2. Electric current converters. 3. Electric power distribution–High tension.
I. Arrillaga, J.
TK2281.S45 2009
621.31'7–dc22

2009023118

A catalogue record for this book is available from the British Library.

ISBN: 978-0-470-74682-0 (H/B)

Typeset in 10/12pt Times by Thomson Digital, Noida, India.
Printed and bound in Great Britain by CPI Antony Rowe, Chippenham, Wiltshire

Contents

Preface

The characteristics of power semiconductors have reached the stage at which they can be used to control the operation of generation, transmission and utilization systems of all types and ratings.

For very high-voltage or very high-current applications, the industry still relies on thyristor-based line-commutated conversion (LCC), which lacks reactive power controllability. However, the ratings of self-commutating switches, such as the IGBT and IGCT, are reaching levels that make the self-commutating technology possible for very high power applications.

The term 'high' requires a reference for its interpretation. In this respect, three rating components are involved, namely voltage, current and power. Of course, high power ratings can only be achieved by correspondingly high current and/or voltage ratings, but not necessarily both. While high power transmission uses high voltage and relatively low current in order to reduce power losses, some industry processes (such as aluminium smelting) require very high current and very low voltage.

Currently, there is a high level of interest in countries such as China, India, Brazil and parts of Africa in generating power from large renewable resource (mainly hydro) plants at remote locations and transmitting this power using ultra high voltage (UHV) to national and/or international load centres. The powers and distances under consideration are typically 6000 MW and 2000 km respectively and the voltage selected by the planners for these projects is $\pm$800 kV DC. Although there is no experience of operating at such voltage level, the general opinion is that they do not represent an unreasonable risk and the manufacturers are ready for the task.

This book reviews the present state and future prospects of self-commutating static power converters for applications requiring either UHV DC (over $\pm$600 kVs), such as required by very large long-distance transmission or ultra high currents (in hundreds of kA), such as those used in aluminium smelters and large energy storing plants.

The authors would like to acknowledge the main sources of information and, in particular, the material reproduced, with permission, from CIGRE, IET and IEEE. They also want to thank Dr Lasantha B. Perera for his earlier contribution to the subject and The University of Canterbury and The University of Inner Mongolia for providing the facilities for their work.

1

Introduction

1.1 Early developments

A variety of electronic valves was tried in the first part of the twentieth century for the conversion of power from AC to DC and vice versa. The mercury-arc valve was the most suitable option for handling large currents, and thus, multiphase grid-controlled mercury-pool cathode valves were developed for industrial and railway applications.

Efficient bulk power transmission, however, requires high voltage rather than current and, thus, the development of a high-voltage DC transmission technology only became possible in the early 1950s, with the invention (by Uno Lamm of ASEA) of the graded-electrode mercury-arc valve [1]. Soon after, with the appearance of the thyristor or silicon controlled rectifier (SCR), the use of power conversion progressed rapidly to higher voltage and power ratings.

The source forcing the commutation process between the converter valves (either mercury-arc or thyristor) was the AC system voltage and thus the converter was said to be line-commutated (LCC). LCC relies on the natural current zeros created by the external circuit for the transfer of current from valve to valve. The commutation is not instantaneous, because of the presence of AC system reactance, which reduces the rate of change of current and, therefore, lengthens the commutation time in proportion to the reactance and the magnitude of the current to be commutated; the duration of the commutation also depends on the magnitude of the instantaneous value of the commutating voltage, which changes with the position of the firing angle. All these variables depend on the operating conditions and, as a result, the prediction and minimization of the commutation angle becomes a difficult problem. This is an important issue for inverter operation, which requires a large firing advance for safe operation, with an increasing demand of reactive power.

By the late 1960s, the successful development of the series-connected thyristor chain had displaced the mercury-arc valve in new high-voltage direct current (HVDC) schemes [2].

SCR-based power conversion technology continues to be used extensively in power transmission (in the form of static VAR compensation (SVC) and HVDC) and in a variety of industry applications. In fact, the power rating capability of present SCR converters is

Self-Commutating Converters for High Power Applications J. Arrillaga, Y. H. Liu, N. R. Watson and N. J. Murray

only limited by the external components attached to the converter, such as the interface transformers.

1.2 State of the large power semiconductor technology

Progress in power semiconductor types and ratings has been such that a review of their current state, important as it is to the subject of this book, will be short lived, and any recommendations on their specific application must be looked at in this context.

Historically, the application of semiconductors to high-voltage applications started with the silicon controlled rectifier (SCR) in the late 1950s. Despite its age, the SCR, though with highly improved current and voltage ratings, is still the most widely used semiconductor in HVDC conversion. However the restricted controllability of the SCR has encouraged the development of alternative power semiconductors of the thyristor and transistor families.

At present the power semiconductor devices available for large power conversion applications are based on the silicon technology and they can be broadly classified in two groups [3].

The first group includes devices with four-layer three-junction monolithic structures, the two early devices in this category being the SCR (silicon controlled rectifier) and GTO (gate turn off thyristor). The devices in this group have low conduction losses and high surge and current carrying capabilities; they operate only as on/off switches with bidirectional voltage blocking capability.

Recent developments in this group are the MCT (MOSFET (metal oxide semiconductor field-effect transistor) controlled thyristor), ETO (emitter turn-off thyristor), MTO (MOS (metal oxide semiconductor) turn-off thyristor) and GCT or IGCT (integrated gate-commutated thyristor). These recent devices were developed to provide fast turn-off capability and low turn-off switching losses.

The majority of commercially available GTOs for providing free current path in voltage source conversion are of the asymmetrical type; they are reversed connected to a fast recovery diode, such that the GTO does not require reverse voltage capability. Asymmetrical GTOs have been used extensively in pulse width modulation (PWM) two- and three-level voltage source converters, active filters and custom power supplies. However, there is little further development of the GTO technology, the interest focusing instead on the GCT, which differs from the GTO by having a turn-off current gain close to unity. This means that, at turn-off, practically all the load current is commutated to the gate circuit for a few microseconds (thus the name gate commutated thyristor). This is achieved by the application of a very strong pulse with a *di/dt* of the order of 3000 A per microsecond. A further development of the GCT is the IGCT (integrated gate commutated thyristor), which, instead of the separate gate drive connected via a lead, uses a gate drive circuit integrated with the semiconductor device, thus achieving very low values of gate inductance. This device has a very short storage time (of about 1 μs), which permits small tolerances (under 0.2 μs) in turn-off times of the different devices, and therefore provides very good voltage sharing as required by the series connection in high-voltage applications.

The IGCT can also be used as an asymmetrical device, in which case a free-wheeling diode with a soft recovery turn-off is needed. IGCTs with blocking voltages up to 6.5 kV are available on the market and IGCTs with 10 kV are under development.

Although the IGCT has overcome many of the problems of the conventional GTO, the gate driver is still complex and a large linear *di/dt* limiting inductor is needed for the anti-parallel diode.

The second group contains devices of three-layer two-junction structure which operate in switching and linear modes; they have good turn-off capability. These are: the BJT (bipolar junction transistor), Darlington transistor, MOSFET, IEGT (injection enhanced transistor), CSTBT (carrier stored trench-gate bipolar transistor), SIT (static induction transistor), FCT (field controlled transistor) and IGBT (insulated-gate bipolar transistor).

There is little manufacturing enthusiasm for developing further some of the devices in the second group, because of the perceived advantages of the IGBT; at present the voltage and peak turn-off currents of the silicon based IGBTs are 6.5 kV and 2 kA.

These devices are mainly designed for use at high PWM frequencies and therefore the switching time must be minimized to reduce losses. This causes high *dv/dt* and *di/dt* and thus requires snubber networks, which result in further losses. Recent advances in the IGBT technology involve the modular and press-pack designs.

A new type of IGBT (referred to as IEGT) has become available that takes advantage of the effect of electron injection from emitter to achieve a low saturation voltage similar to that of the GTO.

1.2.1 Power ratings

As already mentioned, the ratings are changing fast and therefore any comparisons made must indicate the date and source of the information. For instance a 1999 published IEE review [4] of typical maximum ratings (Figure 1.1), showed that the GTO offered the best maximum blocking voltage and turn-off current ratings; these were 6 kV and 4 kA respectively, the switching frequency being typically under 1 kHz. More recently, however, the industry has

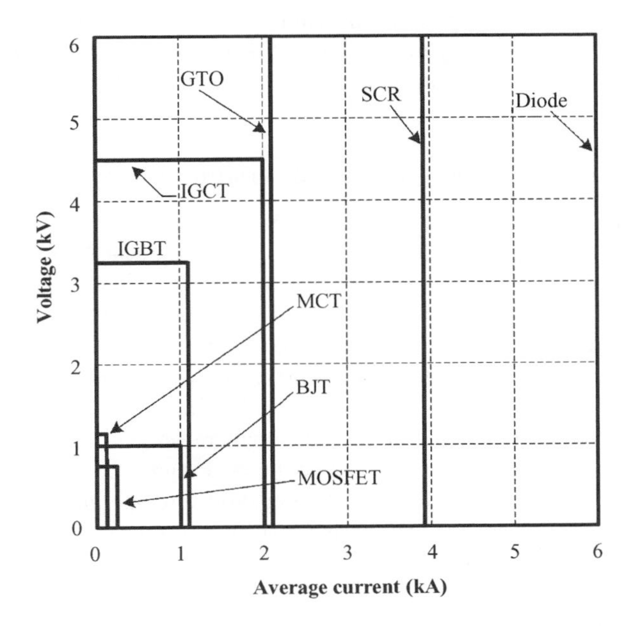

Figure 1.1 Voltage and current ratings of the main power semiconductors. (Reproduced by permission of the IET.)

concentrated in the IGCT development, for which the maximum ratings have already reached 6.5 kV and 6 kA.

The thyristor remains by far the most cost-efficient device for very high power application. The thyristor ratings presently available are typically (12 kV/1.5 kA and 8.5 kV/5 kA).

1.2.2 Losses

The IGBT turn-off losses are lower than those of the SCR and IGCT and so are the turn-on losses in the case of soft-switched IGBTs. The forward voltage drop of the IGBT is, however, much higher than that of a thyristor of comparable voltage rating.

In the assessment of the energy loss of a converter, the most important factor is the frequency used for the switching of the valves, which depends on the type of configuration and control. To illustrate this point a comparison between different alternative converters of the LCC and voltage source conversion (VSC) types has shown the following figures for the power loss of the complete converter station [5]:

- An IGBT based two-level voltage source converter with a PWM frequency of 1950 Hz has a power loss of approximately 3%.
- An IGBT based three-level voltage source converter with a PWM frequency of 1260 Hz has a power loss of approximately 1.8%.
- The loss figure for an SCR based LCC (line-commutated converter) station (including valves, filters and transformers) is 0.8%.

The IGCT has low on-state voltage and low total power losses (about one half of those of the conventional GTO) as shown in Figure 1.2. The IGCT has the lowest total loss (including both device and peripheral circuits) of all present power semiconductors.

1.2.3 Suitability for large power conversion

The main candidates for high power conversion appear to be the IGCT and IGBT. The previous section has already explained that the IGCT offers the best power ratings and lower overall losses, two important factors favouring its use in large power applications.

On the other hand, the IGBT requires much less gate power and has considerably superior switching speed capability, therefore permitting the use of higher switching frequencies (these can be typically 3 kHz for soft-switched devices as compared with 500 Hz for the IGCT). While high switching frequencies have some disadvantages (like switching and snubber losses), they help to reduce the harmonic content and therefore filtering arrangements, reduce machine losses and improve the converter dynamic performance.

The IGBT is more reliable than the IGCT under short-circuit conditions. It is designed to sustain a current surge during conduction and also during turn-on. However, short-circuit faults need to be detected quickly so that turn-off is achieved within 10 μs. The IGCT on the other hand has no inherent current-limiting capability and must be protected externally. The overall reliability of the IGCT for high-power and -voltage application is very impressive, given the reduced voltage stress achieved by the series connection and its potential to eliminate the need for snubbers.

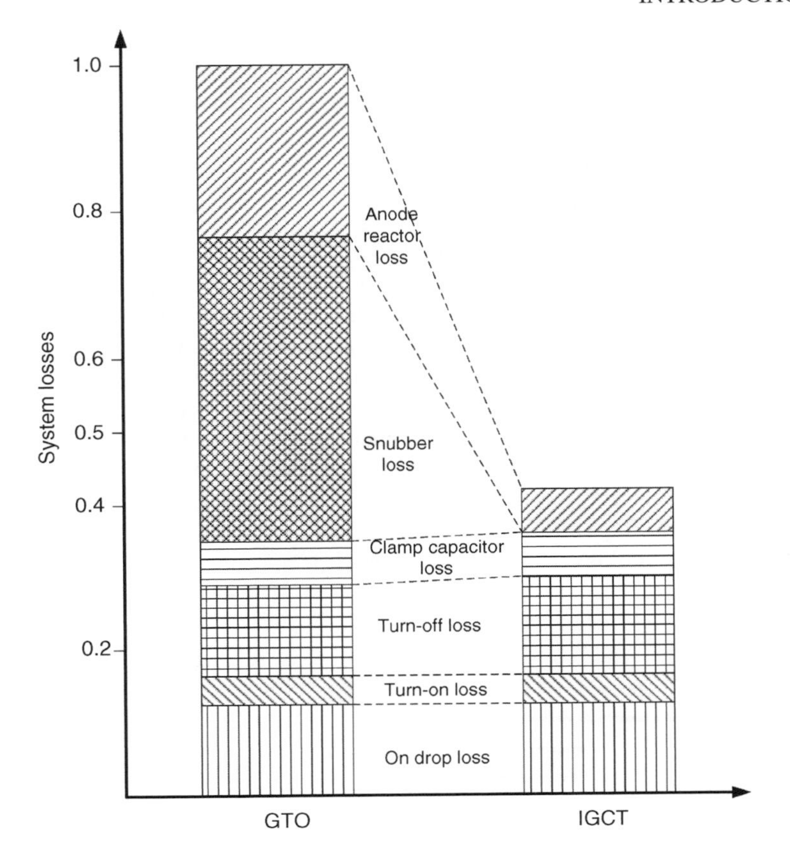

Figure 1.2 Comparison of system loss for a converter using GTO or IGCT (GCT).

For high-voltage applications the converters require high blocking voltage switches, which can only be achieved by the series connection of individual units. Both the IGBT and IGCT are suitable for series connection as the turn-on and turn-off times are relatively small and the switching speeds high. Reliable static and dynamic voltage sharing techniques are now available for thyristor chains (in LCC conversion) and transistor-type switching devices. Parallel sharing resistors are perfectly adequate for static balancing. In the case of the IGBT good dynamic voltage sharing can also be achieved by means of adaptive gate control of the individual units.

The current ratings of present switching devices are sufficiently large for high-voltage application and the use of device paralleling is rarely required. The IGBT technology is also suited for parallel operation, since the high-current IGBT modules themselves consist of many parallel chips.

Heavy investment in the IGBT technology is favouring this device at the expense of the GTO and IGCT alternatives. The availability of press-pack IGBTs at high voltages and currents is strengthening their position in high-voltage applications, where series operation and redundancy of power switches are required.

However, there are still problems with the IGBT for high-voltage applications, due to stray inductances and diode reverse recoveries and, thus, at this stage it is not clear to what extent the fast switching capability should be exploited, as the resulting voltage spikes may exceed the allowable limits.

1.2.4 Future developments

In the future, new wide-band gap (WBG) semiconductor materials such as silicon carbide (SiC) and gallium nitride (GaN), instead of silicon, are likely to increase the power handling capability and switching speed. The best candidate at the moment seems to be the SiC, a device that provides low on-state voltage, low recovery charge, fast turn-on and turn-off, high blocking voltage, higher junction temperature and high power density. In particular this material permits substantial increase in the allowable peak junction temperature, thus improving the device surge capability and reducing the complexity of the cooling system. With the utilization of SiC unipolar power switches it is possible to reduce the power losses by a factor of ten. At present only small chip sizes of SiC are available on the market, but a recent forecast of the future voltage ratings achievable with SiC switching devices (in relation to those of present silicon technology) is shown in Figure 1.3.

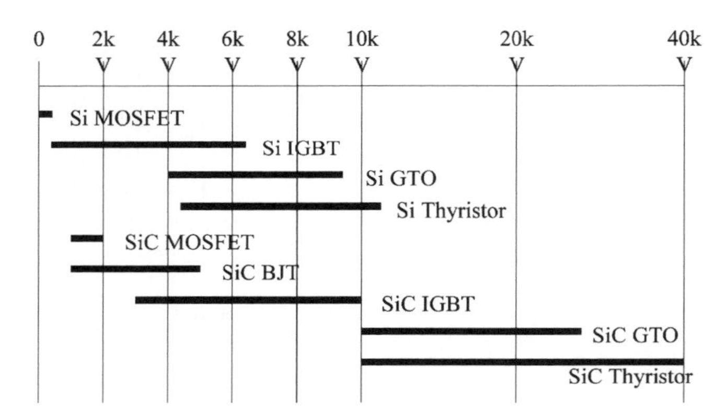

Figure 1.3 Silicon carbide (SiC) switches compared to silicon (Si) switches. (Reproduced by permission of CIGRE.)

Most of the reported development concentrates in raising the device voltage withstand level. There is, however, a place for a low-voltage, high-current device in applications such as aluminium smelters and in superconductive magnetic coils (for fusion reactors and energy storage).

1.3 Voltage and current source conversion

The first consideration in the process of static power conversion is how to achieve instantaneous matching of the AC and DC voltage levels, given the limited number of phases and switching devices that are economically viable. The following circuit restrictions are imposed on a static power converter by the characteristics of the external circuit and of the switching components:

- If one set of nodes (input or output) of the matrix of switches is inductive, the other set must be capacitive so as not to create a loop consisting of voltage sources (or capacitors and voltage sources) when the switches are closed or a cut set consisting of current sources when the switches are opened.
- The combination of open and closed switches should not open-circuit an inductor (except at zero current) or short-circuit a capacitor (except at zero voltage).

For stable conversion some impedance must, therefore, be added to the switching circuit of Figure 1.4(a) to absorb the continuous voltage mismatch that inevitably exists between the two sides.

When the inductance is exclusively located on the AC side (as shown in Figure 1.4(b)), the switching devices transfer the instantaneous direct voltage level to the AC side and, thus, the circuit configuration is basically a voltage converter, with the possibility of altering the DC current by controlling the turn-on and -off instants of the switching devices. A DC capacitor on the DC side and an AC interface inductance on the AC side are the essential components of a

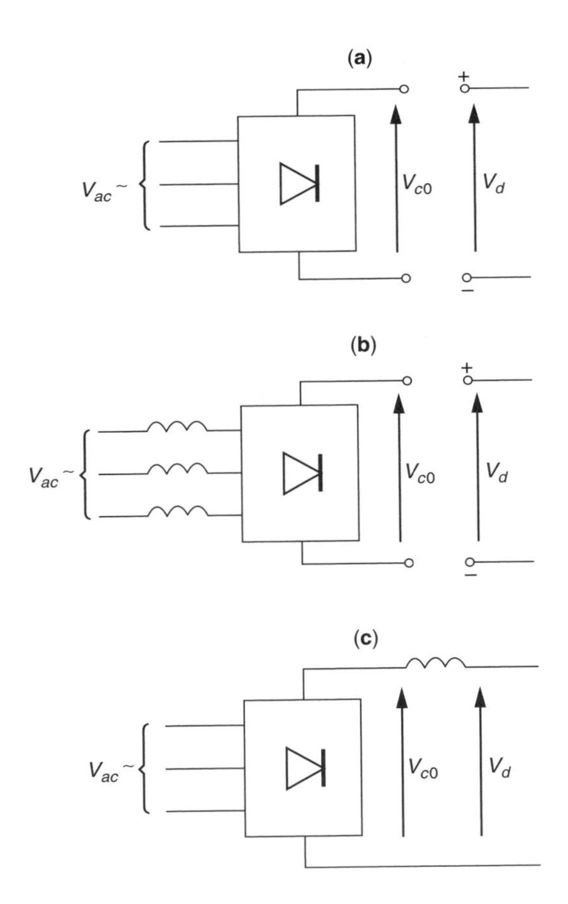

Figure 1.4 AC–DC voltage matching: (a) unmatched circuit; (b) circuit for voltage conversion; (c) circuit for current conversion.

voltage source converter. The designation voltage source converter is used because the function of the voltage source converter is explained by the connection of a voltage source on the DC side in the form of a large capacitor appropriately charged to maintain the required voltage. The AC side inductance serves two purposes: first, it stabilizes the AC current and, second, it enables the control of active and reactive output power from the voltage source converter. The switches must provide free path of bidirectional current, but they are only required to block voltage in one direction. This naturally suits asymmetrical devices like the IGBT or the thyristor type symmetrical ones paralleled with a reverse diode.

If, instead, a large smoothing reactor is placed on the DC side (as shown in Figure 1.4(c)), pulses of constant direct current flow through the switching devices into the AC side. Then, basically a current source converter results. The AC side voltage is then the variable directly controlled by the conversion process. Since the AC system has significant line or load inductance, line-to-line capacitors must be placed on the AC side of the converter. The switches must block voltages of both polarities, but they are only required to conduct current in one direction. This naturally suits symmetrical devices of the thyristor type and, therefore, current source conversion (CSC) constitutes the basis of line-commutated conversion. The asymmetrical type switches are not suited for current source conversion as a diode for sustaining the reverse voltage has to be connected with the asymmetrical switch in series, which causes extra losses.

The valve conducting period in the basic voltage source converter configuration is 180°, i.e. the bridge AC voltage has two levels by the valves and by the AC current when both valves are in the off state (as in this case the AC terminal is floating); whereas in the CSC case the width of the pulse is 120° and, therefore, the output phase (the current in that case) is either $+I_d$, 0 or $-I_d$, i.e. the bridge AC phase current has three levels.

1.4 The pulse and level number concepts

The pulse number (p) is a term commonly used in line-commutated conversion (LCC) and indicates the ratio of the DC voltage ripple frequency to the AC system fundamental frequency. A large number of phases and switches would be needed to produce perfect rectification, that is, a ripple-free DC voltage. This, of course, is not a practical proposition because the AC system consists of three phases only, which limits the number of pulses to six. This pulse number can be derived by the use of a double-star (six-phase) converter transformer, each phase, in series with a single valve, providing one sixth of the DC voltage waveform. However a more efficient alternative in terms of transformer utilization is the three-phase bridge switching configuration, which is the preferred option for other than very low-voltage applications.

The idealized (i.e. with perfect AC and DC waveforms and zero commutation angle) conversion process in a three-phase bridge converter is shown in Figure 1.5.

If on the DC side the rectified voltage is connected to the load via a large smoothing inductor the current will contain practically no ripple (i.e. will be perfect DC) and the converter will inject rectangular shaped currents of 120° duration into the converter transformer secondary phases (positive when the common cathode switches conduct and negative when the common anode switches conduct). On the primary side of the converter transformer the phase current is the AC rectangular waveform shown in Figure 1.6(a). This three-level waveform only applies when there is no phase-shift between the transformer primary and

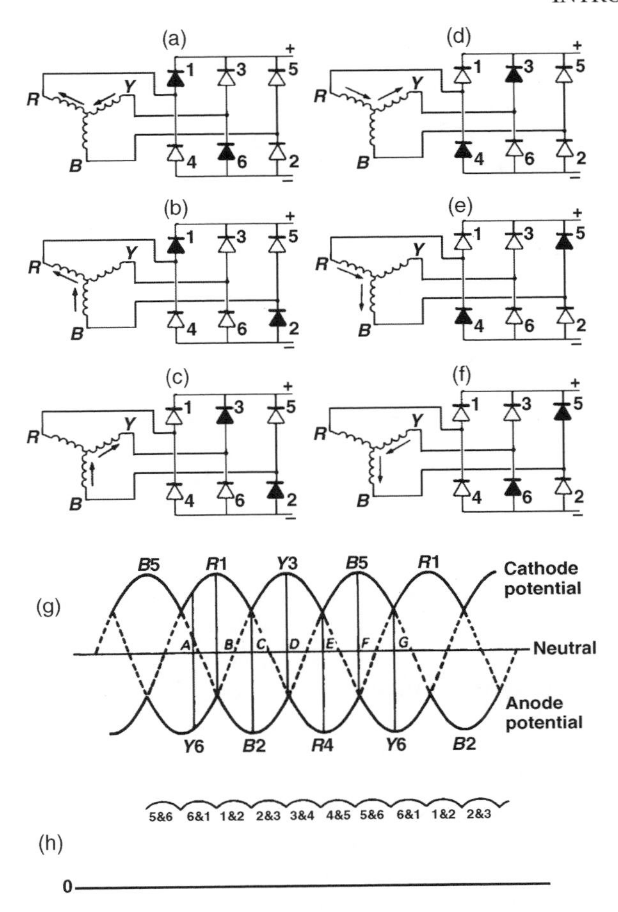

Figure 1.5 Conducting sequence and DC voltage waveforms in a three-phase bridge converter.

secondary phase voltages, i.e. a star–star transformer connection. If instead the converter transformer uses the star–delta or delta–star connection, the primary current, as shown in Figure 1.6(b), has four levels instead of three. So the use of the term level for this purpose is inconsistent.

Instead, the description of the conversion process is normally made in terms of sinusoidal frequency components (i.e. harmonics). In this respect, the three-phase bridge is referred to as a six-pulse configuration. producing *characteristic* voltage harmonics of orders $6k$ (for $k = 1,2,3\ldots$) on the DC side and *characteristic* current harmonics of orders $6k \pm 1$ on the AC side. In practice, any deviations from the assumed ideal AC or DC system parameters will result in other frequency components (though normally of greatly reduced magnitudes), which are referred to as *uncharacteristic* harmonics.

The 'level number' is a term commonly used in self-commutating conversion and indicates the number of voltage levels used by the bridge phase arms. It is normally limited to two or three (in the case of PWM) while higher numbers are used in multilevel configurations.

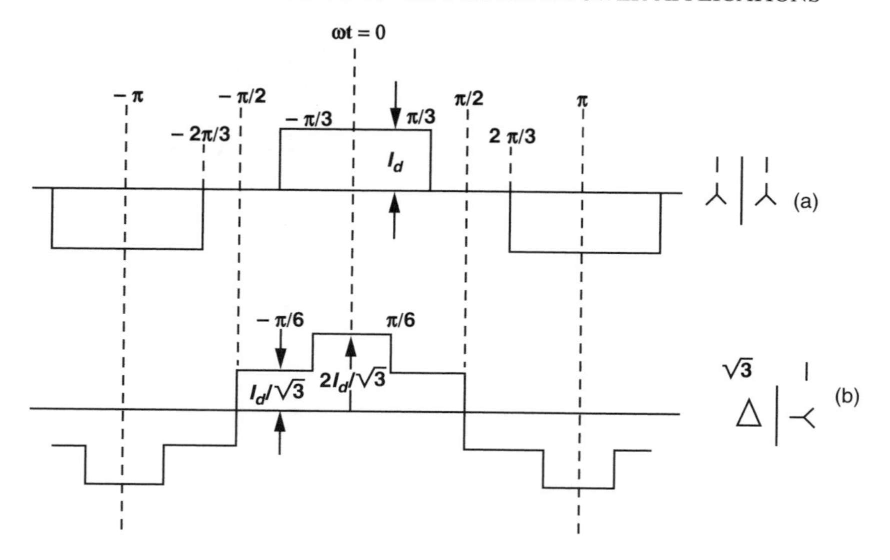

Figure 1.6 Idealized phase current waveforms on the primary side: (a) star–star transformer connection; (b) delta–star transformer connection.

1.5 Line-commutated conversion (LCC)

Three-phase AC–DC and DC–AC converters ratings in hundreds of MW, as required in power transmission and in the metal reduction industry, need to be of high-voltage and high-current designs respectively. These normally consist of complex structures of series or parallel connected power switches, as the ratings of the individual ones are far too low for these purposes. In the series connection the difficulty arises from the need to achieve equal voltage sharing, both during the dynamic and steady states, while in the parallel connection the switches must share the steady state and dynamic currents.

In line-commutated conversion the switchings take place under zero current conditions and snubber circuits are only used to slow down the speed of the individual switches to that of the slowest one in order to achieve good voltage sharing.

Diode rectification (the earliest conversion technology) has no controllability and makes the rectified voltage and current exclusively dependent on the external DC and AC system conditions.

The advent of the silicon controlled rectifier under LCC permited active power conversion control from full negative to full positive rating but at the expense of absorbing varying quantities of inductive reactive power. The DC voltage, however, can be controlled from full negative to full positive rating and the DC current from zero to full rated level. Lack of waveform quality, normally in the form of current harmonic content, is another important problem of SCR–LCC conversion.

The 12-pulse configuration has become the standard configuration in high-voltage thyristor conversion. Higher pulse numbers, such as 48, are commonly used by the metal reduction industry, but in practice they suffer from some low-order harmonic distortion problems. Shunt connected passive filters are, thus, an integral part of the LCC power conversion process and risk causing low-order harmonic resonances with the AC system.

Most of the present HVDC converters, being of the CSC–LCC type, require reactive power for their operation but possess full DC voltage and current controllability. The power and voltage ratings of existing schemes are already in GWs and a $\pm 800\,\mathrm{kV}$ DC technology is currently being introduced. The valves consist of a large number of series-connected thyristors, which are fired synchronously and they are naturally switched off when the current through them reduces to zero; the natural commutation process permits the use of low-cost voltage-balancing snubbers.

1.6 Self-commutating conversion (SCC)

Self-commutation takes place independently from the external circuit source and system parameters and is achieved practically instantaneously. It requires the use of switching devices with turn-on and turn-off capability and the position and frequency of the on and off switching instants can be altered to provide a specified voltage and/or current waveform.

With self-commutation, the switching action takes place under rated voltage and current conditions, and, therefore, a large amount of stored energy is involved that increases the rating and cost of the snubbers.

The following is a wish-list of items for an effective large power conversion:

1. Perfect balancing of the voltages across the series-connected individual switches of the high-voltage valves during the off-state and in the switching dynamic regions; or perfect balancing of the current in the shunt-connected individual switches of the high-current valves during the on-state and in the switching dynamic regions.
2. High-quality output waveforms.
3. Low dv/dt rate across the switches and other converter components to simplify insulation coordination and reduce RF interference.
4. Minimal on-state and switching losses (prefers relatively low switching frequency).
5. Simple structural topology to reduce component costs.
6. Flexibility in terms of active and reactive power controllability.

The self-commutating concepts advanced so far for large power conversion are pulse width modulation (PWM) and multilevel VSC for high-voltage applications and multipulse and multilevel CSC for high-current applications.

1.6.1 Pulse width modulation (PWM) [6]

The ideal properties of the power transistor, in terms of voltage/current characteristics and high-frequency switching capability, permitted the development of PWM in the 1960s, and is still the most flexible power conversion control technique. The PWM principle is illustrated by the output voltage waveform shown in Figure 1.7. As well as providing fundamental voltage control, while maintaining the DC voltage constant, PWM eliminates the low harmonic orders and only requires small filter capacity to absorb the high-frequency components. Such level of flexibility is achieved by modulating the widths of high-frequency voltage pulses. A modulating carrier frequency is used to produce the required (normally

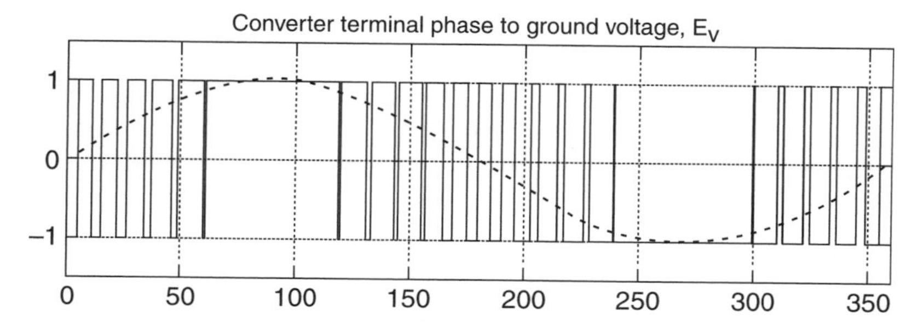

Figure 1.7 Voltage source converter PWM with optimum harmonic cancellation.

voltage) waveform. The PWM concept scores well on items (2), (5) and (6) of the ideal requirements but poorly on items (3) and (4).

PWM, originally used in relatively low-power applications such as the motor drive industry, has recently led to the development of series-connected valves capable of withstanding the high voltages used in power transmission. It was first used in flexible alternating current transmission systems (FACTS) applications, and more recently as the basis of self-commutating HVDC transmission systems. At the time of writing, however, it is uncertain whether this technology will continue developing to meet the needs of very large-power, long-distance transmission. In this respect there are important difficulties to overcome in terms of efficiency and reliability of operation. Besides the high switching losses, the use of two or three DC voltage levels subjects the valves and all the surrounding system components to very high *dv/dt*'s following every switching event, increasing the conversion losses and complicating the system insulation coordination.

The power rating capability of PWM is also limited at present by its reliance on cable transmission. To extend the use of VSC–PWM to overhead line transmission will also require improvements in the control of DC line fault recovery.

1.6.2 Multilevel voltage source conversion

The main object of multilevel conversion is to generate a good high-voltage waveform by stepping through several intermediate voltage levels, i.e. the series-connected devices are switched sequentially at the fundamental frequency, producing an output waveform in steps. This eliminates the low-order harmonics and reduces the *dv/dt* rating of the valves by forcing them to switch against a fraction of the DC voltage.

This concept scores well on items (2), (3) and (4) of the ideal requirements list but poorly on items (5) (the number of auxiliary switches increasing approximately with the square of the

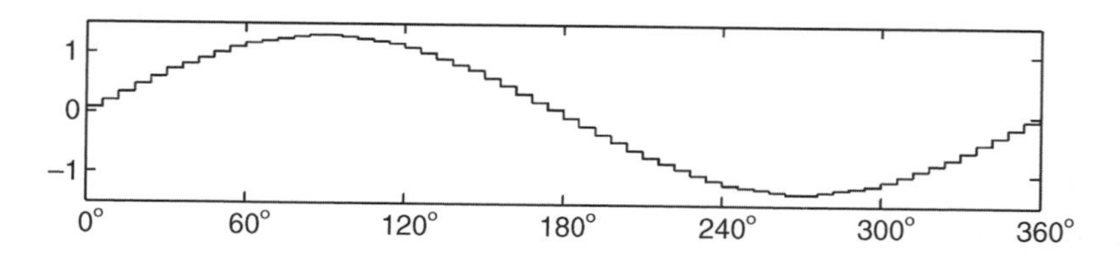

Figure 1.8 Typical output voltage waveforms of a multilevel VSC configuration.

level number in most of the proposed topologies) and (6) (lacking independent reactive power control at the terminals of DC interconnections).

The use of the robust thyristor-type switches (such as the GCT and IGCT), makes the multilevel alternative better suited to the needs of very large power conversion, but the structural complexity and limited control flexibility has so far discouraged the use of the multilevel configurations.

The original multilevel configurations are of the VSC type, which provides free current paths without the need to absorb substantial electromagnetic energy.

Modular multilevel converter (MMC) is a recent Siemens concept [7] that, in common with other multilevel configurations, provides a fine gradation of the output voltage, thus reducing the harmonic content, the emitted high-frequency radiation and the switching losses. The important advantage of this configuration over earlier multilevel proposals is that it permits four-quadrant power controllability. This alternative scores well on item (6).

Finally, a structurally simpler multilevel concept is multilevel reinjection [8,9], applicable to both VSC and CSC, suited to very large power ratings; this scores well in items (1) to (6).

1.6.3 High-current self-commutating conversion

At present most of the perceived application of large power self-commutating semiconductors appears to be in the high-voltage area. In comparison there seems to be little interest in the development of low-voltage, high-current self-commutating switching. However, the latter is important for applications such as aluminium smelters and superconductive magnetic energy storage, both of which are likely to benefit from a more flexible and cost-effective power electronic technology.

In self-commutating CSC the forced commutation from rated current to zero involves large electromagnetic energy (stored in the AC system inductance). Thus interfacing forced-commutated CSC with the AC system requires the provision of costly high capacitors. An acceptable self-commutated CSC configuration should be able to provide a continuously varying AC current. It will be shown that CSC multilevel current reinjection can achieve that target, as well as scoring well in the other items of the wish-list.

1.7 Concluding statement

Only two decades ago it would have been considered science fiction to talk about a transistor-based 300 MW HVDC transmission link, and yet this is now a reality. Trying to predict the final state of the static conversion technology is a very difficult task.

The next five chapters of the book describe the present state of the large power conversion technology and the concepts and configurations likely to influence the design of large static power converters in the next decade. These concepts and converter configurations are used in the last six chapters, with reference to the most likely applications requiring high power and/or current conversion.

References

1. Lamm, U. (1964) Mercury-arc valves for high voltage dc transmission. *IEE Proceedings*, **3** (10), 1747–53.

2. Lips, H.P. (1997) Semiconductor power devices for use in HVdc and FACTS controllers. *International Colloquium on HVDC and FACTS*, Johannesburg, South Africa, 1997, paper 6.8.
3. Lips, H.P. (1996). Semiconductor power devices for use in HVDC and FACTS controllers. Conference Internationale des Grandes Reseaux Working Group (CIGRE WG) 14.17.
4. Shakweh, Y. (1999) New breed of medium voltage converters. *IEE Power Engineering Journal*, **13** (2), 297–307.
5. CIGRE Study Committee B4-WG 37 (2005) VSC Transmission. CIGRE, Paris.
6. Arrillaga, J., Liu, Y.H. and Watson, N.R. (2007) *Flexible Power Transmission – The HVDC Options*. John Wiley & Sons Ltd, London.
7. Dorn, J., Huang, H. and Retzmann, D. (2007) *Novel voltage source converters for HVDC and FACTS applications*. CIGRE Annual Conference, Osaka (Japan), November.
8. Perera, L.B., Liu, Y.H., Watson, N.R. and Arrillaga, J. (2005) Multi-level current reinjection in double-bridge self-commutated current source conversion. *IEEE Transactions on Power Delivery*, **20** (2), 984–91.
9. Liu, Y.H., Arrillaga, J. and Watson, N.R. (2004) Multi-level voltage reinjection – a new concept in high power voltage source conversion. *IEE Proceedings, Generation, Transmission and Distribution*, **151** (3), 290–8.

2

Principles of Self-Commutating Conversion

2.1 Introduction

The conventional line-commutated large power converters normally use the current source conversion principle. Both during rectification and inversion they absorb varying quantities of reactive power in their normal operation and inject substantial harmonic currents into the AC system. These conditions require complex and expensive converter stations and cause voltage and harmonic interactions with the AC system to which they are connected. Although the experience of many years has managed to reduce, if not completely eliminate, these problems, the extra cost involved is very substantial and the scheme's reliability is reduced by the number of components.

The progressive development of the self-commutated conversion techniques permits the design of high-power converter systems without the need of passive filters and shunt reactive power compensators; this results in substantial cost reduction and elimination of potential harmonic interactions between the converter and the power system.

Although the progress already made with self-commutation has been great, we are still speculating on the extent to which self-commutating based technologies should be developed to exploit the great potential available in the future.

For power transmission application, the voltage needs to be high to reduce losses and, thus, to match the high voltage rating the converter valves consist of many series-connected switches. Similarly, for industry applications requiring very high currents, many switching devices need to be connected in parallel to match the high current rating.

The dynamic and steady state voltages and currents of these configurations are now well understood and the design of very large power conversion systems presents no special problem.

This chapter introduces the principles of self-commutating conversion with emphasis on VSC, which is at present the preferred option. In later chapters, however, it will be shown that for high-power applications there is considerable potential for a self-commutating CSC technology.

2.2 Basic VSC operation

Figure 2.1 displays the switching circuit for one phase of the basic two-level converter and its corresponding voltage output waveform, with the midpoint of the DC capacitors used as a reference for the AC output voltage. The two switches are turned on–off in a complementary way (one on and the other off) to generate a certain output of discrete two voltage levels $(+V_{dc}/2$ or $-V_{dc}/2)$.

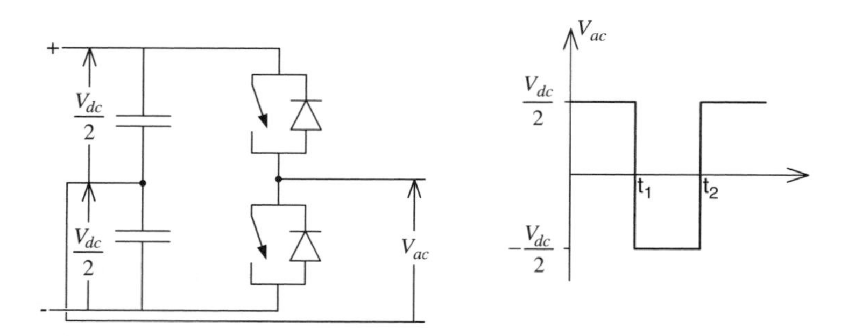

Figure 2.1 Two-level single phase voltage source converter. (Reproduced by permission of CIGRE.)

Since the conduction in a solid state switch is unidirectional, an anti-parallel diode needs to be connected across to form a switch pair, to ensure that the bridge voltage only has one polarity, while the current can flow through the pair in both directions.

When the two main switches are blocked, the anti-parallel diodes form an uncontrolled rectifier. In this condition the application of an external AC voltage charges the DC upper and lower capacitors, via the uncontrolled rectifiers, to the peak value of the AC voltage connected across them. Once the DC capacitor is charged and the external source connected, the voltage source converter is ready for operation.

The main switches can be switched on and off in any desired pattern; however, immediately before one switch is turned on, the opposite switch must be turned off, as their simultaneous conduction would create a short circuit of the DC capacitors. This action causes a small blanking period (of a few microseconds) where none of the switches are on and the current path is via the freewheeling diodes. To explain this condition let us start at a point, after $t = 0$ in Figure 2.1, where the upper switch is on and, thus, the AC terminal is connected to the plus terminal of the DC capacitors via the main switch and freewheeling diode pair; this allows the current to flow through the pair in either direction.

When the upper switch is turned off (at instant t_1), the reactance of the AC circuit maintains its present current and the diodes in parallel with the upper and lower switches provide the current path. If the current in the upper pair at instant t_1 is from the plus terminal of the DC capacitors to the AC terminal, the lower freewheeling diode turns on, and thus the AC output voltage changes from plus to minus $V_{dc}/2$, i.e. the polarity reversal has been initiated by turning off the upper switch. Similarly, when the current in the upper pair at instant t_1 is from the AC terminal to the plus terminal of the DC capacitors, the upper freewheeling diode is on and the AC output voltage remains at plus $V_{dc}/2$, i.e. the polarity is unchanged by turning off the upper switch.

Thus there is no need to turn on the lower switch immediately following the turn off of the upper one. When, after the blanking time the lower switch is turned on, the lower switch pair takes over the current in either direction. During the blanking time the AC terminal voltage polarity depends on the current direction in the switch pair to be turned off, while the AC side reactance (the transformer leakage reactance and phase reactor as well as the AC system) maintains the current unchanged.

A similar procedure, via the opposite switch pair of the main switch and freewheeling diode, can be used to explain the transfer from the negative to positive output voltage waveform. This process applies to all VSC configurations irrespective of the fundamental power flow operating condition (i.e. in the four quadrants).

To generate the required AC voltage waveform, a repeated switching sequence is used as follows:

- In a certain time interval ($T = t_n - t_0$), the number of switching actions (n) and switching times ($t_1, t_2, \ldots, t_n$), as well as the sequence start time (t_0) form the switching pattern.
- The parameter ($T = t_n - t_0$) controls the AC output voltage frequency, the n switching times ($t_1, t_2, \ldots, t_n$) control the fundamental amplitude (which is related to the half cycle average area) and the sequence start time (t_0) controls the phase angle.

Therefore the frequency, amplitude and phase angle components of the voltage source converter fundamental voltage are controlled by the gate signals. That means that at the fundamental frequency, the voltage source converter functions as a controllable voltage source; its frequency and phase angle are directly determined by the firing pulse pattern, while its magnitude is determined by the DC voltage and the firing pulse pattern together. The power implications are discussed in the following section.

2.2.1 Power transfer control

If the voltage source converter is connected to a passive network on the AC side, the power flows unidirectionally from the DC input side to the passive AC load. When connected to an active AC system, the power can be made to flow in either direction by making the phase angle of the converter AC output voltage positive or negative with respect to that of the AC system voltage.

The two switching pairs block the unidirectional DC voltage and permit bidirectional current; this permits bidirectional power flow between the voltage source converter and the AC system to which the voltage source converter is connected.

To simplify the analysis, the voltage source converter voltage (V_2) is connected to an AC voltage source (V_1) by an inductance (X). Figure 2.2(a) shows the fundamental frequency phasor representation of the voltage source converter operating as an inverter and supplying active and reactive power to the AC system. In this operating condition the diagram shows that the voltage source converter output voltage (V_2) has a larger amplitude and is phase-advanced with respect to the AC system voltage (V_1).

The active power exchange (P) between two active sources is expressed as:

$$P = V_1 V_2 \sin(\delta)/X \tag{2.1}$$

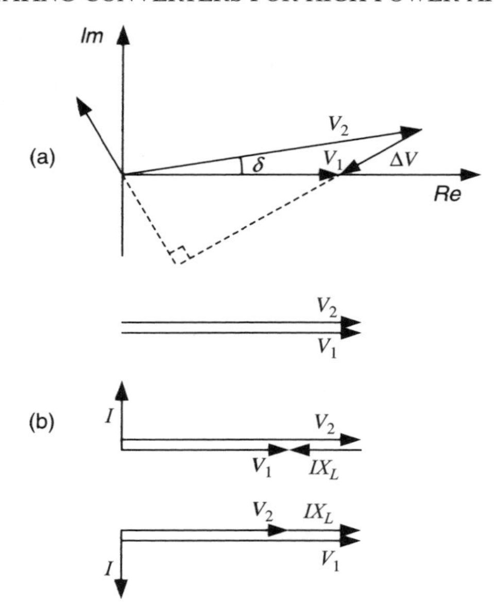

Figure 2.2 Operating modes of the voltage source converter: (a) phasor diagram when providing active and reactive power; (b) phasor diagram when operating as a reactive power controller.

and the reactive power (Q) measured at the voltage source converter terminal is

$$Q = V_2\,[V_1\cos(\delta) - V_2]/X \tag{2.2}$$

where δ is the phase difference between V_1 and V_2.

In this operating condition the inductive reactive power is defined as positive.

If the converter is used only as a reactive power compensator, then there is no need for an active DC system, and the converter is terminated by a DC capacitor. However, the capacitor size should be sufficient to permit the charge and discharge operations caused by the switching sequence of the converter valves without exceeding the specified ripple.

Figure 2.2(b) illustrates the case when the voltage source converter operates purely as a reactive power compensator, i.e. the converter (V_2) and supply system (V_1) voltages are in phase with each other. When both voltages are of the same magnitude there is no exchange of reactive power between the converter and the system (i.e. the current is zero); when V_2 is larger than V_1 the current leads the voltage by 90°, i.e. the converter behaves as a capacitor and, thus, generates reactive power; when V_2 is smaller than V_1 the current lags the voltage by 90°, i.e. the converter behaves like an inductor and, thus, absorbs reactive power.

In this mode of operation the voltage source converter is analogous to an ideal rotating synchronous compensator (and is usually referred to as a STATCOM) [1]; however the voltage source converter has very small inertia (related to the firing pattern change time) and its response is very close to instantaneous. Thus the voltage source converter acts as an AC voltage source, controlled to operate at the same frequency as the AC system to which it is connected.

The voltage/current characteristics of the voltage source converter operating in this condition, shown in Figure 2.3, illustrate that this type of converter can provide full capacitive support for voltages down to approximately 0.15 per unit.

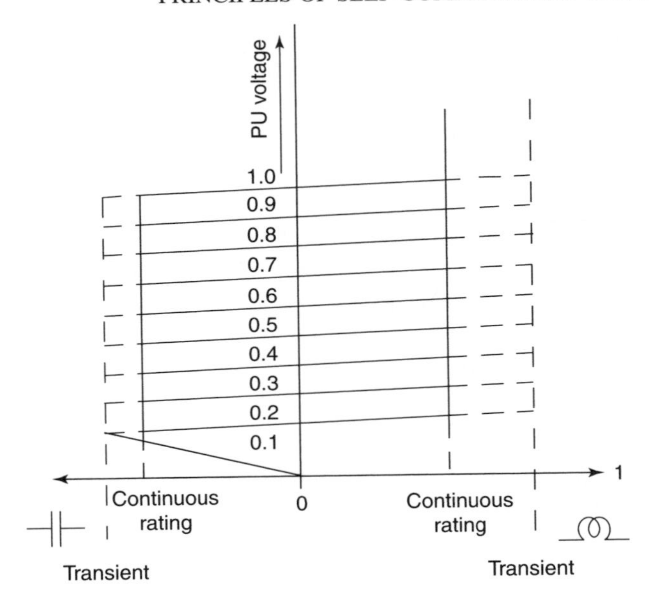

Figure 2.3 Voltage/current characteristics of a voltage source converter operating as a static compensator (STATCOM).

If the converter is connected to an active DC system (e.g. to another converter), the AC current and voltage can have any phase relationship, and the converter can act as a rectifier or an inverter, and with leading or lagging reactive power (i.e. four quadrant operation is possible). Thus the voltage source converter can be controlled to operate at any point within a circle, the radius of which represents the converter MVA rating.

There are, of course, active and reactive power limits determined by the maximum allowable valve current and maximum allowable DC voltage respectively on the storage capacitor. For a given AC system voltage, the DC voltage rating is determined by the maximum AC output voltage that the converter must generate to provide the maximum required reactive power.

The *PQ* characteristics depend on the converter AC output voltage and, therefore, the interface transformer ratio can be used to optimize them at the design stage. Also, the self-commutated voltage source converter can be designed to control the DC voltage with a small percentage over and below the AC voltage peak level; moreover, the voltage source converter continuously provides a wide range of AC output voltage amplitude control to maximize the steady state power capability and can minimize the power losses in the AC power transmission system without the need of an on-load tap changer (OLTC).

2.3 Main converter components

As explained in the introductory chapter, voltage source converter power conversion implies the presence of a voltage source on the DC side. The voltage source maintains a prescribed voltage across its terminals irrespective of the magnitude or polarity of the current flowing through the source.

The circuit diagram of Figure 2.4 [2] shows the basic components of a VSC substation, which consist of one or more converter units including the high-voltage valve chains, DC capacitors, reactors, transformers and filters, as well as (not shown) control, monitoring, protection and auxiliary equipment.

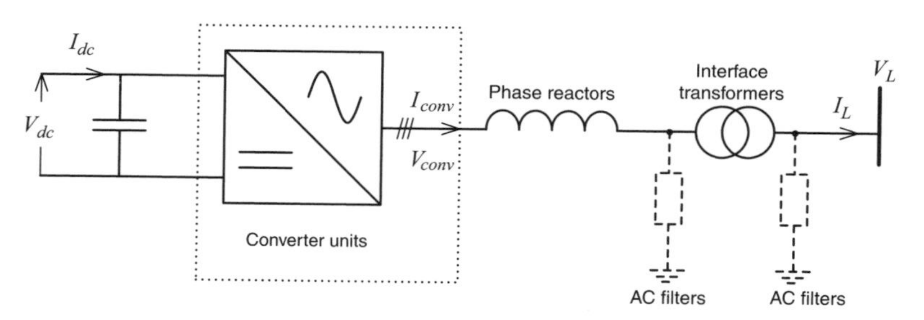

Figure 2.4 Basic diagram of a voltage source converter substation. (Reproduced by permission of CIGRE.)

2.3.1 DC capacitor

In a voltage source converter the DC side is strongly capacitive, and thus voltage stiff. In this respect, the DC capacitor serves the purpose of stabilizing the DC voltage. Consequently the voltages are well defined and are normally considered independent of the converter switching operation. The DC average voltage can be controlled by charging or discharging the capacitor, which is achieved by active power exchanges between the AC and DC sides.

In practice, voltage source converter operation produces some harmonic currents in the DC circuit. These currents, due to the presence of DC circuit impedance, give rise to corresponding harmonic voltages, and therefore to DC voltage ripple. They also affect the ripple on the capacitors of other stations connected to the same DC system. The DC voltage ripple is mainly influenced by the following factors:

- the voltage source converter valve switching strategy;
- the capacitance of the DC capacitor;
- AC system unbalance and distortion.

The size of the DC capacitor reduces with increasing switching frequencies, because every time the valves are switched the DC current in the capacitor changes direction, thus reducing the effective ripple.

2.3.2 Coupling reactance

The voltage source converter is not directly connected to the AC supply when the latter is a strong system (i.e. has a large short-circuit capacity compared to the converter rating). In such a case, some coupling reactance is needed between the system busbar and the converter terminals. As well as reducing the fault current, this coupling reactance stabilizes the AC current, helps to reduce the harmonic current content and enables the control of active and reactive power from the voltage source converter. This component includes the inductance of the interface transformers and any extra phase reactor possibly added for these purposes. The interface reactors and transformers must withstand any high-frequency voltage stresses imposed by the voltage source converter operation. Thus, depending on the converter configuration, it may be economical to place a high-frequency blocking filter between the converter and the interface transformer.

2.3.3 The high-voltage valve

For high-voltage applications the limited voltage rating of individual switching devices requires their connection in series to form the high-voltage rating valve, and voltage balance across the individual switches is essential for the valve reliability.

The self-commutated switching devices can be categorized into two types: thyristor type (GTO, IGCT, MCT, etc.) and transistor type (IGBT, MOSFET, etc.). The gate controllability of the transistor-type semiconductor switches can be used for voltage balancing across the switches during the on/off switching dynamics.

All semiconductor switches experience a change in the turn-on delay time with temperature, which can vary from device to device. The turn-on time variation is also affected by rise time, i.e. the slower turning-on devices will have a greater share of the total voltage. They also experience turn-off variation times, dependent on temperature and commutated current levels, which vary from device to device, i.e. the quicker turning-off devices will have a greater share of the total voltage.

The turn-on and turn-off time difference between the individual switches in the series chain causes the dynamic voltage unbalancing. The leakage current difference causes the steady voltage unbalance during the blocking state.

Good voltage sharing is critical to ensure that all the devices in a series chain experience similar operating conditions. The dynamic balance is achieved by placing snubber circuits in parallel, or (in the case of transistor switches) using active gate control during switching dynamics.

The snubber circuit usually consists of passive capacitors, inductors, resistors and possibly fast recovery diodes. The use of a snubber capacitor helps to achieve dynamic voltage sharing (due to spread in turn-off delays) while a di/dt inductor helps the switches to share voltage (due to spread in turn-on delays) and diode voltage sharing under reverse recovery conditions. A di/dt limiting inductor reduces the turn-on losses substantially. Although the contribution of the turn-off snubber capacitor to loss reduction is less significant it is still justifiable. The snubber shifts the switching losses from the switch to a low-cost passive component where it can be dissipated away or recovered into the supply or load.

Adequate static voltage sharing of series-connected switches is achieved by employing a parallel-connected resistor per switching module. The resistor function is to provide a leakage current which is dominant over the switches, thus forcing sharing to occur by potential divider action. The resistor current is typically ten times the switch leakage current, in order to swamp all the variations of leakage currents under different operating and temperature conditions.

For easy series assembly a typical switch building block consists of a single switching device and a freewheeling diode, snubber components and the gate drive. The gate board is equipped with overvoltage, undervoltage and overcurrent protection. High reliability is provided by mounting the switch on a separate heat-sink, thermally decoupled from the snubber resistor. For diagnostic purposes the switches are connected by fibre optics to a pulse distribution board.

As a MOS-controlled transistor-type device with a high impedance gate and capability of blocking, saturating and amplifying, the IGBT requires low energy to switch the device and maintain power flow. This facilitates the series connection by providing good voltage distribution even at high switching frequencies (in kHz). Moreover, the gate power for the series-connected semiconductors can be provided via the voltage dividing circuits across the individual IGBTs.

In high-voltage applications, many IGBTs are connected in series, as shown in Figure 2.5. For instance the valves used in the original 150 kV HVDC light schemes had up to 300 series-connected IGBTs. The IGBT valve will, in principle, experience the same requirements as the thyristor valve of the conventional line-commutated HVDC converter.

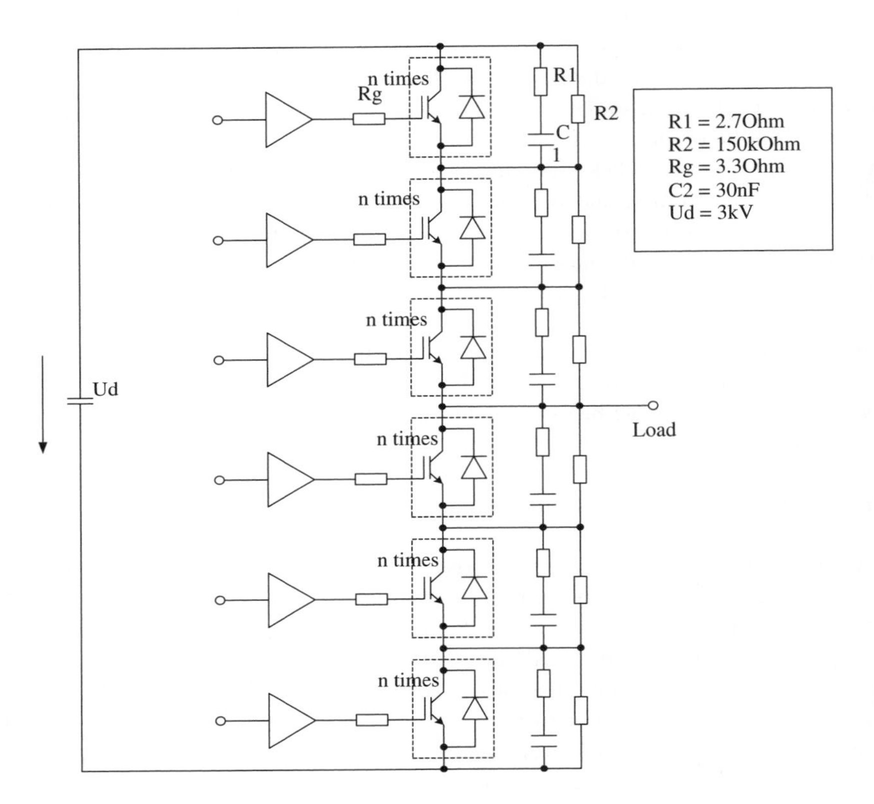

Figure 2.5 Circuit diagram of an IGBT chain phase module.

Moreover, the number of series-connected switches required to sustain the converter voltage rating must be increased to provide the voltage source converter valve with redundancy to continue operating following the failure of individual devices. Also, for the valve to continue operating, a faulty switch must not create an open circuit, i.e. the faulty device in the series chain must remain in the short-circuit state and be capable of conducting current continuously until the next scheduled maintenance period.

A faulty thyristor-type switch can remain in the short-circuiting state, but a faulty standard IGBT module does not have the required short-circuit capability, as the IGBTs use bond wires that open the circuit upon failure. Therefore specially developed press pack designs need to be used in a series chain, as reported for DC transmission applications [3–5].

In the IGBT series chain, small differences in the range of some 10 ns are critical to voltage sharing and, thus, active voltage control of the gate driver circuit is often used to ensure voltage balancing, both in the static blocking state and in the switching dynamic period. Moreover, in addition to active voltage control, a small RC snubber for damping oscillations is normally employed, as shown in Figure 2.5.

The use of snubber circuits adds to the complexity of the valve and increases the power losses. It is possible to design the gate control of the IGBT valve to maintain acceptable voltage sharing without snubbers, provided that the spread in the device's characteristic switching times, switching transient behaviour and blocked state leakage currents are sufficiently small. The gate unit must maintain the device operation safe during short-circuit conditions. The IGBT is better behaved in this respect.

In contrast with the thyristor valves of line-commutated conversion, the individual switches (GTOs, IGCTs or IGBTs) in VSC transmission schemes have to withstand high voltage for a substantial part of the operation time and, therefore, the probability of an incident particle initiating a destructive current avalanche during the blocking stage is greatly increased.

While the current source converters used in power transmission normally operate in stable conditions (i.e. the power changes occur at moderate rates), the higher controllability of voltage source converters puts heavier demands on their use as adaptive system controllers.

As well as the rated current, the valves have to withstand peak current due to ripple and transient overcurrent levels and need to be provided with an additional protection margin; they also need to be capable of turning off the current following a short circuit close to the valve itself.

2.3.4 The anti-parallel diodes

The anti-parallel diodes in the voltage source converter constitute an uncontrolled bridge rectifier and, thus, unlike the main switches (which can be switched off in a few microseconds during transients) the diodes have to be designed to withstand the fault-created stresses.

A DC short-circuit fault creates a current path through the diodes and the fault current is only limited by the AC system and converter impedances. This fault current needs to be cleared by a circuit breaker on the AC side, which will require a few cycles to operate.

The diode will also experience inrush current (as well as an overvoltage on the DC bus) if the voltage source converter is energized by the circuit breaker under zero voltage on the DC side. This overcurrent must be limited by external components (such as pre-injection breaker resistors). Alternatively, the DC capacitor can be charged independently prior to converter energization. In this case a DC blocking voltage will appear after the initial current surge.

2.4 Three-phase voltage source conversion

2.4.1 The six-pulse VSC configuration

The basic six-pulse VSC configuration, shown in Figure 2.6, consists of a three-phase bridge connecting the AC source to a predominantly capacitive DC system.

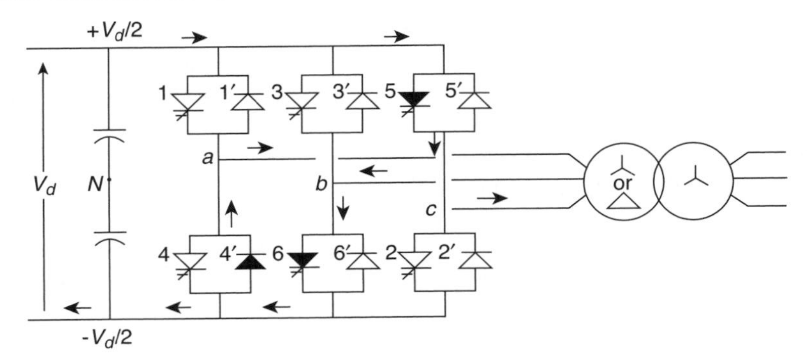

Figure 2.6 Three-phase full-wave bridge voltage source converter. (Reproduced by permission of CIGRE.)

The bridge valve unit is an asymmetric turn-off device (shown in Figure 2.6 as a GTO) with a reverse-connected diode in parallel. The turn-off device carries out the instantaneous inverter function, while the diode is needed to provide a path for the transfer of energy from the AC to the DC side (i.e. rectifier operation) to charge the capacitor. During each cycle there will be periods of rectification and inversion determined by the phase angle, and, therefore, the average current will determine the net power flow direction (i.e. rectification or inversion). If the converter operates as a rectifier with unity power factor, only the diodes conduct the current, whereas during inverter operation with unity power factor only the turn-off devices conduct. By appropriate control of the turn-on and turn-off switchings, a three-phase AC waveform is produced at the AC output.

For fast transients the DC capacitor can be regarded as a perfect voltage source. Thus in the short time sequence in which the switching devices are controlled to interconnect the DC and AC terminals, the DC voltage remains practically constant and the voltage source converter produces a quasi-square wave AC voltage.

The (self-commutating) voltage and (line-commutated) current source conversion processes can be treated as dual and, thus, the analysis of a three-phase voltage source converter follows similar lines to that of a three-phase LCC if current is substituted by voltage. There are, however, two important differences between CSC and VSC. One is the duration of the valve conducting period, which in the case of the voltage source converter is 180° instead of the 120° generally adopted for the CSC configuration. The 180° conduction is needed in the voltage source converter case to avoid the condition in which both arms of the bridge leg are in the off-state; this condition would occur if a 120° period was used, which would cause uncertainty in the output voltage. The other important difference is the absence of commutation overlap, which makes the voltage source converter operation more predictable and easier to analyse. As in the case of the LCC CSC, the valves in the self-commutating VSC circuit of Figure 2.6 are controlled to commutate only at the power system frequency.

Figure 2.7 shows the following waveforms:

- The phase voltages v_{aN}, v_{bN}, v_{cN} with respect to the capacitor mid-point (not with respect to the transformer neutral, as these two points are not directly connected).

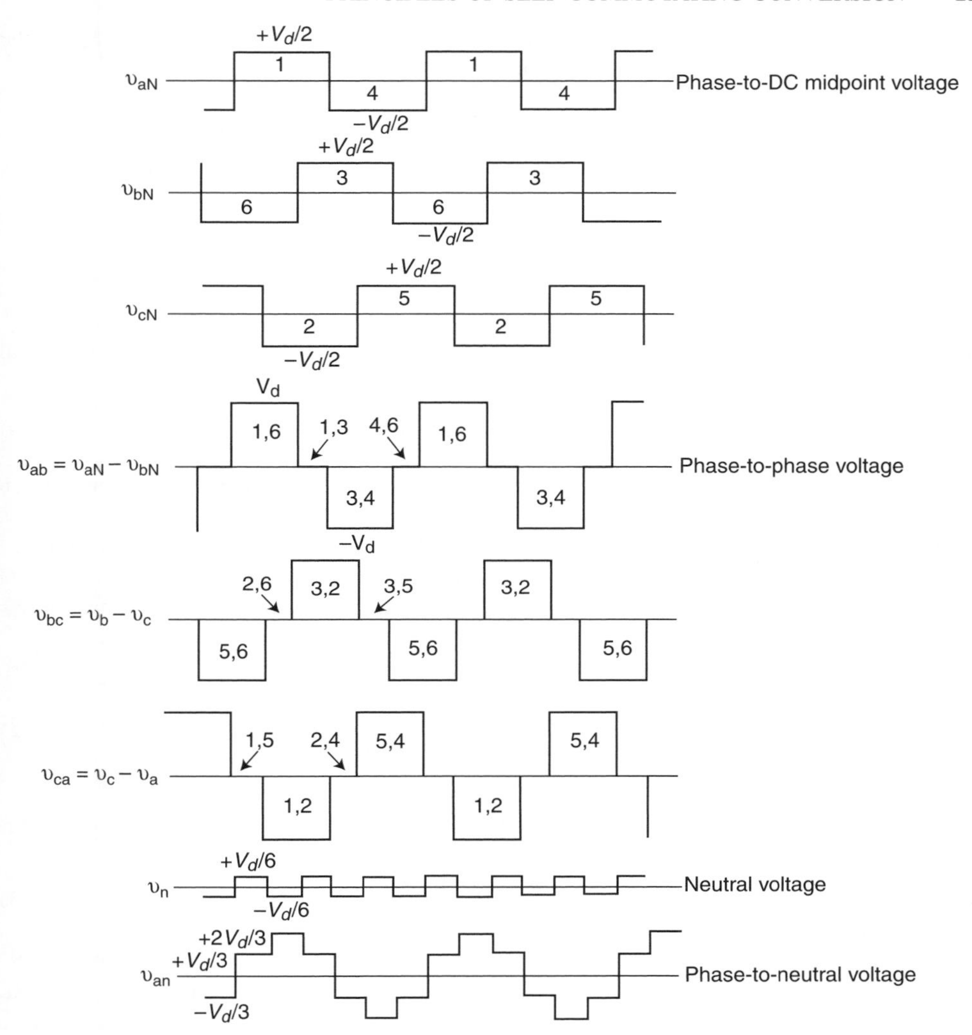

Figure 2.7 AC voltage waveforms of the six-pulse voltage source converter. (Reproduced by permission of CIGRE.)

- The line voltages on the converter side (i.e. v_{aN}, v_{bN}, v_{cN}). Again, as in the line-commutated case, the AC side has only three-phase conductors (i.e. the transformer neutral is floating) and, therefore, these voltage waveforms are of 120° duration.
- The voltage of the floating neutral in the secondary of the star connected transformer v_n, with respect to the midpoint N. This potential is the sum of the instantaneous potentials of the three phases (v_{aN}, v_{bN}, v_{cN}) and is a square AC waveform varying at three times the fundamental frequency, with a magnitude equal to one sixth of the DC voltage.
- The phase voltage v_{an} across the star-connected transformer secondary.

A Fourier analysis of the phase voltage waveforms shows the following sequences for the fundamental and harmonic frequency components:

- The fundamental component is a symmetrical set of positive sequence.
- The triplen harmonics are all of zero sequence (i.e. their values are the same in the three phases at all instants).
- The fifth harmonic is of positive sequence.
- The seventh harmonic is of negative sequence, etc.

Subtracting the neutral voltage v_n from v_{aN} (refer to Figure 2.7) gives the phase to neutral voltage (v_{an}), i.e.

$$v_{an} = (2V_d/\pi)\,[\cos(\omega t) + (1/5)\cos(5\omega t) - (1/7)\cos(7\omega t) - (1/11)\cos(11\omega t)\ldots] \quad (2.3)$$

which is a three-level voltage of magnitudes 0, $V_d/3$ and $2V_d/3$ and is free from triplen harmonics. It should be noted that v_{aN} and v_{an} are in phase.

Voltages v_{bn} and v_{cn} are the same but phase shifted by 120° and 240° respectively from v_{an}.

The phase-to-phase voltage (v_{ab}) consists of three levels, i.e $\pm V_d$and zero; its fundamental component is phase-shifted by 30° with respect to v_{an} and its amplitude is $\sqrt{3}$ times the v_{an} amplitude.

The time domain expression of the 120° phase-to-phase voltage is:

$$V_{ab} = (2\sqrt{3}/\pi)V_d\,[\cos(\omega t) - (1/5)\cos(5\omega t) + (1/7)\cos(7\omega t)\ \ldots] \quad (2.4)$$

which shows that the triplen harmonics have now been eliminated.

The total *rms* value of V_{ab} is $0.816V_d$, and the fundamental component (V_c) of the converter voltage:

$$\mathrm{V}_c = (\sqrt{6}/\pi)V_d = 0.78V_d \quad (2.5)$$

Thus in the stiff voltage source converter the AC voltage output is a function of the DC voltage only and, therefore, to alter the former requires a corresponding change in the latter. This can be achieved by charging or discharging the DC capacitor either from a separate source or from the AC system.

The harmonic voltages are related to the fundamental frequency voltage by:

$$V_n = V_c/n \quad (2.6)$$

The DC current waveform contains a direct component and a series of harmonics of orders n = 6k. The direct component is equal to:

$$I_d = \frac{3\sqrt{2}}{\pi} I\cos(\theta) \quad (2.7)$$

where I is the AC phase current *rms* and θ the firing angle, which in self-commutation is also the power factor angle.

Current I_d changes from $+1.35I$ to $-1.35I$ and vice versa as the firing angle changes from full rectification to full inversion.

The ratio of the nth harmonic to the peak value of the instantaneous DC current has a minimum value of

$$\sqrt{2}/(n^2-1), \text{ when the power factor is unity,}$$

and a maximum of

$$\sqrt{2}n/(n^2-1), \text{ when the power factor is zero.}$$

All these equations show that the higher the value of n (the harmonic order) the lower is the harmonic amplitude, which explains the importance of eliminating the low orders. However, during unbalanced conditions the second harmonic and other low-frequency orders reappear.

In the above development it is assumed that the AC current is perfectly sinusoidal. In practice the phase currents will contain small contents of other harmonics that may be present in the AC system and converter voltage waveforms.

2.4.2 Twelve-pulse VSC configuration

The combination of two bridges with their AC voltage waveforms phase-shifted by means of the star and delta interface transformer connections constitutes a twelve-pulse converter. The difference between the CSC and VSC configurations is in their AC side primary winding transformer connections, which must be in parallel (in the CSC case) and series (in the VSC case). The 12-pulse structure and voltage waveforms of the VSC configuration are illustrated in Figure 2.8.

2.5 Gate driving signal generation

The gate signal generation unit constitutes the basic part of the converter control system. Self-commutated converters are driven by the gate signals to control their AC and DC output voltages and currents for the electrical power exchange between the AC and DC power sources.

The fundamental component of the power converter AC output is required for active and reactive power control. Its control flexibility, as well as the steady and dynamic performance determined by the gate signals in real time, govern the converter functions and performance. However, the switching control nature of the converter, as well as producing the required fundamental, can cause unexpected harmonics.

Therefore the gate signal generation strategy is to design a firing pattern, with which the converter provides a flexible control of the fundamental component with the required response performance in real time, and minimize or suppress the harmonic components.

As there are innumerable possible firing patterns, there is plenty of scope for research and optimization.

2.5.1 General philosophy

The common feature of all VSC configurations is the generation of a controllable fundamental frequency component needed to control the electrical power exchange between the AC and DC voltage sources. This is achieved by gate driving signals defined by firing patterns.

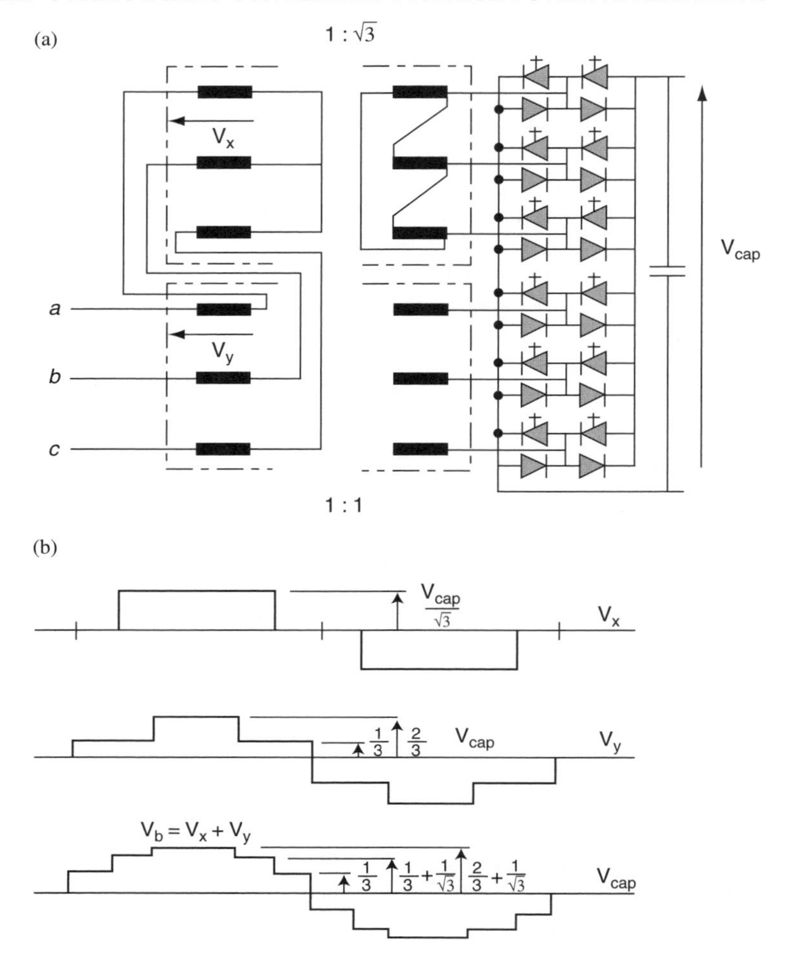

Figure 2.8 Twelve-pulse voltage source converter: (a) circuit structure; (b) voltage waveforms.

The three defining parameters of a sinusoidal waveform are the frequency (F), magnitude (M) and phase angle (δ), and the firing pattern used to generate the desired fundamental component AC output is related to these three parameters. Moreover, the unwanted harmonic components generated by the discontinuous switching operation of the voltage source converter have to be reduced to acceptable levels.

The firing pattern is a virtual image of the voltage source converter AC output and, thus, to eliminate the DC component in the AC output the firing pattern must be an odd symmetry function $[f(t - t_0) = -f(-t + t_0)]$ referenced to a zero crossing point (t_0); to eliminate the even harmonics the firing pattern has to be a half-wave symmetry function $[f(t) = -f(1/(2F) + t)]$.

The fundamental magnitude of the voltage source converter AC output voltage is related to the DC voltage and the half-cycle average area of the firing patterns. For the two-level voltage source converter, the switching instants and the individual durations of the two levels determine the fundamental and harmonic components. The technique used for fundamental magnitude control (by adjusting the individual level duration) is referred to as pulse width modulation (PWM), when the firing pattern has more than one pulse in every fundamental half cycle. If there is only one pulse in every fundamental half-cycle, the switching strategy is referred to as fundamental switching.

For the two-level voltage source converter configuration the fundamental switching pattern cannot control the AC magnitude of the fundamental component; in this case the fundamental magnitude control has to be achieved by regulating the DC voltage within a limited margin. In contrast, if the DC voltage is sufficiently high the PWM firing pattern has the freedom to control frequency, phase angle and magnitude.

A critical factor in the choice of the modulation principle is the ratio of the modulation frequency to the output frequency

$$p = f_p/f$$

This factor determines the harmonic spectrum for a certain degree of control and for a given modulation pattern. A sufficiently large value of this ratio will reduce all the low-order harmonics to be within the specified limits. However, a high frequency ratio will cause high switching losses. Moreover the achievable voltage–time area will be reduced and with it the fundamental component at full output voltage.

Two other commonly used factors are:

- The control ratio γ, defined as the ratio of the fundamental components of the modulated (V_I) and unmodulated ($V_{1(OM)}$) waveforms

$$\gamma = \frac{V_1}{V_{1(OM)}} \tag{2.8}$$

- The utilization ratio, which is the measure of how well the modulation principle utilizes the maximum available voltage–time area at full output voltage,

$$k = \frac{V_{1(MAX)}}{V_{1(OM)}} \tag{2.9}$$

The switching pattern start time (or phase angle related to a reference signal) is an important controllable variable for active power control as described earlier in the chapter. If the converter is connected to an active AC system, the frequency of the firing pattern is synchronous with that of the AC power system; if the converter is connected to a passive system the firing pattern frequency has to be controlled independently.

The firing patterns for a three-phase system can be categorized into symmetrical and independent phase patterns. The former takes the three phases as a whole and, thus, the waveforms of the three phases are always in symmetry; in the independent pattern each individual phase is independently controlled and, thus, the three-phase waveforms can be either symmetrical or asymmetrical.

As the self-commutated converter functions as an electrical power controller in real time, the converter firing pattern parameters (F, M, δ) have to be changed in real time to control the active and reactive power. This is an essential characteristic of the converter to achieve fast dynamic response.

The firing pattern can be implemented by an electronic hardware circuit unit or a software package run on a digital processor to generate the required gate signals. The hardware circuit design and the software programming based on the specific firing pattern should be introduced in electronic circuit design and programming techniques. In the following section several well developed firing pattern techniques are introduced for the two-level voltage source converter configuration.

2.5.2 Selected harmonic cancellation

To control the fundamental and harmonic voltages simultaneously, the phase potentials must be reversed a number of times during each half-cycle at predetermined angles of the square wave.

Generally, at any fundamental switching frequency, each chop (or reversal) provides one degree of freedom, which permits either cancelling a harmonic component or controlling the magnitude of the fundamental voltage [6, 7]. Thus for m chops per half-cycle one chop must be utilized to control the fundamental amplitude and so $m - 1$ degrees of freedom remain. The $m - 1$ degrees of freedom may be utilized to eliminate completely $m - 1$ specified low-order harmonics or to minimize the total harmonic distortion.

By way of example, Figure 2.9 shows three waveforms with different numbers of voltage reversals. The notches are placed symmetrically about the centre line of the half cycle. With the reversal angles shown $(\alpha_1, \alpha_2, \alpha_3, \ldots \alpha_p)$ the following *rms* value of the nth voltage results:

$$V_n = \frac{1}{\sqrt{2}}\frac{4}{\pi}\frac{V_d}{2}\left[-\int_0^{\alpha_1}\sin(n\alpha)d\alpha + \int_{\alpha_1}^{\alpha_2}\sin(n\alpha)d\alpha - \int_{\alpha_2}^{\alpha_3}\sin(n\alpha)d\alpha \ldots + \int_{\alpha_p}^{90}\sin(n\alpha)d\alpha\right]$$

$$= \frac{\sqrt{2}}{\pi n}V_d\left\{[\cos(n\alpha)]_0^{\alpha_1} - [\cos(n\alpha)]_{\alpha_1}^{\alpha_2} + [\cos(n\alpha)]_{\alpha_2}^{\alpha_3} \ldots - [\cos(n\alpha]_{\alpha_p}^{90}\right\}$$

$$= \frac{\sqrt{2}}{\pi n}V_d\{2[\cos(n\alpha_1) - \cos(n\alpha_2) + \cos(n\alpha_3 + \ldots] - 1\} \tag{2.10}$$

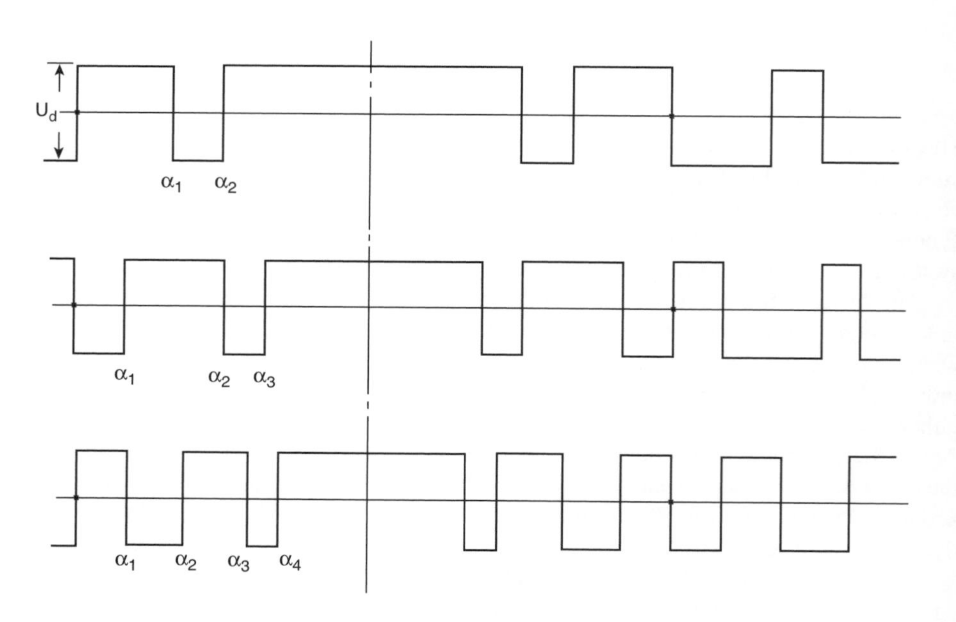

Figure 2.9 Voltage reversals.

For instance, to control the fundamental component and eliminate the fifth and seventh harmonics will need three values of α, which can be determined from the equations:

$$\frac{\pi V_1}{\sqrt{2}V_d} + 1 = 2(\cos\alpha_1 - \cos\alpha_2 + \cos\alpha_3) \tag{2.11a}$$

$$1 = 2[\cos(5\alpha_1) - \cos(5\alpha_2) + \cos(5\alpha_3)] \tag{2.11b}$$

$$1 = 2[\cos(7\alpha_1) - \cos(7\alpha_2) + \cos(7\alpha_3)] \tag{2.11c}$$

The remaining harmonic orders, i.e. 11th, 13th, 17th, etc. can then be calculated from

$$V_n = \frac{\sqrt{2}}{\pi n} V_d\{2[\cos(n\alpha_1) - \cos(n\alpha_2) + \cos(n\alpha_3)]\} \tag{2.12}$$

where $\alpha_1, \alpha_2, \alpha_3$ are the values obtained from equations 2.11. The results of this analysis are plotted in Figure 2.10.

At any fundamental frequency, elimination of the lower-order harmonics from the phase waveforms will cause the portion of the *rms* which was provided by the eliminated harmonics to be spread over the remaining harmonic magnitudes. However, the integrating filter characteristic of the system impedance is more effective in reducing the current harmonics at higher orders. The complete elimination of selected low-order harmonics will increase the magnitudes of the first uncancelled harmonics.

An alternative solution to selective harmonic cancellation is the minimization of the total harmonic content, or of a range of selected harmonics [8, 9]. These techniques, however, do not respond quickly to transient conditions, because the pulses do not occur at fixed intervals and closed loop control is implemented cycle by cycle.

Another possible solution is the use of a hysteresis band modulator, which calculates the error between the specified output and the measured output; the state of the switches is then changed in response to the magnitude of the error. This closed loop technique provides a faster transient response. However, the variable nature of the switching period produces a continuous spread of the output spectra which includes sub-harmonics and is thus impractical for low switching frequencies.

2.5.3 Carrier-based sinusoidal PWM

In this case the reversal instants, instead of being fixed, are determined by the intersections of a sinusoidal reference voltage of fundamental frequency and an amplitude (or modulation index) M, with a saw-tooth modulating waveform of a (carrier) frequency pf (where p is the frequency ratio) [10]. The dominant harmonic of this alternative is that of order p.

For low-frequency ratios, p should be an integer, to ensure that the two intersecting waves are synchronous, thus avoiding discontinuities and fluctuations. The basic method controls the line-to-line voltage from zero to full voltage by increasing the magnitude of the saw-tooth or the sine-wave signal, with little regard to the harmonics generated; for instance the output voltage waveform of Figure 2.11 contains even harmonics.

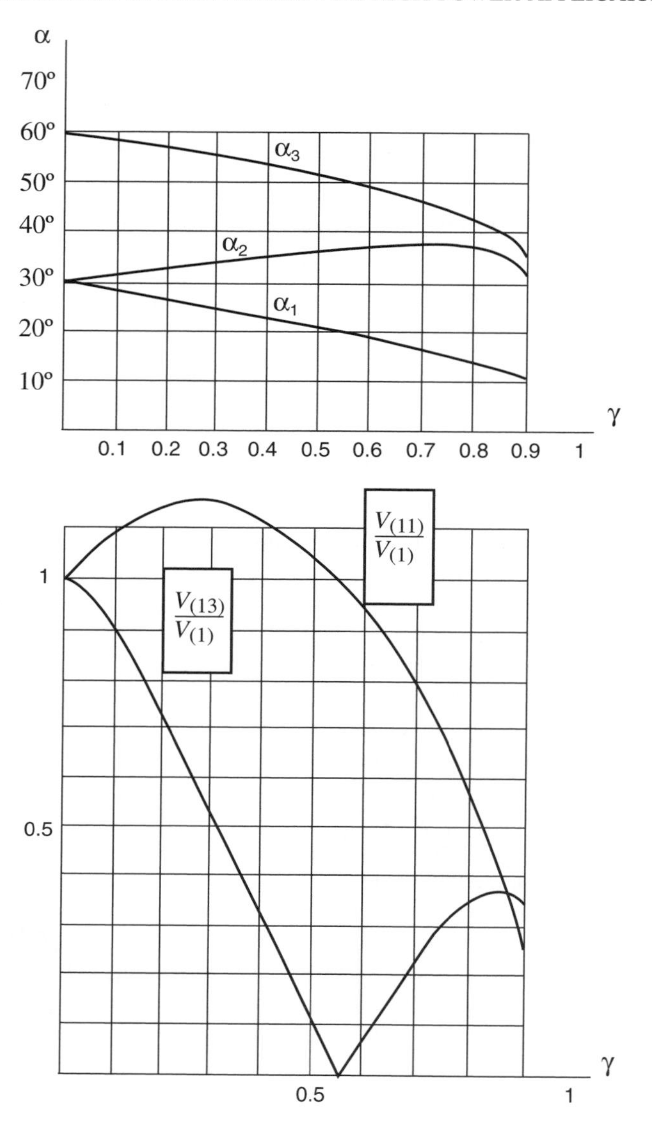

Figure 2.10 Elimination of the fifth and seventh harmonics by means of voltage reversals.

The carrier frequency should preferably be an odd multiple of three ($n = 3, 9, 15 \ldots$). This alternative, shown in Figure 2.12, provides half and quarter wave symmetries, which eliminates the even harmonics from the carrier spectrum and allows symmetrical three-phase voltages to be generated from a three-phase sine-wave set and one saw-tooth waveform.

For high values of p (i.e. over 20) the value of p can be increased continuously as the output frequency decreases. This means that the fixed triangular wave will be compared with a sinusoid of variable frequency and amplitude.

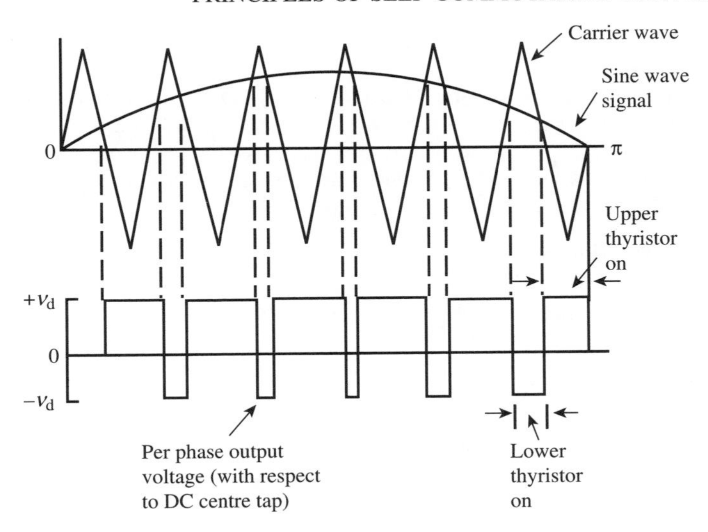

Figure 2.11 Principle of PWM DC–AC conversion.

A computer simulation of the harmonic content will first find the values of the intersecting points $(\alpha_1, \alpha_2, \alpha_3, \ldots)$ from the equations

$$M \sin(\alpha_i) = (-1)^i \left(\frac{2p}{\pi}\alpha_i - 2i\right) \quad \text{for} \quad 1 \leq i \leq \frac{p-1}{2} \tag{2.13}$$

With those values the *rms* content of the harmonics can be calculated from equation (2.10).

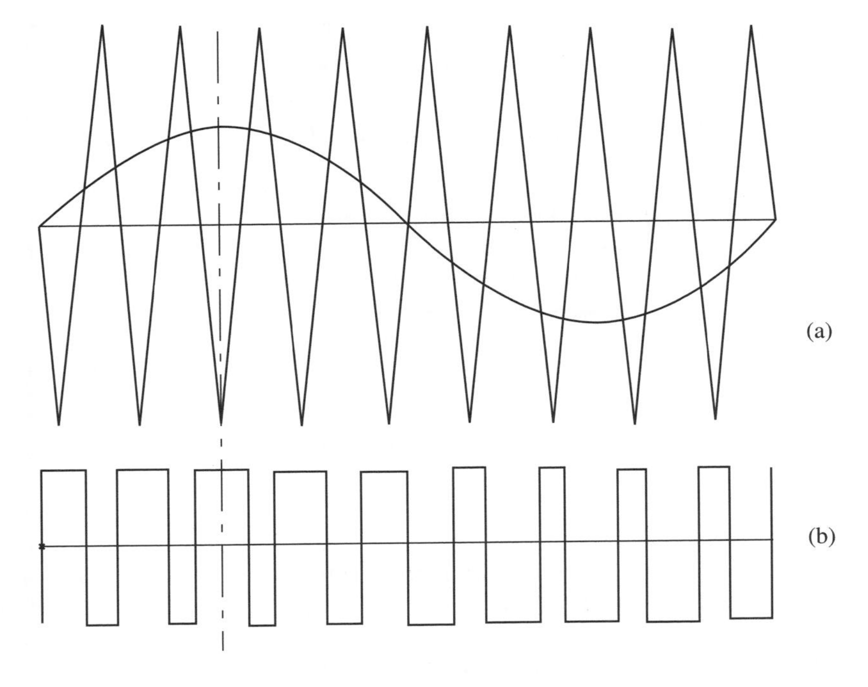

Figure 2.12 Sinusoidal PWM.

The following comments apply when $M<1$ and p is an odd integer greater than 6:

- Even harmonics, odd harmonics of orders $n < (p-2)$ and some higher harmonics are eliminated.
- Harmonics of an order equal to p or in its neighbourhood, as well as the multiples of p, are amplified. (Figure 2.13 shows these effects.)
- The ideal utilization ratio is $k_i = \frac{\pi}{4}$.

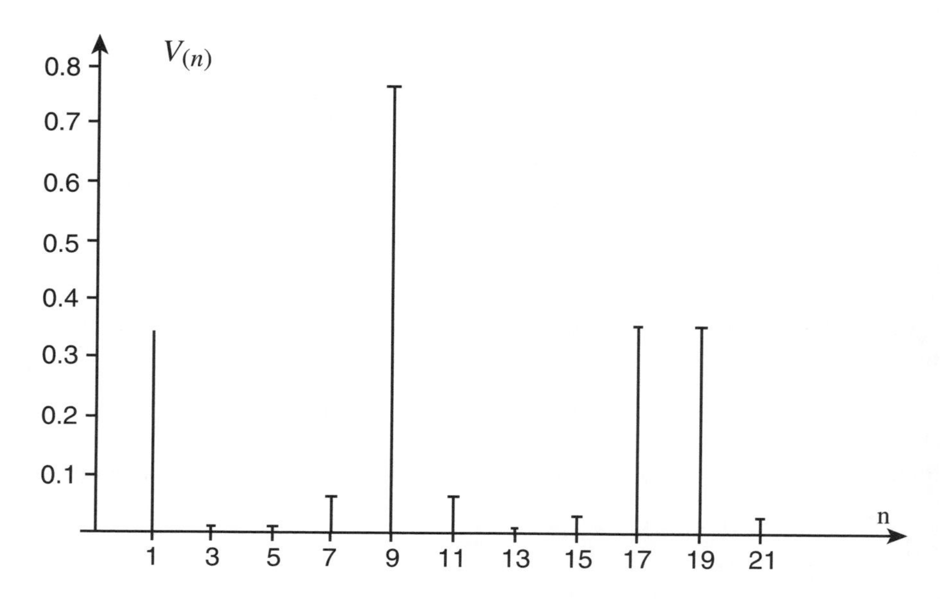

Figure 2.13 Example of sinusoidal PWM for $p=9$ and $M=0.5$.

2.6 Space-vector PWM pattern

For the three-phase two-level voltage source converter shown in Figure 2.6 the three-phase pole-switching functions (S_a, S_b, S_c) are defined as:

$S_i = 0$, when the upper switch is off and the lower on, the AC terminal potential being $-V_d/2$,

$S_i = 1$, when the upper switch is on and the lower off, the ac terminal potential being $+V_d/2$,

where $i = a, b, c$.

Based on the complementary switching rule, there are eight switch states $\vec{S}_i = (S_a \quad S_b \quad S_c)$, $i = 0, 1, \ldots, 7$ (as shown in Figure 2.14(a)). The three-phase AC output voltages of the voltage source converter are determined by these eight switch states. Eight voltage vectors, $\vec{U}_0 = [000], \ldots, \vec{U}_7 = [111]$ are defined, corresponding to the switch states $\vec{S}_0, \ldots, \vec{S}_7$,

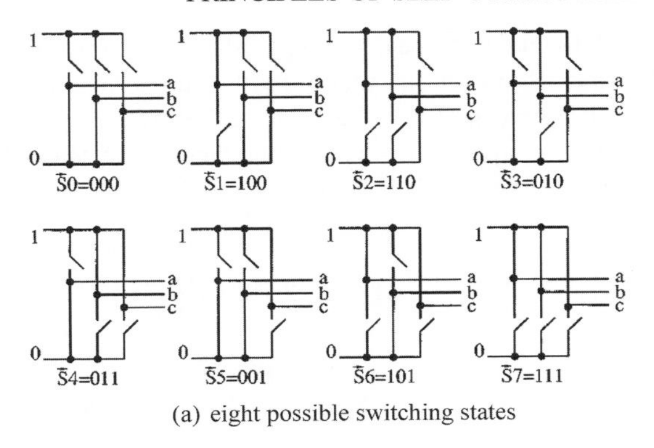

(a) eight possible switching states

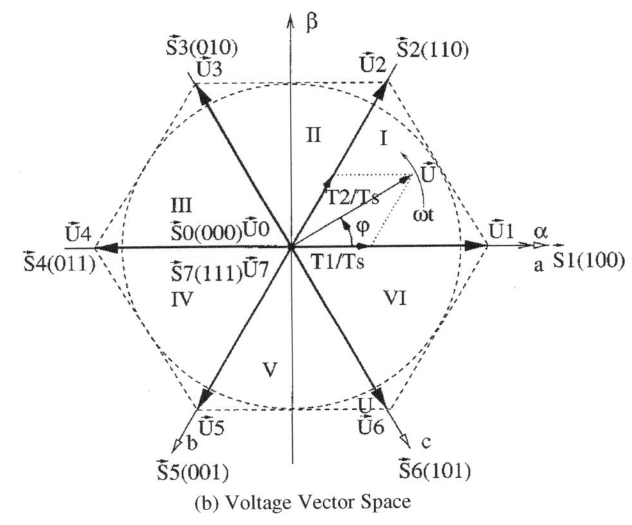

(b) Voltage Vector Space

Figure 2.14 Switching states and voltage vector space: (a) eight possible switching states; (b) voltage vector space.

respectively. The lengths of vectors $\vec{U}_1, \ldots, \vec{U}_6$are unity (i.e. $V_d/2$) and the lengths of $\vec{U}_0$ and $\vec{U}_7$ are zero, and these eight vectors form the voltage–vector space displayed in Figure 2.14(b). The voltage–vector space is divided by these vectors into six sectors.

In the vector space the following operation rules are obeyed:

$$\vec{U}_1 = -\vec{U}_4 \quad \vec{U}_2 = -\vec{U}_5 \quad \vec{U}_3 = -\vec{U}_6 \quad \vec{U}_0 = \vec{U}_7 = \vec{0} \quad \vec{U}_1 + \vec{U}_3 + \vec{U}_5 = \vec{0} \tag{2.14}$$

In a sampling interval, the average output voltage vector can be written as

$$\vec{U}(t) = \frac{t_0}{T_s}\vec{U}_0 + \frac{t_1}{T_s}\vec{U}_1 + \ldots + \frac{t_7}{T_s}\vec{U}_7 \tag{2.15}$$

where $t_0, t_1, \ldots, t_7 \geq 0$ and $t_0 + t_1 + \ldots + t_7 = T_s$ are the turn-on time of the vectors $\vec{U}_0, \vec{U}_1, \ldots, \vec{U}_7$, and T_s is the switching cycle time.

According to equations 2.14 and 2.15, the decomposition of $\vec{U}(t)$ into $\vec{U}_0, \vec{U}_1, \ldots, \vec{U}_7$, can be done in many possible ways. However, in order to reduce the number of switching actions and make full use of active turn-on time for space vectors, the vector $\vec{U}(t)$ is commonly split into the two possible and nearest adjacent voltage vectors and zero vectors $\vec{U}_0$ and $\vec{U}_7$ in an arbitrary sector. For example, in sector I, in a switching cycle interval, vector $\vec{U}(t)$ can be expressed as

$$\vec{U}(t) = \frac{t_1}{T_s}\vec{U}_1 + \frac{t_2}{T_s}\vec{U}_2 + \frac{t_0}{T_s}\vec{U}_0 + \frac{t_7}{T_s}\vec{U}_7 \tag{2.16}$$

where $T_s, t_1, t_2 > 0$, and $T_s - t_1 - t_2 = t_0 + t_7 \geq 0$, $t_0, t_7 \geq 0$.

Let the average amplitude of $\vec{U}(t)$ be $m \times V_d/2$ for controlling the fundamental component, the following relations are then found:

$$\frac{m}{\sin(120^o)} = \frac{t_1}{T_s}\frac{1}{\sin(60^o - \phi)} = \frac{t_2}{T_s}\frac{1}{\sin(\phi)} \tag{2.17}$$

Thus, in sector I the on intervals of the switching states S_1 and S_2 are determined by:

$$\begin{aligned} \frac{t_1}{T_s} &= \frac{2}{\sqrt{3}} m \sin(60^\circ - \phi) \\ \frac{t_2}{T_s} &= \frac{2}{\sqrt{3}} m \sin(\phi) \\ t_0 + t_7 &= T_s - t_1 - t_2 \end{aligned} \tag{2.18}$$

for $0^\circ \leq \phi \leq 60^\circ$.

Similarly in sector II the on intervals of the switching states S_2 and S_3 are determined by:

$$\begin{aligned} \frac{t_2}{T_s} &= \frac{2}{\sqrt{3}} m \sin(120^\circ - \phi) \\ \frac{t_3}{T_s} &= \frac{2}{\sqrt{3}} m \sin(\phi - 60^\circ) \\ t_0 + t_7 &= T_s - t_2 - t_3 \end{aligned} \tag{2.19}$$

for $60^\circ \leq \phi \leq 120^\circ$

In sector III the on intervals of the switching states S_3 and S_4 are determined by:

$$\begin{aligned} \frac{t_3}{T_s} &= \frac{2}{\sqrt{3}} m \sin(180^\circ - \phi) \\ \frac{t_4}{T_s} &= \frac{2}{\sqrt{3}} m \sin(\phi - 120^\circ) \\ t_0 + t_7 &= T_s - t_3 - t_4 \end{aligned} \tag{2.20}$$

for $120^\circ \leq \phi \leq 180^\circ$

In sector IV the on intervals of the switching states S_4 and S_5 are determined by:

$$\frac{t_4}{T_s} = \frac{2}{\sqrt{3}} m \sin(240° - \phi)$$
$$\frac{t_5}{T_s} = \frac{2}{\sqrt{3}} m \sin(\phi - 180°) \tag{2.21}$$
$$t_0 + t_7 = T_s - t_4 - t_5$$

for $180° \leq \phi \leq 240°$

In sector V the on intervals of the switching states S_5 and S_6 are determined by:

$$\frac{t_5}{T_s} = \frac{2}{\sqrt{3}} m \sin(300° - \phi)$$
$$\frac{t_6}{T_s} = \frac{2}{\sqrt{3}} m \sin(\phi - 240°) \tag{2.22}$$
$$t_0 + t_7 = T_s - t_5 - t_6$$

for $240° \leq \phi \leq 300°$

In sector VI the on intervals of the switching states S_6 and S_1 are determined by:

$$\frac{t_6}{T_s} = \frac{2}{\sqrt{3}} m \sin(360° - \phi)$$
$$\frac{t_1}{T_s} = \frac{2}{\sqrt{3}} m \sin(\phi - 300°) \tag{2.23}$$
$$t_0 + t_7 = T_s - t_6 - t_1$$

for $300° \leq \phi \leq 360°$

Once the control input variables F, M ($M = m \times V_d/2$) and δ are determined in real time $kT_s \leq t < (k+1)T_s$, and the switching cycle time T_s is chosen, the switching vectors for the next switching cycle can be determined by selecting a value of j that places ϕ in the region:

$$0 \leq \phi = \frac{360°(2k+1)FT_s}{2} + \delta - 360° \times j < 360° \tag{2.24}$$

and then using one of equations 2.18 to 2.23 to determine the right switching vectors and calculate the on intervals of the chosen vectors.

Equations 2.18 to 2.23 indicate that there is a degree of freedom in the choice of the switching vectors $\vec{S}_0$ and $\vec{S}_7$ within one switching cycle (i.e. either only one or both of them).

Unlike the carrier-based sinusoidal PWM, there are no separate modulation signals in each of the three phases for the space vector PWM pattern. Instead, a voltage vector is processed as a whole for the three phases. This forces the three phase waveforms to be in symmetry, but without the freedom of independent control of the individual phases.

Figures 2.15 and 2.16 show the three-phase voltage waveforms and their spectra for switching patterns 1 and 2 respectively. The space vector switching pattern 1 uses the

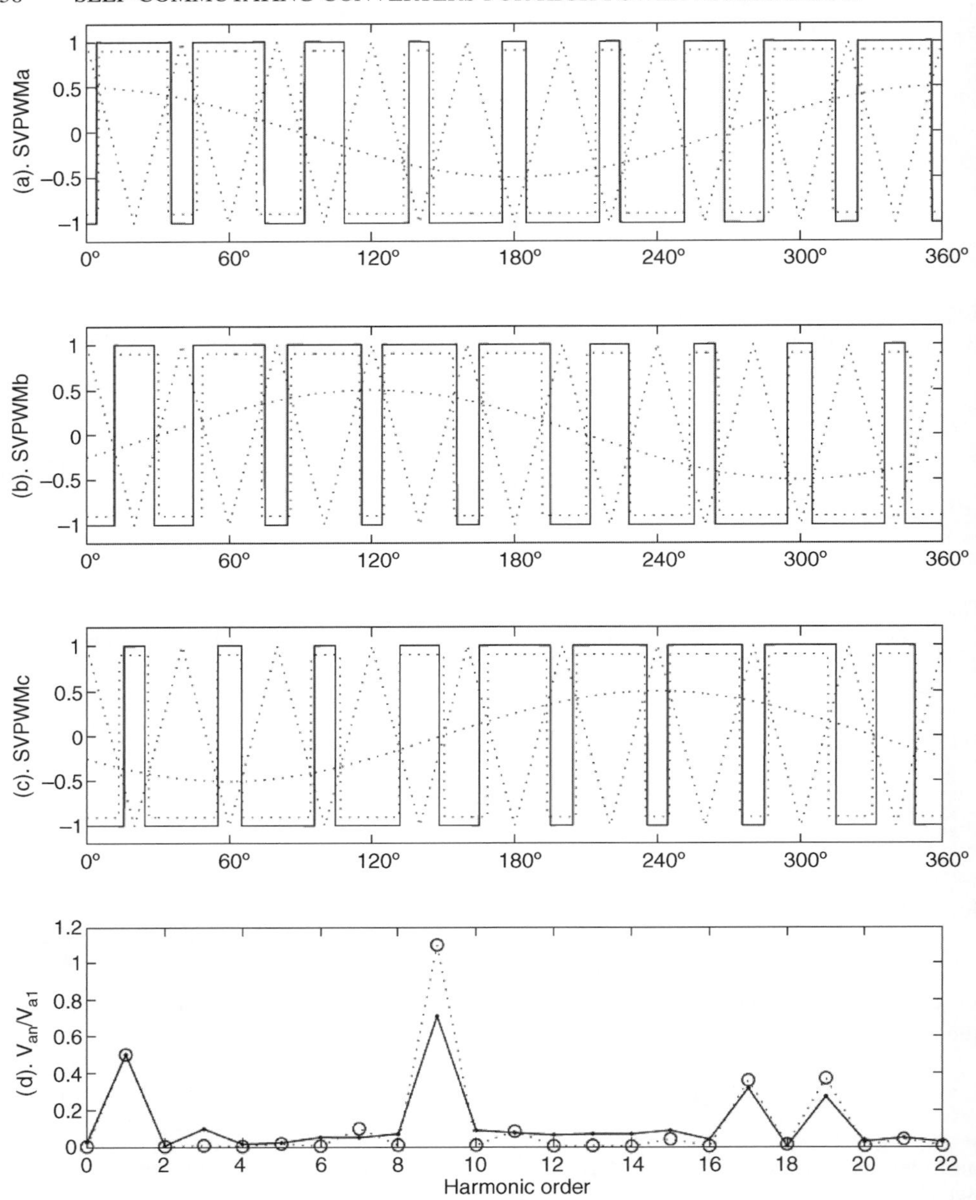

Figure 2.15 Voltage waveforms and spectrum of switching pattern 1.

zero switching vectors $\vec{S}_0$ and $\vec{S}_7$ in each switching cycle. Once the switching vector durations (t_0, t_7, t_i and t_{i+1}) are determined, the switching vectors are arranged as follows:

$$||S_0|S_i|S_{i+1}|S_7|S_{i+1}|S_i|S_0||$$

$$\left|\left|\leftarrow\frac{t_0}{2}\rightarrow\right|\leftarrow\frac{t_i}{2}\rightarrow\left|\leftarrow\frac{t_{i+1}}{2}\rightarrow\right|\leftarrow t_7\rightarrow\left|\leftarrow\frac{t_{i+1}}{2}\rightarrow\right|\leftarrow\frac{t_i}{2}\rightarrow\left|\leftarrow\frac{t_0}{2}\rightarrow\right|\right|$$

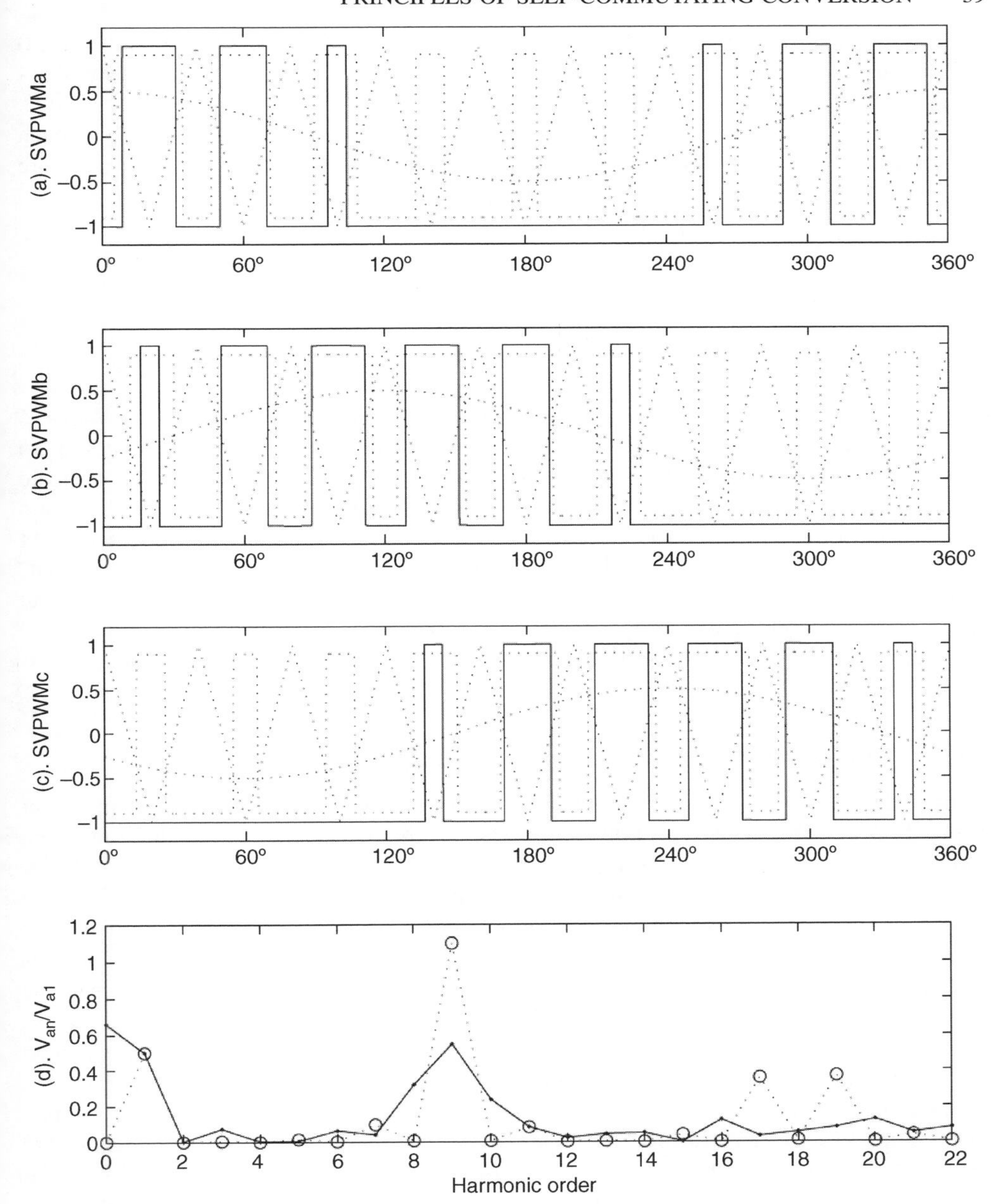

Figure 2.16 Voltage waveforms and spectrum of switching pattern 2.

The space vector PWM (SVPWM) pattern 1 has the same number of switching actions in a fundamental cycle, compared to the triangle carrier-based sinusoidal PWM (SPWM), if the SVPWM switching cycle time is equal to the SVPWM carrier cycle time. The waveforms shown in Figure 2.15(a), (b) and (c) indicate that there are not significant differences between the two switching strategies SVPWM and SPWM (in dashed lines). The spectra in Figure 2.15(d) indicate that the SVPWM waveforms include more triplen harmonics, although they will not appear in the line voltage waveforms, as they are all in phase.

The space vector switching pattern 2 only uses the zero switching vectors $\vec{S}_0$ in each switching cycle (i.e. $t_7 \equiv 0$). Once the switching vector durations (i.e. t_0, t_i and t_{i+1}) are determined, the switching vectors are arranged as follows:

$$|| \ \vec{S}_0|\vec{S}_i|\vec{S}_{i+1}|\vec{S}_{i+1}|\vec{S}_i|\vec{S}_0||$$

$$\left\| \leftarrow \frac{t_0}{2} \rightarrow \middle| \leftarrow \frac{t_i}{2} \rightarrow \middle| \leftarrow \frac{t_{i+1}}{2} \rightarrow \middle| \leftarrow \frac{t_{i+1}}{2} \rightarrow \middle| \leftarrow \frac{t_i}{2} \rightarrow \middle| \leftarrow \frac{t_0}{2} \rightarrow \right\|$$

The SVPWM pattern 2 has less switching actions in a fundamental cycle, as compared with the triangle carrier-based SPWM, if the SVPWM switching cycle time is equal to the SVPWM carrier cycle time. The waveforms shown in Figure 2.16(a), (b) and (c) indicate that there are significant differences between the two switching strategies SVPWM and SPWM (in dashed lines). The spectra in Figure 2.16(d) indicate that the SVPWM waveforms include not only triplen harmonics but also even harmonics and a DC component, therefore these voltage waveforms are only suitable to supply symmetrical loads. The substantial advantage of the SVPWM pattern 2 is the reduced number of the switching actions (66.7% of the SVPWM pattern 1) with a corresponding switching loss reduction.

2.6.1 Comparison between the switching patterns

Semiconductor converters are used for electrical power control in a wide range of power ratings (kW to GW); the power rating dominates the topological structures of the converters, and also determines the gate firing strategies to be used. For high-power applications, efficiency is a high priority and, thus, the use of a low switching frequency is preferred to reduce the switching losses. Also, high waveform quality (power conversion performance) is required to keep the filter cost and losses low. The fundamental component controllability is also related to the gate switching patterns. A compromise has to be made between the switching frequency, waveform quality and controllability.

The fundamental switching strategy uses the lowest switching frequency to generate square waveforms with odd order harmonics; the six-pulse voltage source converter shown in Figure 2.6 and its fundamental switching pattern shown in Figure 2.7 are a typical example. The multipulse converters use more than one six-pulse voltage source converter to increase the power rating and improve the waveform quality by using the fundamental switching pattern for each six-pulse voltage source converter; the 12-pulse configuration in Figure 2.8 is commonly used in many applications.

Under the fundamental switching frequency gate control, the voltage source converter AC output voltage amplitude cannot be regulated to change within a wide range, as the PWM gate control does. This is not a serious problem when the converter is connected to a power grid, because under normal operating conditions the grid bus voltage is maintained within a narrow margin. Therefore, the voltage source converter AC output voltage magnitude is only required to vary within a small range, which can be achieved by controlling the DC voltage; however, during a serious fault condition the grid bus voltage could change substantially, a situation that the fundamental switching voltage source converter, without a large range of magnitude controllability, cannot handle.

For high-power applications the converter will require more switching devices and will normally consist of more than two fundamental switching voltage source converter units; this provides the opportunity to arrange their firing control to achieve a magnitude controllable AC output voltage; such a technique will be used in subsequent chapters.

The switching pattern based on the method of selected harmonic cancellation uses lower switching frequency to generate the controllable fundamental component and without the unexpected lower-order harmonics. The complete elimination of selected low-order harmonics will increase the magnitudes of the first uncancelled harmonics. However, the integrating filter characteristic of the system impedance is more effective in reducing the current harmonics at higher orders.

A concern with the derivation of the switching pattern is the time consumed to solve the sinusoidal function equations, as these need to be solved in real time for any change in the fundamental magnitude. This problem can be overcome with the use of the fast and powerful digital signal processor (DSP).

The switching pattern of the carrier-based SPWM uses the independent phase waveforms to be modulated and a common carrier waveform for all the phases to generate the controllable fundamental components for each phase independently. While the frequency is common to the three phases, the magnitude and phase angle of each phase can be set independently from the others so as to improve the control flexibility of the voltage and current. These features are available to converters connected to the power grids, where the equivalent source voltages and impedances of the three phases are symmetrical or with negligible differences for normal conditions, but there is serious asymmetry for faulty conditions.

For three-line output converter systems the zero sequence harmonics can be added to the fundamental waveform to improve the DC voltage utilization [11].

As the independent phase waveforms to be modulated determine the converter system response performance, to obtain a quick response the delay time involved in the derivation of the waveform variations along the fundamental frequency, magnitude and phase angle orders has to be minimized.

The space–vector PWM patterns are constructed via a voltage vector processed as a whole for the three phases. Such processing is always in symmetry, but lacks the freedom of independent control of the individual phases.

To minimize the number of switching actions in each switching cycle, the two zero-length switching vectors are inserted at the beginning and stopping ends and the middle of a switching cycle.

The SVPWM pattern 1 has the same number of switching actions in a fundamental cycle, as the triangle carrier-based SPWM. There are not significant differences between the SVPWM and SPWM. However, the SVPWM pattern 1 includes more triplen harmonics than the SPWM pattern.

The SVPWM pattern 2 has fewer switching actions in a fundamental cycle, as compared with the SPWM pattern. The SVPWM waveforms include not only triplen harmonics but also some low even harmonics and a DC component, therefore these voltage waveforms are only suitable for the supply to symmetrical loads. The substantial advantage of the SVPWM pattern 2 is the reduction in the number of switching actions.

2.7 Basic current source conversion operation

Current source conversion (CSC) requires a large inductance on the DC side to make the DC current well defined and slow to change. The AC side voltage is then the variable directly controlled by the conversion process. Since the AC system has significant line or load inductance, line-to-line capacitors must be placed on the AC side of the converter. The switches must block voltages of both polarities, but they are only required to conduct current in one direction. This naturally suits symmetrical devices of the thyristor type and, therefore, CSC constitutes the basis of the conventional line-commutated conversion process.

As the switches in CSC only carry the current unidirectionally, there is no need for the anti-parallel diode. In the CSC configuration shown in Figure 2.17(a), the roles of the DC capacitor and AC inductance are interchanged (with respect to the voltage source converter alternative). In VSC the DC capacitor facilitates the rapid transfer of current from the outgoing valve to its opposite in the same phase leg, irrespective of the direction of the AC current; the capacitor must therefore be large enough to handle alternate charging and discharging with little change in DC voltage. In the self-commutating CSC configurations, although the valves can also be turned off at will, they also require an alternate path, otherwise the current turn-off will have to dissipate a large amount of energy in an inductive circuit. Therefore AC capacitors need to be connected between the phases to facilitate the rapid transfer of current.

The current source converter injects an AC current into the system or load with the necessary AC voltage behind to force the current injection; therefore, the DC current source must be capable of driving such a current.

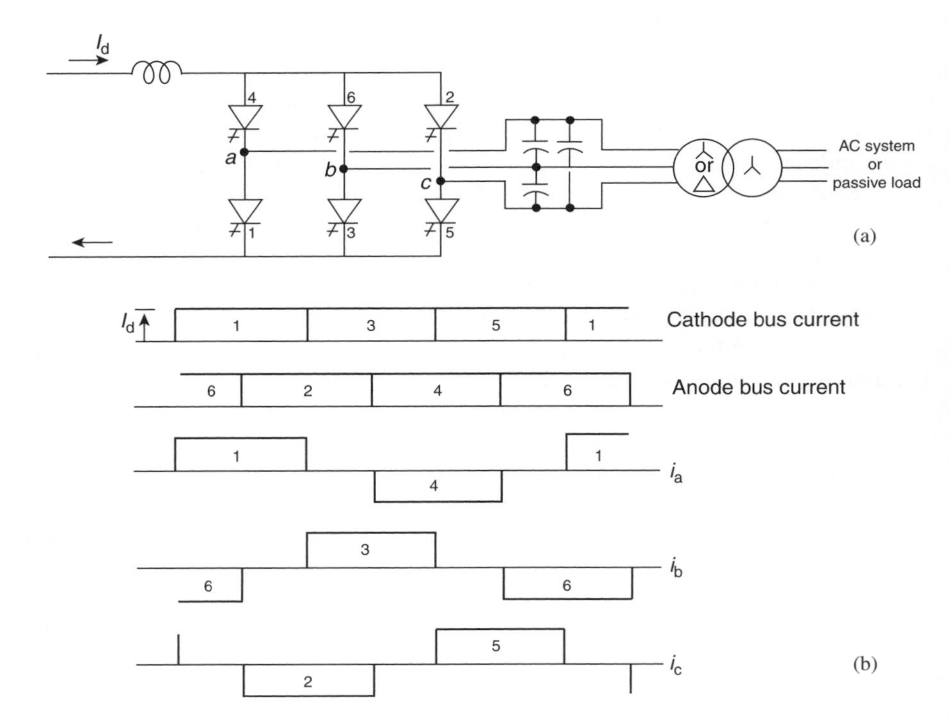

Figure 2.17 Current source converter: (a) circuit diagram; (b) current waveforms.

Similarly to the voltage source converter, the current source converter is capable of operating with leading power factor and the self-commutation of the valves permits the converter to operate as an inverter feeding power into a passive, as well as an active, load or system.

The introduction of thyristor-type self-commutating switches practically eliminates the main disadvantages of LCC and makes the robust thyristor-based conversion an alternative worth considering for HVDC application.

2.7.1 Analysis of the CSC waveforms

The valve conduction sequence is the same as for the line-commutated conversion process, but without the commutation overlap. The three-phase injected currents are as shown in Figure 2.17(b).

In the time domain the phase current is expressed by

$$I = (1/\pi)2\sqrt{3}\, I_d[\cos(\omega t) - (1/5)\cos(5\omega t) + (1/7)\cos(7\omega t) \ldots] \tag{2.25}$$

The harmonic currents are of orders $6k \pm 1$ (for k=1,2 ...) and their *rms* values are obtained from the expression:

$$I_n = (1/\mathrm{n})(\sqrt{6}/\pi)I_d \tag{2.26}$$

Thus the *rms* value of the fundamental component is

$$I_1 = (\sqrt{6}/\pi)\, I_d = 0.78\, I_d \tag{2.27}$$

The harmonic voltage content on the DC side is obtained from the expressions of the line-commutated conversion but in the absence of commutation overlap, i.e. with $u = 0$.

As a percentage of the maximum average DC voltage ($V_{c0} = 3(\sqrt{2})V_c/\pi$)these are:

$$\sqrt{2}/(n^2 - 1) \approx \sqrt{2}/n^2 \quad \text{for} \quad \alpha = 0$$

and

$$\sqrt{2}n/(n^2 - 1) \approx \sqrt{2}/n \quad \text{for} \quad \alpha = \pi/2$$

So the higher harmonics increase faster with the firing angle.

2.8 Summary

From the operational viewpoint the use of PWM adds considerable flexibility to the voltage source converter. The modulation process develops independent fundamental and harmonic frequency voltage control from a constant DC source; this in turn provides four-quadrant power controllability and avoids the use of low-order harmonic filters.

With the addition of PWM, VSC is becoming a competitive technology for high-voltage applications, such as HVDC transmission for power ratings of several hundred MW.

In HVDC CSC schemes, a change in the power direction requires DC voltage reversal, a condition that makes the control of multiterminal interconnections difficult. PWM based VSC can achieve that purpose without altering the DC voltage magnitude and polarity. The ability to control the transfer of active power as well as the terminal voltage has made VSC PWM conversion (based on IGBT switching) attractive for HVDC transmission. The question is whether such a high level of flexibility is necessary and affordable for very high power ratings.

From the design viewpoint, however, the frequent and rapid switching of the AC bus voltage between two levels results in repetitive high-frequency transient stresses, which affects the design of the voltage source converter components and, unless appropriately filtered, will cause electromagnetic interference. Wound components are particularly vulnerable to repetitive, high-frequency transient stresses and the dielectrics of insulation materials causes premature ageing.

Among the advantages of increased switching frequency are simpler filtering arrangements, smaller DC ripple and smaller size of VSC substations. An additional benefit of PWM is that the size of the DC capacitor can be reduced. The size of the capacitor is decided to ensure that the voltage ripple is between 2 and 10%. Every time the valves are switched the current in the DC capacitor changes direction and this effect reduces the voltage ripple.

The main disadvantages of PWM are the higher converter losses, reduced power capability for each device and increased voltage (dv/dt) stresses. These problems can be reduced by the multilevel converters described in later chapters.

References

1. Hingorani, N.G. and Gyugyi, L. (2000) *Understanding FACTS. Concepts and Technology of Flexible AC Transmission Systems*. IEEE Press.
2. CIGRE (2004) VSC Transmission. *Document B4-WG 37*. CIGRE, Paris.
3. Fujii, T., Yoshikawa, K., Koga, T., Nishiura, A., Takahashi, Y., Kakiki, H., Ichijyou, M. and Seki, Y. (2000) 4.5kV-2000A Power Pack IGBT (Ultra High Power Flat-Packaged Pt Type RC-IGBT). *Proceedings of the 12th International Symposium on Power Semiconductor Devices and ICs*, 22–25 May 2000, pp. 33–36.
4. Kaufmann, S., Lang, T. and Chokhawala, R. (2001) Innovative Press Pack Modules for High Power IGBTs. *Proceedings of the 13th International Symposium on Power Semiconductor Devices and ICs*, 4–7 June 2001, pp. 59–62.
5. Gunturi, S., Assal, J., Schneider, D. and Eicher, S. (2003) Innovative metal system for IGBT press pack modules. *Proceedings of the 15th International Symposium on Power Semiconductor Devices and ICs*, 14–17 April 2003, pp. 110–113.
6. Patel, H.S. and Hoft, R.G. (1973) Generalised technique of harmonic elimination in voltage control in thyristor inverters. Part 1. Harmonic elimination. *IEEE Transactions on Industry Applications*, **9** (3), 310–17.
7. Patel, H.S. and Hoft, R.G. (1974) Generalised technique of harmonic elimination and voltage control in thyristor-inverters. Part 2. Voltage control techniques. *IEEE Transactions on Industry Applications*, **10** (5), 666–73.

8. Goodarzi, G.A. and Hoft, R.G. (1987) GTO Inverter Optimal PWM waveform. *Proceedings of the IEEE Industry Applications Society Annual Meeting*, **1**, 312–16.
9. Enjeti, P.N., Ziogas, P.D. and Lindsay, J.F. (1990) Programmed PWM techniques to eliminate harmonics: a critical evaluation. *IEEE Transactions on Industry Applications*, **26** (2), 302–16.
10. Holtz, J. and Beyer, B. (1992) Optimal PWM for AC servos and low cost industrial drive. *Proceedings of IEEE Industry Application Meeting*, **1** 1010–17.
11. Keliang, Z. and Wang, D. (2002) Relationships between space–vector modulation and three-phase carrier-based PWM: a comprehensive analysis. *IEEE Transactions on Industrial Electronics*, **49** (1), 186–96.

3

Multilevel Voltage Source Conversion

3.1 Introduction

As explained in the first chapter, both the transistor and thyristor types of self-commutating power switches can now be combined in series to form reliable high-voltage valves. The series-connected power switches can be fired synchronously or asynchronously (with switch voltage clamping assistance). Synchronous control, used in two-level VSC, causes static and dynamic voltage sharing problems as well as high dv/dt.

Possible alternatives to two-level conversion for high-voltage applications are the multi-pulse and multilevel topologies. Increasing the pulse number has been traditionally achieved in current source conversion by the series or parallel connection of bridges, their respective voltage waveforms being phase-shifted with respect to each other by appropriate connections of the interface transformers. Applying the multibridge concept to self-commutating voltage source conversion improves the converter output waveforms without the assistance of high-frequency switching. However, for high-pulse conversion the increased number of converter transformers required makes this solution unattractive.

A more effective alternative for high-voltage application is the multilevel concept with asynchronous firing control; this improves the dynamic voltage balancing of the valves, while the steady state voltage sharing is achieved by means of clamping devices.

Like in the case of PWM, multilevel converters can vary the phase position of the converter fundamental frequency voltage with respect to the AC system voltage waveform; however, their effect on the fundamental frequency voltage magnitude is very different. In the PWM solution the magnitude of this voltage can be varied independently from the DC voltage, whereas in the multilevel alternative the voltage magnitude is fixed by the DC voltage. The main object of multilevel conversion is to generate a good high-voltage waveform by stepping through several intermediate voltage levels, i.e. the series-connected devices are switched sequentially producing an output waveform in steps. This eliminates the low-order harmonics and reduces the dv/dt rating of the valves by forcing them to switch against a fraction of the DC voltage.

Self-Commutating Converters for High Power Applications J. Arrillaga, Y. H. Liu, N. R. Watson and N. J. Murray

In multilevel conversion the main valves are switched at the fundamental frequency, thus eliminating the need for higher-frequency components typical of PWM. However, the level number is limited by the increasing structural converter complexity with higher numbers and, thus, some PWM may still be needed when the level number is low.

Even though conventional HVDC and SVC schemes use line-commutated current source conversion (CSC), CSC has so far not been considered a viable alternative in self-commutating conversion.

VSC and CSC are dual topologies and thus the description of multilevel VSC is applicable to CSC, by considering the converter output in terms of current instead of voltage. The following multilevel VSC concepts are introduced in this chapter:

1. PWM/multibridge conversion;
2. diode clamped conversion;
3. flying capacitor conversion;
4. cascaded H-bridge conversion;
5. modular multilevel conversion.

3.2 PWM-assisted multibridge conversion [1]

Rather than increasing the frequency of the PWM pattern to reduce the harmonic content of the output voltage, a multibridge configuration can be subjected to a phase-shifted carrier. Considering n units, each carrier is shifted by T/n, where T is the period of the carrier waveform. Thus the voltage harmonic components of the individual units are shifted with respect to each other and can be designed to be cancelled when the outputs of the various bridges are added.

Each bridge of a transformer-connected multibridge converter is modulated independently and the overall output can be described by the same equations presented in the earlier sections. Figure 3.1 illustrates the operating principle with reference to a four-bridge configuration. In Figure 3.1(a), the three sine waves of frequency f_1 intersect with four triangular carriers of frequency f_c to produce twelve PWM waveforms. The four PWM control waveforms associated with phase A are shown in Figure 3.1(b) and the resultant PWM waveform for this phase is shown in Figure 3.1(c). Finally, when the latter is combined with the corresponding waveform in phase B (not shown), their difference produces a nine-pulse phase-to-phase waveform. It must be remembered, however, that each of the component bridges is still a two-level converter and, thus, the higher output pulse number should not be confused with the high-level output of the multilevel configurations discussed in later chapters.

The phase of the spectral components of the natural and uniformly sampled PWM implementations are $n\phi_1 - m\phi_c$ and $n\phi_1 - m\phi_c - (m + n\omega_1/\omega_c)\pi/2$ respectively. In both cases, however, the phase of the modulator signal term f_1 is not affected by changes in the carrier phase ϕ_c (since $m = 0$); therefore, the input signal frequency component will add regardless of the phase relationship of the carriers used for the individual bridges.

Changing the phase of the carrier signal affects the phase of the carrier terms and their sidebands (since $m \neq 0$). However, choosing ϕ_{ci} for each modulator such that $\sum m\phi_{ci} = 0$ causes all carriers and sideband terms of order m to sum to zero. Consider as an example the case

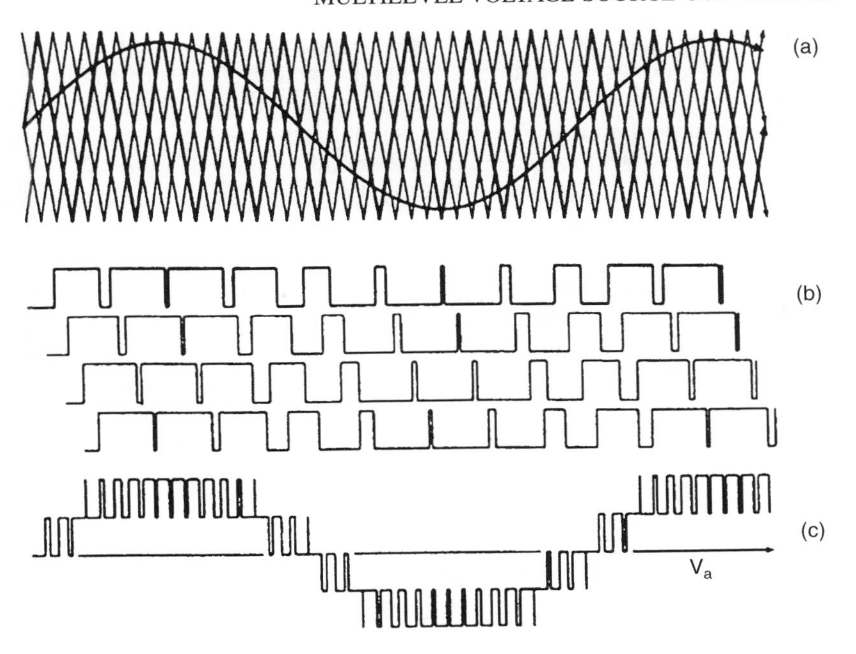

Figure 3.1 Five-level PWM waveforms: (a) triangular carriers and modulating wave; (b) individual PWM waveforms; (c) output unfiltered voltage.

of a three-phase three-bridge four-level converter. The phases of the triangular carriers of the individual phases are $-2\pi/3$, 0 and $2\pi/3$, and thus the phasor sum of the harmonic terms will be zero for $m = 1,2,4,5,7,8,\ldots$. Only the spectral components associated with $m = 0,3,6,\ldots$ will appear in the PWM output, i.e. for a switching frequency of 450 Hz, the lowest carrier and sidebands are centred around $3 \times 450 = 1350$ Hz. A reported high-power application [8] used 48 modules with 120 Hz carrier frequency to synthesize a 60 Hz sinusoid. Despite the low pulse number (N = 2) used, the first uncancelled carrier appears at $48 \times 120 = 5760$ Hz.

The PWM control strategies can, to a large extent, be developed independently of the converter topology. Thus the results discussed in this section with reference to the transformer summed multibridge converter are generally applicable to the multilevel configurations to be described in later chapters.

3.3 The diode clamping concept

The neutral clamped three-level concept was introduced by Navae [2] following the development of the GTO. However, the three-level configuration does not provide sufficient waveform quality. Complemented by PWM this configuration has been extensively used in the drives industry and more recently in several HVDC transmission schemes.

3.3.1 Three-level neutral point clamped VSC

In the two-level voltage source converter, each phase leg of the three-phase converter consists of a switch pair (which for a high-voltage application requires a number of series-connected switches); the DC side capacitor is shared by the three phase legs. The switch pairs are gated in a complementary (bipolar) way, such that the output is either connected to the positive or negative potential of the capacitor. The main drawbacks of this configuration are static and dynamic voltage-sharing problems (requiring complex balancing techniques) and the high dv/dt generated by the synchronous commutation of all the switches.

These problems are substantially reduced in the three-level neutral point clamped topology, where the converter unit AC terminals are switched between three discrete voltage levels. This scheme has the advantage of 'effectively' doubling the switching frequency as far as the output harmonics are concerned, compared to the bipolar voltage switching scheme. Also, the output voltage steps are halved with respect to those of the two-level configuration. The three-phase unit, shown in Figure 3.2 for a GTO-based converter, has an extra terminal on the DC side connected to the centre point of the equally split DC source. The switching configuration has two sets of valves in series, with their intermediate point connected to the DC supply centre tap via extra diodes. For a given valve voltage rating, however, the total DC supply voltage is doubled so that the output voltage per valve remains the same as in the two-level configuration.

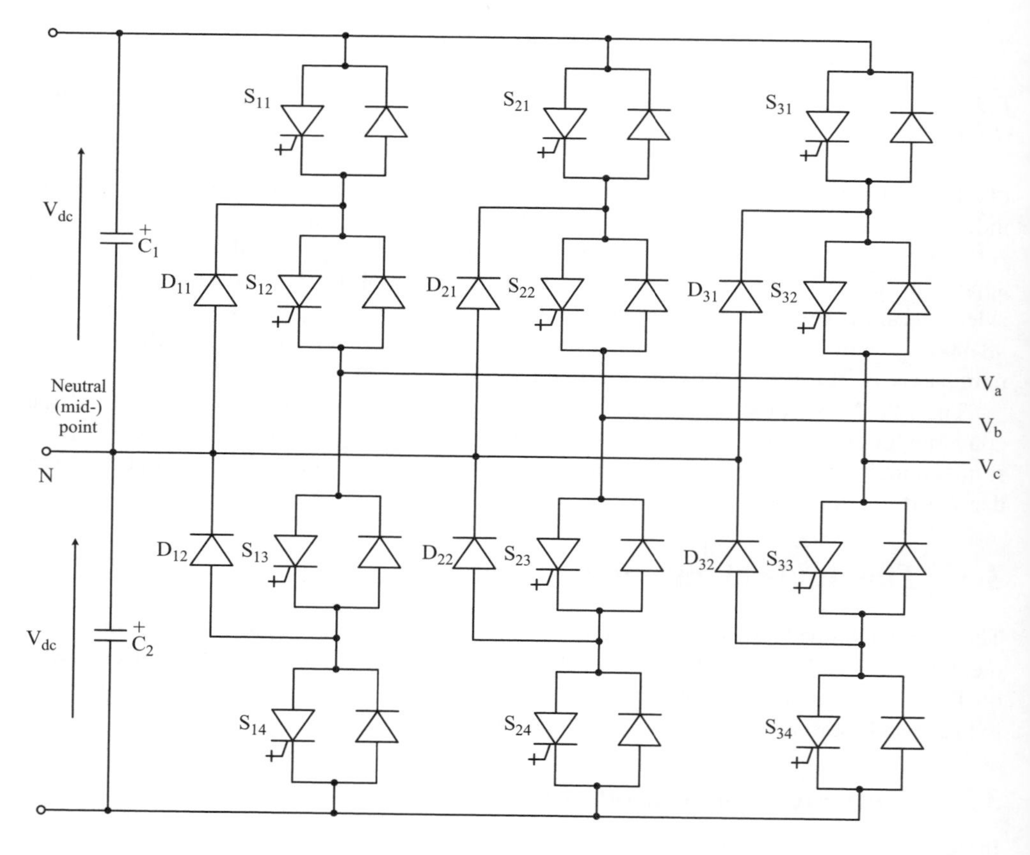

Figure 3.2 Diagram of a three-phase NPC converter.

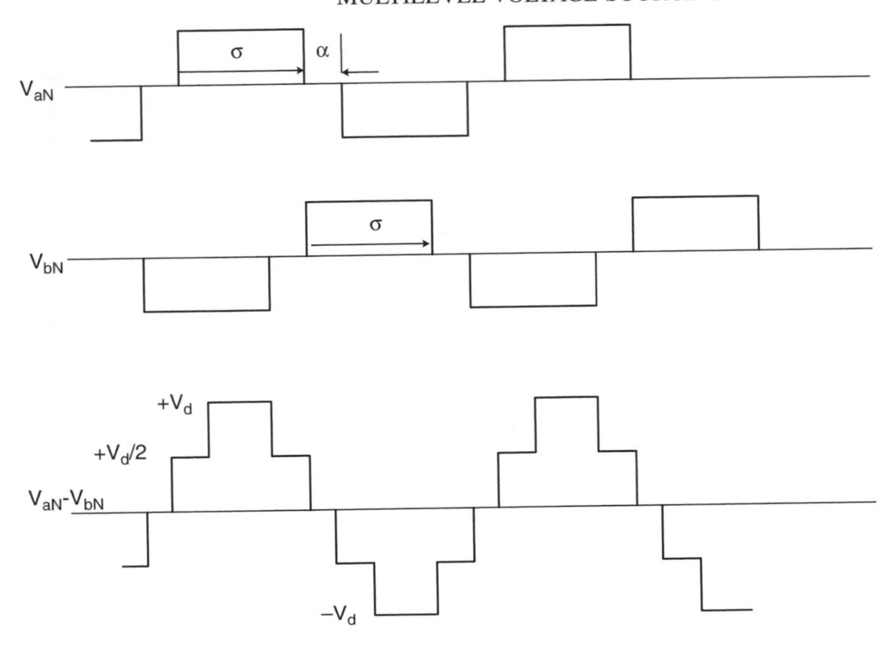

Figure 3.3 AC voltage waveforms for a three-level voltage source converter.

The phase-to-neutral and phase-to-phase output voltage waveforms are shown in Figure 3.3 with respect to the midpoint of the DC capacitor. The phase-to-neutral voltage consists of three levels, i.e. positive, negative and zero. The positive level is produced by switching on the two series-connected upper valves of the phase unit; similarly, the negative level is produced by switching the two series-connected lower valves; the zero level is produced by switching the upper and lower middle valves, thus connecting the centre tap of the DC supply to the output via the two extra diodes. In the zero voltage output region the current continues to flow via the upper middle GTO device and upper centre tap diode (when positive), or the lower middle GTO and the lower centre tap diode (when negative). This process, however, increases the flow of triplen harmonics ripple current through the midpoint of the DC supply.

The frequency components of the phase to N (the capacitor neutral or midpoint) voltage are given by:

$$v_{aN} = \frac{2V_d}{\pi}\left[\sin\left(\frac{\sigma}{2}\right)\sin\left(\omega t+\frac{\sigma}{2}\right)-\frac{1}{3}\sin\left(\frac{3\sigma}{2}\right)\sin\left(\omega t+\frac{\sigma}{2}\right)+\frac{1}{5}\sin\left(\frac{5\sigma}{2}\right)\sin\left(\omega t+\frac{\sigma}{2}\right)-\cdots\right] \tag{3.1}$$

and the *rms* value of the voltage harmonic of order *n* is

$$V_n = \frac{\sqrt{2}V_d}{\pi n}\sin\left(\frac{n\sigma}{2}\right) \tag{3.2}$$

The relative durations of the positive, negative and zero regions are a function of the control angle σ, which defines the conduction interval of the top upper and the bottom lower valves. Therefore this parameter controls the magnitude of the fundamental *rms* component of the output voltage, which according to equation 3.2 reaches a maximum of $\sqrt{2}V_d/\pi$ at $\sigma = 180°$ and becomes zero at $\sigma = 0°$. Thus an important advantage of the three-level configuration is its capability to control the magnitude of the output voltage without changing the number of valve switchings per cycle. Another advantage is that, with judicious choice of the zero voltage time (α in Figure 3.3), selected harmonic components of the output waveform can be eliminated. For example, for α equal to 30° the positive and negative intervals are 120° and 60°, respectively, and that means that the waveform is free of triplen harmonics. Figure 3.4 shows the variation of the fundamental and harmonic components of the output voltage (as a ratio of the maximum value) versus the pulse width.

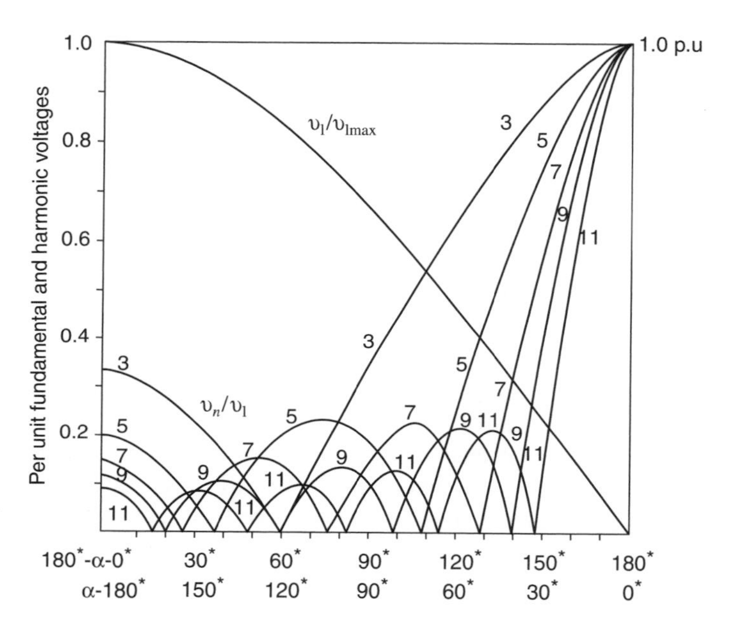

Figure 3.4 Fundamental and harmonic voltages for a three-level voltage source converter.

The operating advantages of this scheme are realized at the expense of some increase in the circuit complexity and a more rigorous control of the current transfers between the four valves, with constrained voltage overshoot. Another requirement is the need to accommodate the increased triplen harmonic content flowing through the midpoint of the DC supply; thus some extra DC storage capacitor will be needed to reduce the fluctuation of the midpoint voltage.

The three-level configuration reduces the dv/dt across the valves during switchings, as well as the frequency requirements of the PWM control and, therefore, the switching losses. To illustrate this point, a comparison of converter losses (including both the valves and converter transformers) to achieve the same harmonic reduction, shows that the losses in the three-level converter are of the order of 2%, those of a two-level converter 3.5%, while the corresponding figure for the line-commutated converter is 0.8%.

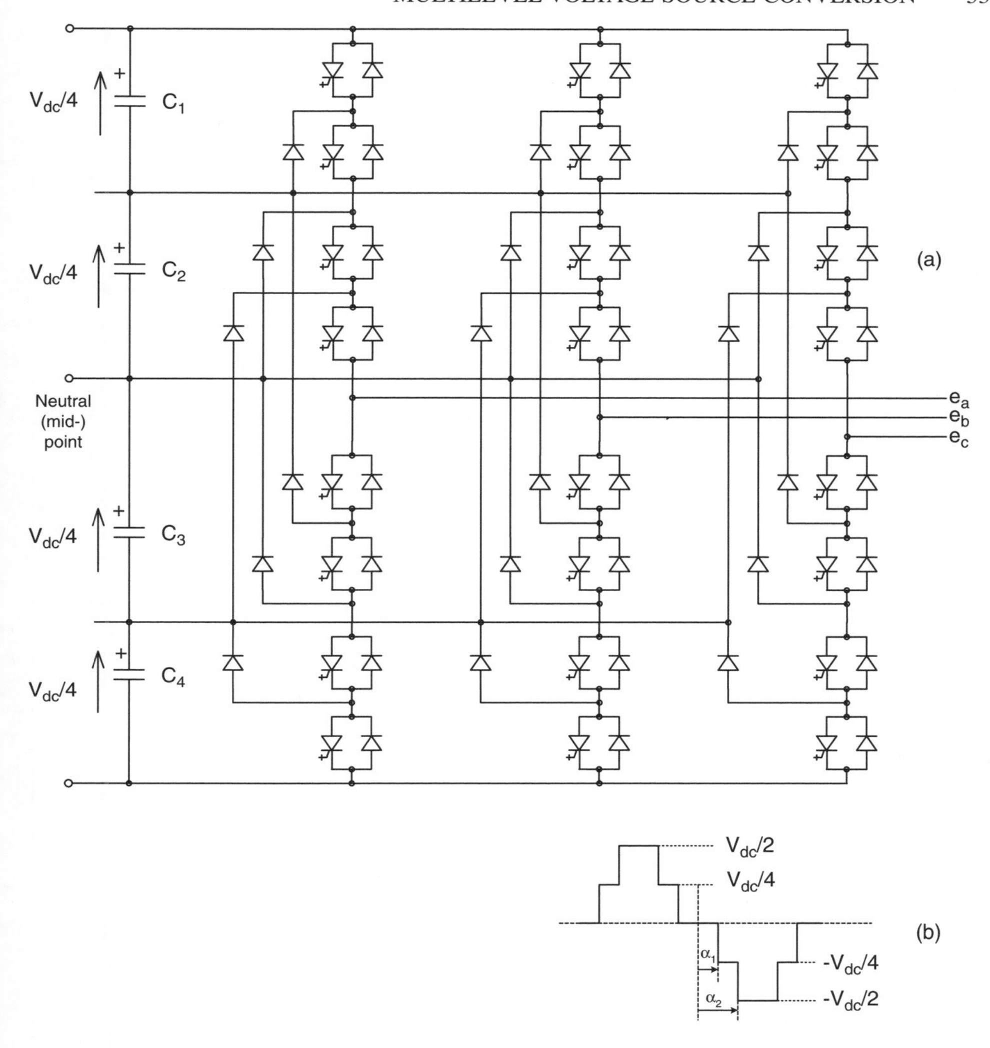

Figure 3.5 Diagram of a three-phase five-level diode clamped converter (a) and associated AC voltage (one phase) waveform (b).

3.3.2 Five-level diode-clamped VSC

Figure 3.5(a) shows a three-phase five-level diode clamped configuration, where the DC bus consists of four capacitors, C_1, C_2, C_3, C_4, and the voltage across each capacitor is nominally $\frac{V_d}{4}$. Thus the voltage stress across each switching device is limited to $\frac{V_d}{4}$ and this is achieved by the clamping diode. The output voltage waveform of the five-level pole is shown in Figure 3.5(b), with the capacitor midpoint used as a reference.

Let us now consider one phase of the five-level configuration, shown in Figure 3.6 [3]. If the negative DC rail (instead of the capacitor midpoint) is used as a reference (i.e. zero voltage) the switching pattern shown in Table 3.1 (where the symbols '1' and '0' indicate

Table 3.1 Switching patterns of the five-level diode-clamped voltage source converter

Output	Switch state							
V_{a0}	S_{a1}	S_{a2}	S_{a3}	S_{a4}	S'_{a1}	S'_{a2}	S'_{a3}	S'_{a4}
$V_5 = V_{dc}$	1	1	1	1	0	0	0	0
$V_4 = 0.75V_{dc}$	0	1	1	1	1	0	0	0
$V_3 = 0.50V_{dc}$	0	0	1	1	1	1	0	0
$V_2 = 0.25V_{dc}$	0	0	0	1	1	1	1	0
$V_1 = 0$	0	0	0	0	1	1	1	1

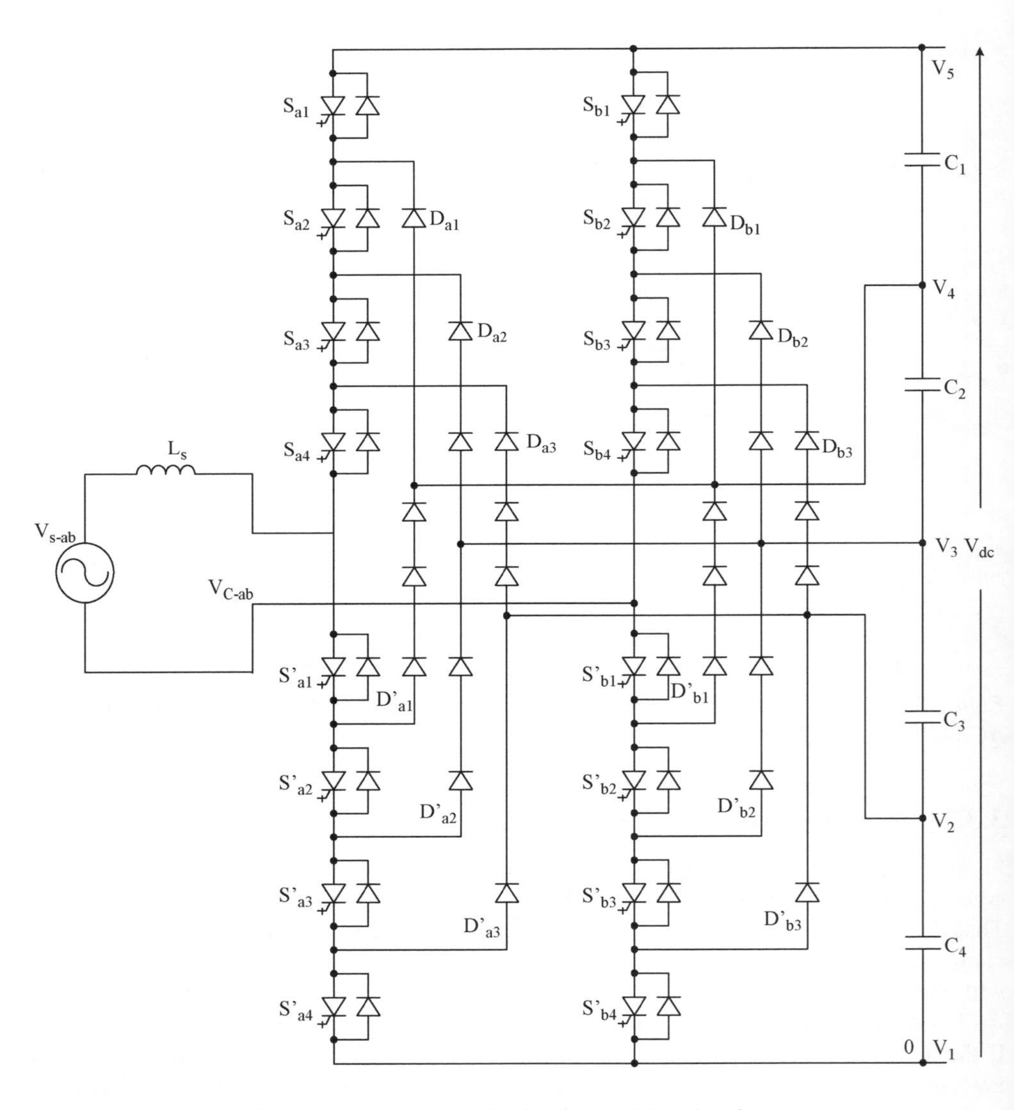

Figure 3.6 A single-phase diode-clamped five-level converter.

on and off states respectively) is used to synthesize the five-level waveform across the phase 'a' leg:

- Turning on all the upper switches (S_{a1}–S_{a4}) produces $V_{a0} = V_d$.
- Turning on three upper switches (S_{a2}–S_{a4}) and one lower switch ($S_{a'1}$) produces $V_{a0} = \frac{3V_d}{4}$.
- Turning on two upper switches (S_{a3}–S_{a4}) and two lower switches ($S_{a'1}$–$S_{a'2}$) produces $V_{a0} = \frac{V_d}{2}$.
- Turning on one upper switch (S_{a4}) and three lower switches ($S_{a'1}$–$S_{a'3}$)($S_{a'1}$–$S_{a'3}$) produces $V_{a0} = \frac{V_d}{4}$.
- Turning on all the lower switches ($S_{a'1}$–$S_{a'4}$) produces $V_{a0} = 0$.

The same switching pattern applies across the phase 'b' leg (if a is replaced by b) but the phase is shifted by 180° for the single-phase configuration (in the three-phase configuration the shifts between the phases will be 120°). It is clear from Figure 3.6 that switch S_{a1} conducts only for a small part of the half-cycle, while switch S_{a4} conducts almost for the entire cycle; thus, if all the switches have the same rating, the outer switches will be oversized and the inner switches undersized.

An important problem of this configuration is achieving capacitor voltage balancing, which is highly dependent on the displacement between the voltage and current fundamental components (or, ignoring waveform distortion, the power factor). When the converter operates at zero power factor (i.e. as a VAR compensator) the capacitor voltages are balanced by the equal charge and discharge in each half-cycle, as shown in Figure 3.7(b). At the other extreme,

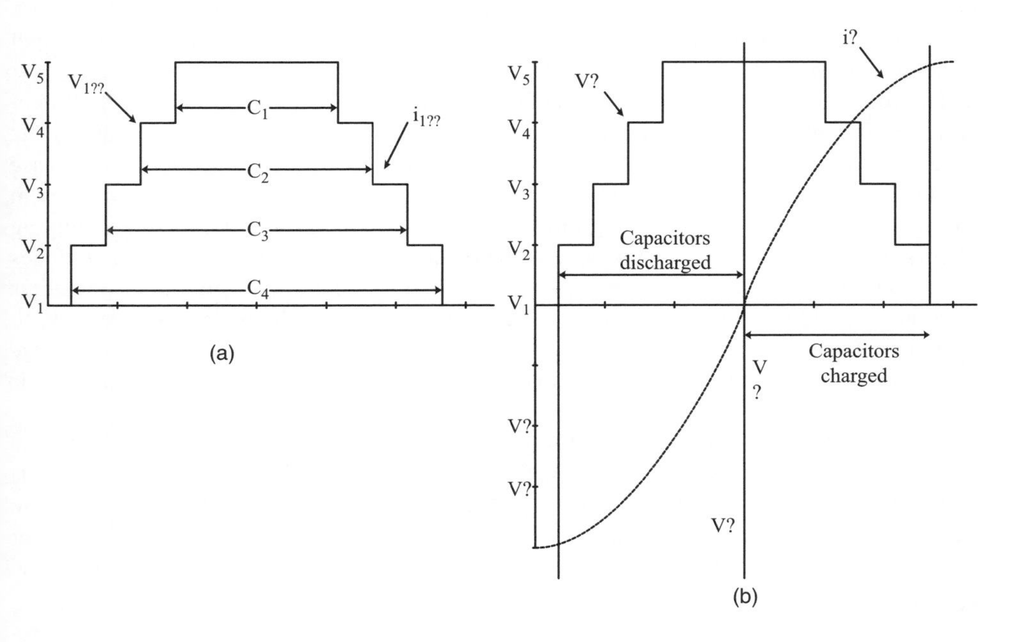

Figure 3.7 Capacitor charge profiles: (a) voltage and current in phase; (b) voltage and current out of phase by 90°.

when operating at unity power factor, Figure 3.7(a) shows that the charging (for rectifier operation) and discharging (for inverter operation) times of the capacitors are different. Thus additional circuitry, such as an isolated supply for each of the levels, is needed to solve the capacitor balancing problems.

Also, apart from the two outer switches, i.e. S_1 and S'_4, the rest are not directly clamped to the DC capacitors. Therefore, depending on the stray inductances of the structure, the indirectly clamped switches may be subjected to a larger proportion of the nominal blocking voltage during their off state. Indirect clamping is also a problem in the diode structures, the blocking voltage of each diode depending on its position in the structure. This normally requires the use of more diodes in series; however diversity in the diodes' switching characteristics as well as stray parameters call for large resistor–capacitor (RC) snubbers, which add to the cost and volume of the structure.

Moreover, the three inner DC rails carry bidirectionally controlled current, which prevents the use of polarized turn-on snubbers, as the latter will worsen the static overvoltage problem of the inner devices.

3.3.3 Diode clamping generalization

In high-voltage source conversion the DC voltage has to be shared by a number of series-connected capacitors. Also the power converter valves require the use of switching devices in series to withstand the high voltage (which is much higher than the individual switch voltage rating). There are, therefore, multiple intermediate nodes available to implement the multilevel conversion, though additional components are required in the form of clamping diodes.

An increase in the level number provides a corresponding reduction in the harmonic content. It also improves the steady state and dynamic voltage balancing of the valves, because groups of switches are clamped to capacitors with equal voltage and, therefore, the output voltage waveform is controlled to change level by level.

The multilevel diode-clamped (MLDC) configuration is a generalization of the three-level neutral-clamped converter concept described in section 3.2.1. Figure 3.8 [4] shows one phase structure of the generalized (m-level) MLDC voltage source converter. In this case $(m-1)$ clamping units are needed with every clamping node, because the maximum blocking voltage across the clamping diodes is related to its position in the diode clamping network. In Figure 3.8 the clamping node with potential V_i is connected by a group of diodes to the node between valves G_{ui} and $G_{u(i+1)}$, and the maximum and minimum potentials are V_i and V_0, thus the maximum voltage across the path from the node at potential V_i to the node between G_{ui} and $G_{u(i+1)}$ is $V_{m-1}-V_i = (m-1-i)V_L$; also the maximum voltage across the path from the node at potential V_i and the node between G_{di} and $G_{d(i+1)}$ is $V_i-V_0 = iV_L$.

Thus the number of clamping diodes connected with every level node is $(m-1)$, and the clamping path from the node at potential V_i to the main upper switches is formed by $(m-1-i)$ clamping diodes; similarly, the clamping path from the node at potential V_i to the main down switches is formed by i clamping diodes. An m-level converter needs $(m-2)$ inner nodes to be clamped and, therefore, the total number of clamping diodes is given by the expression:

$$N_{CD} = (m-1)(m-2)$$

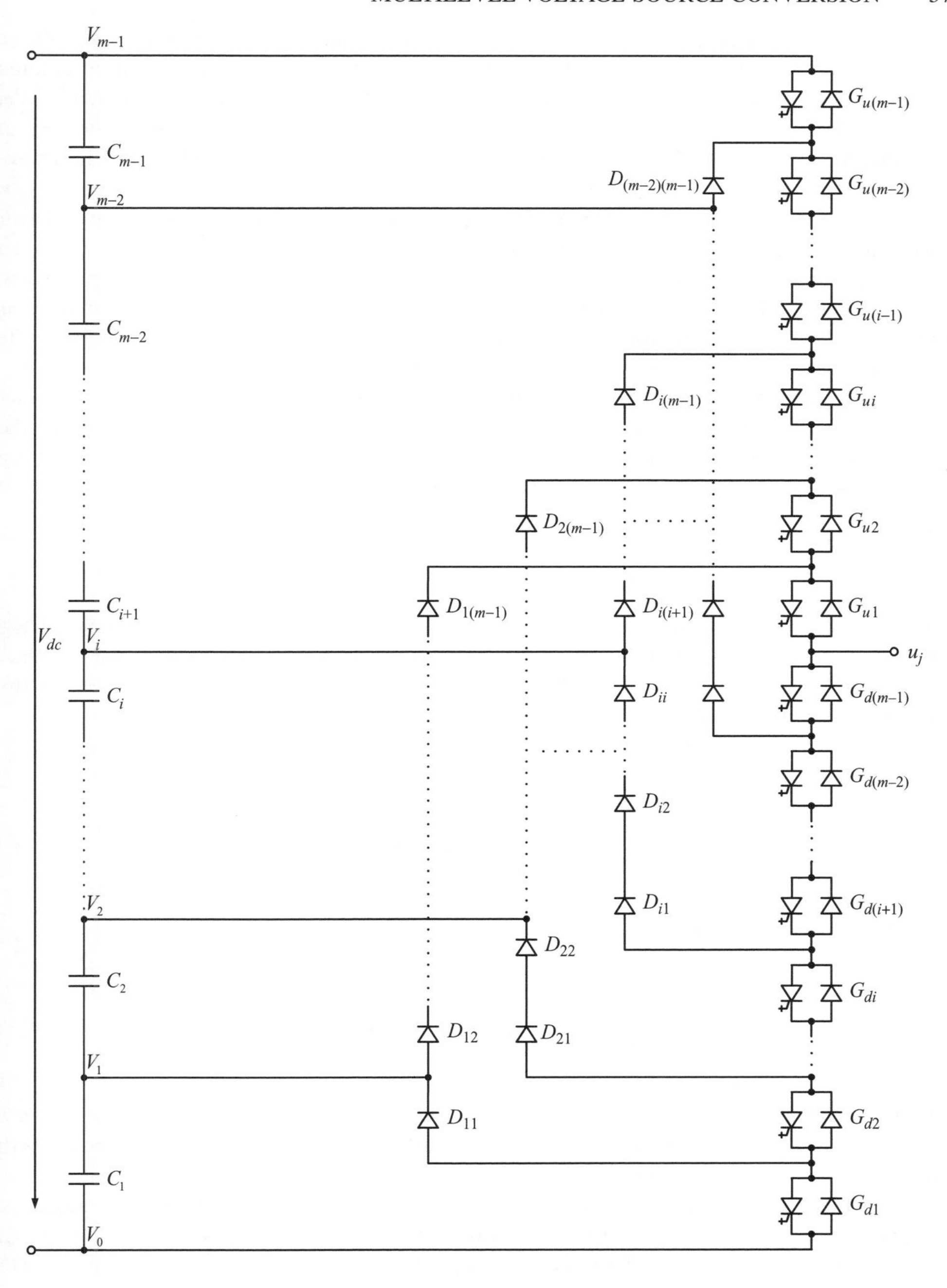

Figure 3.8 One phase of a generalized MLDC voltage source converter.

The required number of main switches is easier to determine. Since the output voltage u_j in Figure 3.8 is varied from V_0 to $V_{(m-1)}$, for an m-level converter, the maximum voltage across the path from the output terminal to the lowest level node is $V_{(m-1)}-V_0 = (m-1)V_L$, and the maximum voltage across the path from the highest level node to the output terminal is also $V_{(m-1)}-V_0 = (m-1)V_L$. The number of main switches is, therefore, $2(m-1)$.

In principle, the use of MLDC-VSC may be justified in high-voltage reactive power controllers (such as the STATCOM), that require many series-connected switches; it needs a relatively small DC capacitance, has low switch blocking voltages and low switching losses. However the application of MLDC conversion to high-voltage active power control (as required in HVDC transmission) is less attractive because of the difficulty of maintaining the individual capacitor voltages balanced, particularly for high-level numbers.

Raising the level number beyond five is impractical because the number of clamping diodes required increases in proportion to the square of the level number, which makes the capacitor balancing system more complex, introduces unequal power ratings for the switching devices (causing design and maintenance difficulty) and makes it difficult to add redundant switches.

3.3.3.1 Analysis of the waveforms [5]

The multilevel diode clamping waveforms are described here under ideal conditions, i.e. all the DC capacitances are of infinite size, the on state switches volt-drop is zero, their off state impedance is infinite and the free-wheeling diodes are ideal. Also the voltages across the capacitors are constant and of the same value ($V_L = V_d/(m-1)$).

The Fourier components of the output voltage are given by the expression:

$$V_n = \frac{4V_d}{n\pi(m-1)} \sum_{i=1}^{(m-1)/2} \cos(n\vartheta_i) \tag{3.3}$$

for odd level numbers and

$$V_n = \frac{4V_d}{n\pi(m-1)} \left\{ \frac{1}{2} + \sum_{i=1}^{(m/2)-1} \cos(n\vartheta_i) \right\} \tag{3.4}$$

for even level numbers, where ϑ_i is the ith switching angle, m the number of levels and n the harmonic order (which in the simulation results presented here is limited to the 49th order).

The fundamental component and the individual harmonics of the phase voltage waveform, for a given number of levels, is calculated for the sets of switching angles which eliminate all possible low-order harmonics. There is no unique solution for the switching angles required to eliminate a given number of harmonics and, thus, the results shown relate to the switching set that gives minimum total harmonic distortion (THD).

Using a transformer reactance of 15%, Figure 3.9 shows the variation of the AC line current THD with the number of levels. The results are given in normalized form for 1 per

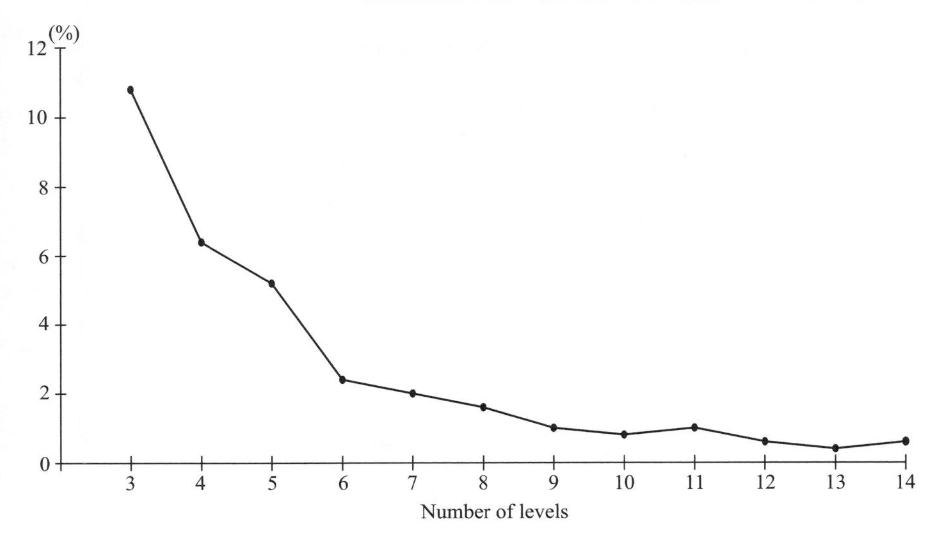

Figure 3.9 Total harmonic distortion in the AC line current.

unit leading current operation, which is the worst case in terms of harmonic performance and device utilization. The study shows that from three to six levels there is significant improvement, their respective THDs being 11% and 2.5%. For 14 levels this figure is reduced to about 0.6% and there is no significant reduction in the THD above 12 levels. The current waveform is also analysed assuming 1 per unit leading operation and quarter wave symmetry (i.e. $0 < \vartheta < \pi/2$).

In an interval $(\vartheta_i - \vartheta_{i+1})$, where the converter phase voltage remains constant the following relationship applies:

$$L\frac{di}{dt} = \sqrt{2}V\sin(\omega t) - v_I \tag{3.5}$$

where V is the *rms* value of the ac system voltage referred to the converter side, v_I is the instantaneous converter AC side voltage with respect to the AC neutral and L is the transformer leakage inductance (per phase).

Replacing the integration variable ωt by ϑ and using the terminal condition that the current is equal to zero at $\vartheta = \frac{\pi}{2}$, the current becomes:

$$i(\vartheta) = -\frac{\sqrt{2}V\cos(\vartheta) + v_I(\vartheta - \frac{\pi}{2})}{\omega L} \tag{3.6}$$

Figures 3.10(a) and 3.10(b) show respectively the nominal current waveform together with the corresponding normalized converter voltage output for the five- and eight-level configurations.

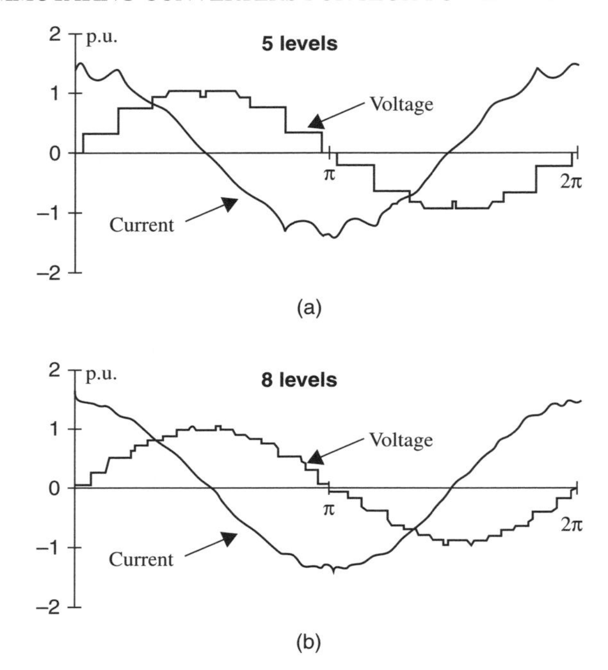

Figure 3.10 Leading AC current and normalized converter phase voltage: (a) for five levels; (b) for eight levels.

3.3.3.2 Capacitor voltage balancing

The three-phase m-level converter has $(m-1)$ different DC side capacitors $(C_1 - C_{m-1})$, which are periodically charged and discharged to control the DC voltage and, in turn, the AC fundamental voltage component. Each capacitor charging current is controlled by the switching strategy used and the AC side current. The voltage change in each capacitor results from the combined effect of the capacitor size and the charging current.

In normal operation, with balanced converter voltages and currents, the size of each capacitor can be decided to ensure that the voltage across the different capacitors change equally when the total DC side voltage is adjusted [6]. Capacitor voltage equality is a basic requirement for maintaining the MLDC conversion normal. To keep the capacitor voltages balanced, the average current into each intermediate node has to be equal within a specific period to ensure that the voltage across each capacitor increases or decreases equally.

Even in the three-level configuration the capacitor voltages can become unequal in the presence of unbalanced currents. However, capacitor unbalance is more of a problem for higher level numbers and operating conditions requiring power angles different from 90°. The cause of the problem has already been explained in section 3.2.2 with reference to the five-level configuration.

When the capacitors are not properly balanced, the voltage stress on the switching devices increases and the converter voltage waveform becomes more distorted. This condition can be avoided by using an adaptive control over the individual capacitor voltages, as shown in Figure 3.11 [7] for one of the capacitors. It can be achieved by generating an error signal

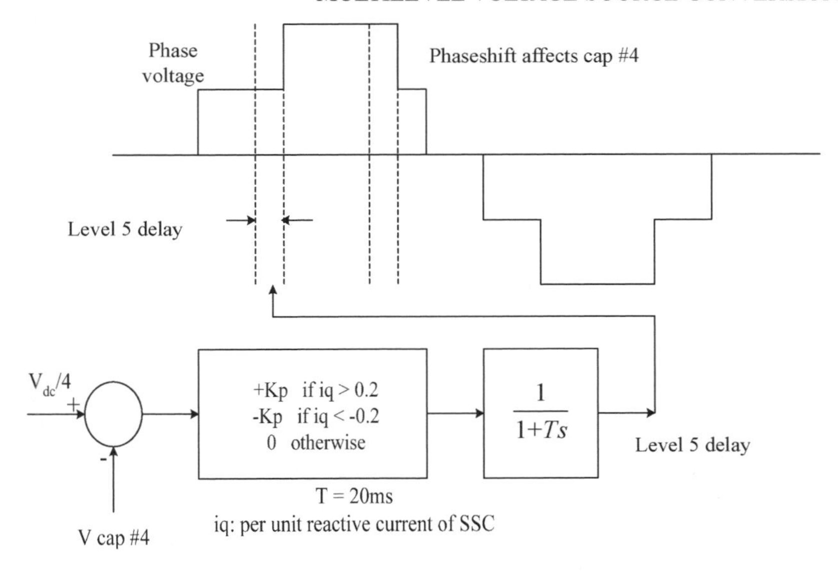

Figure 3.11 Balancing of the individual capacitor voltages.

when the individual capacitor voltage differs from its nominal share of the total DC voltage and this signal is used to phase-shift the related portion of the output voltage wave. This solution is applicable to reactive power compensation applications but not in active power control applications, because in these cases the voltage waveform cannot be seriously distorted.

3.4 The flying capacitor concept [8–10]

3.4.1 Three-level flying capacitor conversion

Figure 3.12 shows the flying capacitor circuit (also known as floating capacitor or imbricated cell conversion), as well as the corresponding output voltage waveform. The latter is identical to that of the neutral point clamped converter shown in Figure 3.1. In the three-level flying capacitor solution, the additional voltage step is obtained by the use of a separate DC capacitor in each phase; this capacitor is pre-charged to one half of the total DC voltage across the converter bridge.

The converter valves switch the AC buses between the three voltage levels by directing the current path through the DC capacitors, adding to or subtracting from the voltages of the individual capacitors in any desired manner. The switches in each phase arm are arranged in two pairs (A_1, B_1) and (A_2, B_2). Within each pair, the switches need always to be in complementary states. Thus to create the $-V_{dc}$ level of the phase output waveform, switches A_1 and B_2 must be turned on; similarly switches A_2 and B_1 need to be turned on to produce the $+V_{dc}$ level. The intermediate DC capacitor is bypassed for part of the fundamental frequency cycle.

The total capacitor power rating is considerably larger for this topology than for the two-level or three-level neutral point clamped topologies. This is partly due to the need for

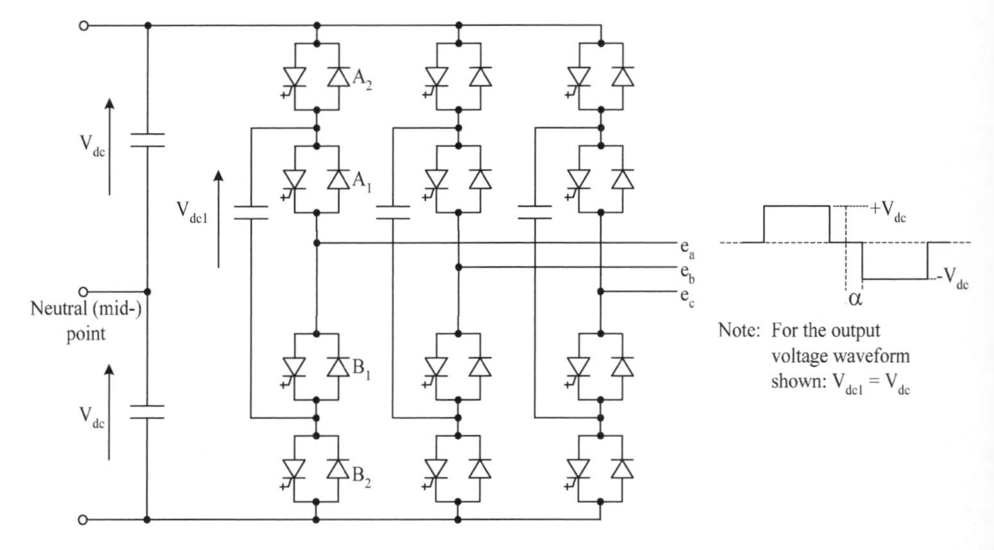

Figure 3.12 Diagram of a three-phase three-level flying capacitor converter and associated AC voltage (one phase) waveform.

extra separate capacitors in each phase and partly because the DC capacitors carry a significant ripple current.

3.4.2 Multi-level flying capacitor conversion

The three-level flying capacitor concept can be extended to provide any number of voltage levels, by adding further capacitors and subdividing the VSC valves. The size of the voltage increment between two capacitors defines the size of the voltage steps in the output voltage waveform. As an example, Figure 3.13 shows one arm of a five-level configuration, as well as the output voltage waveform which is exactly the same as that of the five-level diode-clamped alternative of Figure 3.5.

The valves switch the AC buses between the different voltage levels by directing the current path through the DC capacitors as explained for the three-level case. The flying capacitor has greater flexibility than the diode-clamped converter. It is possible to create the same output voltage using alternative connections of series capacitors allowing current to flow in the direction required for recharging. For instance, in Figure 3.13 the output voltage level V_{dc1} can be produced by:

1. Turning on switches A_1, B_2, B_3, B_4.
2. Turning on switches A_2 and B_1 (thus connecting $V_{dc2} - V_{dc1}$ to the load); this reverses the current flow through the V_{dc1} capacitor.
3. Turning on switches A_3, B_1, B_2 (thus connecting $V_{dc3} - V_{dc2}$ to the load).
4. Turning on switches A_4, B_1, B_2, B_3 (thus connecting $V_{dc4} - V_{dc3}$ to the load).

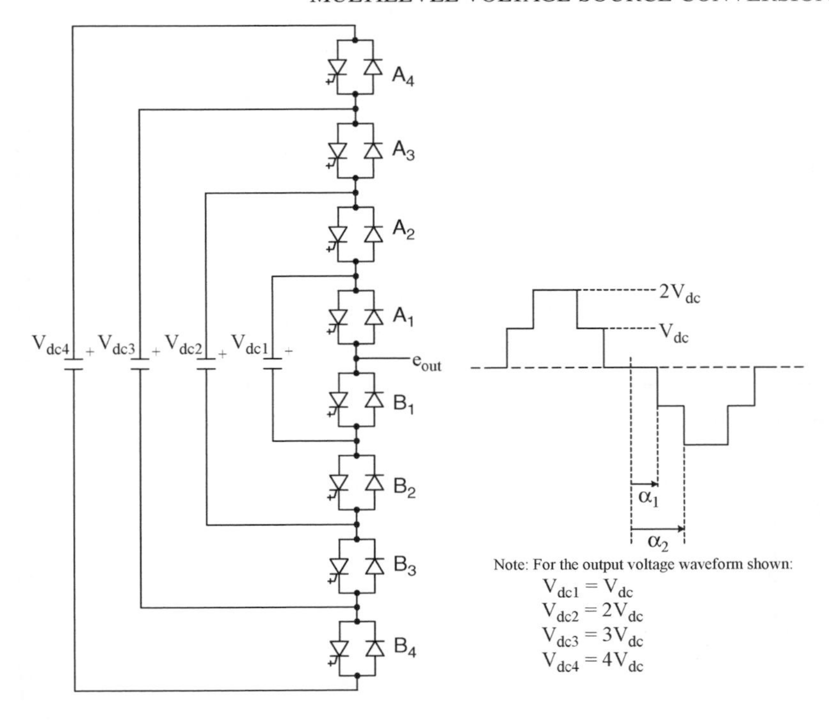

Figure 3.13 Five-level 'flying capacitor' type pole.

The required number of flying capacitors for an m-level three-phase converter, provided that the voltage rating of each capacitor used is the same as that of the main power switches is given by:

$$3[1+2+\cdots+(m-2)]+(m-1)$$

which, for the case of $m=5$, results in 22 capacitor units.

Naturally, as the number of flying capacitors increases, so does the cost of equipment and the size of the converter station. However, compared to the three-level neutral point diode-clamped topology, the individual valves are switched approximately half the number of times in the five-level topology and the harmonic performance is still superior.

However, as indicated earlier, the control of a multilevel topology is complicated and higher frequency is required to balance each capacitor voltage.

3.4.2.1 Three-phase configuration

In the three-phase configuration shown in Figure 3.14 all capacitors have the same voltage rating and, thus, a series connection of capacitors is used to indicate the voltage level between the clamping points. The capacitors are directly connected to the output through the converter switches and the AC voltage waveform is very much the same as that produced

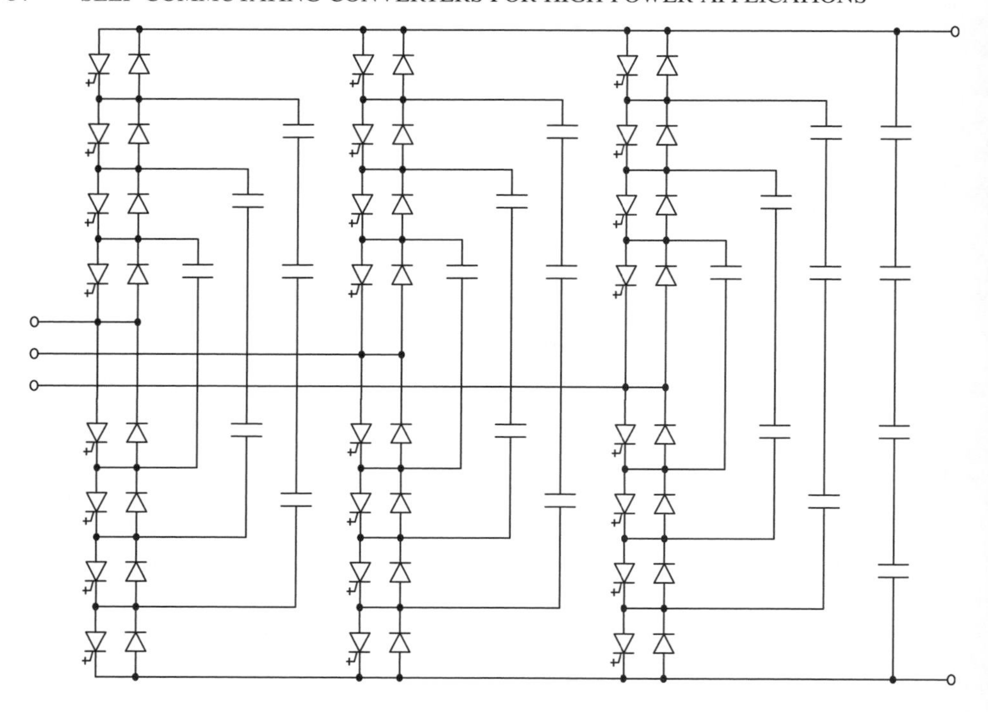

Figure 3.14 Three-phase flying capacitor multilevel configuration.

by the diode clamping scheme. Thus, in general for an m-level converter, the phase voltage has m steps and the line voltage (for the single-phase configuration), $(2m-1)$. For a three-phase unit, the main DC capacitors are shared by the three phases but the floating capacitors are not.

Capacitor voltage control through the switching pattern becomes complex, however, as the output is extended to three phases. Besides the difficulty of balancing the voltage in real power conversion, this configuration requires large numbers of storage capacitors, their size being largely proportional to the square of their nominal voltages and inversely proportional to the switching frequency.

If the voltage rating of each capacitor is the same as that of a main power switch, an m-level converter will require a total of $(m-1)(m-2)/2$ auxiliary capacitors per phase leg in addition to $(m-1)$ main DC bus capacitors. To balance the capacitors' charge and discharge, it is possible to use two or more switch combinations for the middle voltage levels $\left(\text{i.e.} \frac{3V_d}{4}, \frac{V_d}{2} \text{ and } \frac{V_d}{4}\right)$ in one or several fundamental cycles. Thus by appropriate selection of the switch combination, this configuration can be used for real power conversion; however, in this case the switch combination is rather complicated and the switching frequency needs to be higher than the fundamental frequency.

As the load current passes through the clamping capacitors, the current rating of these capacitors is higher. Also special complex pre-charging systems are required to maintain capacitor balance. On the other hand, all the semiconductor switches have the same duty, it has a lower switch blocking voltage and offers the lowest (of the multilevel schemes) converter switching losses.

3.5 Cascaded H-bridge configuration

The basic building block of the H-bridge configuration, shown in Figure 3.15, is a single-phase full bridge link consisting of four switches, S_1, S_2, S_3, S_4, connected to an isolated capacitor. The H-link can generate three voltage levels, i.e. $+V_{dc}, 0, -V_{dc}$. Turning on S_1 and S_4 gives $-V_{dc}$, turning on S_2 and S_3 gives $+V_{dc}$ and bypassing the cell (via S_1, S_2 or S_3, S_4) gives zero voltage.

Several links connected in series (cascaded) constitute a phase of the converter, and the phase output voltage waveform is the sum of the voltages of the activated links. For instance, Figure 3.16 shows the development of the phase voltage waveform for a chain of four links.

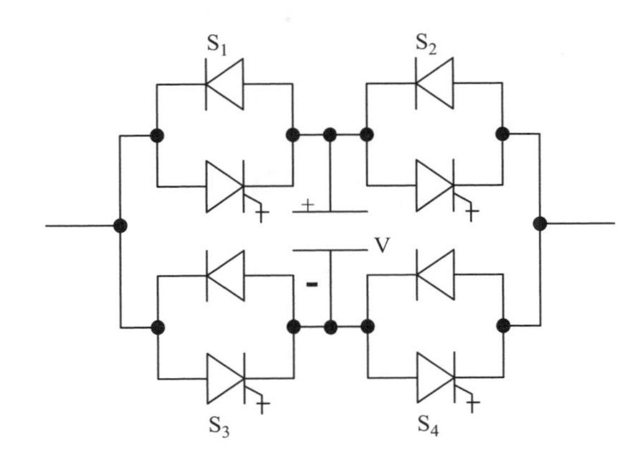

Figure 3.15 One link of the H-bridge configuration.

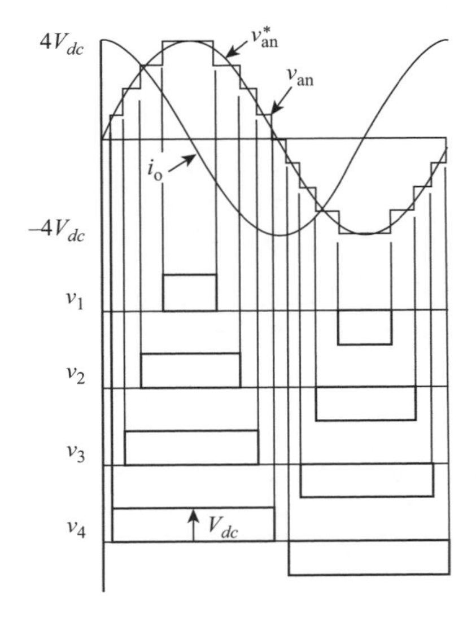

Figure 3.16 Waveforms of the five-level H-bridge converter.

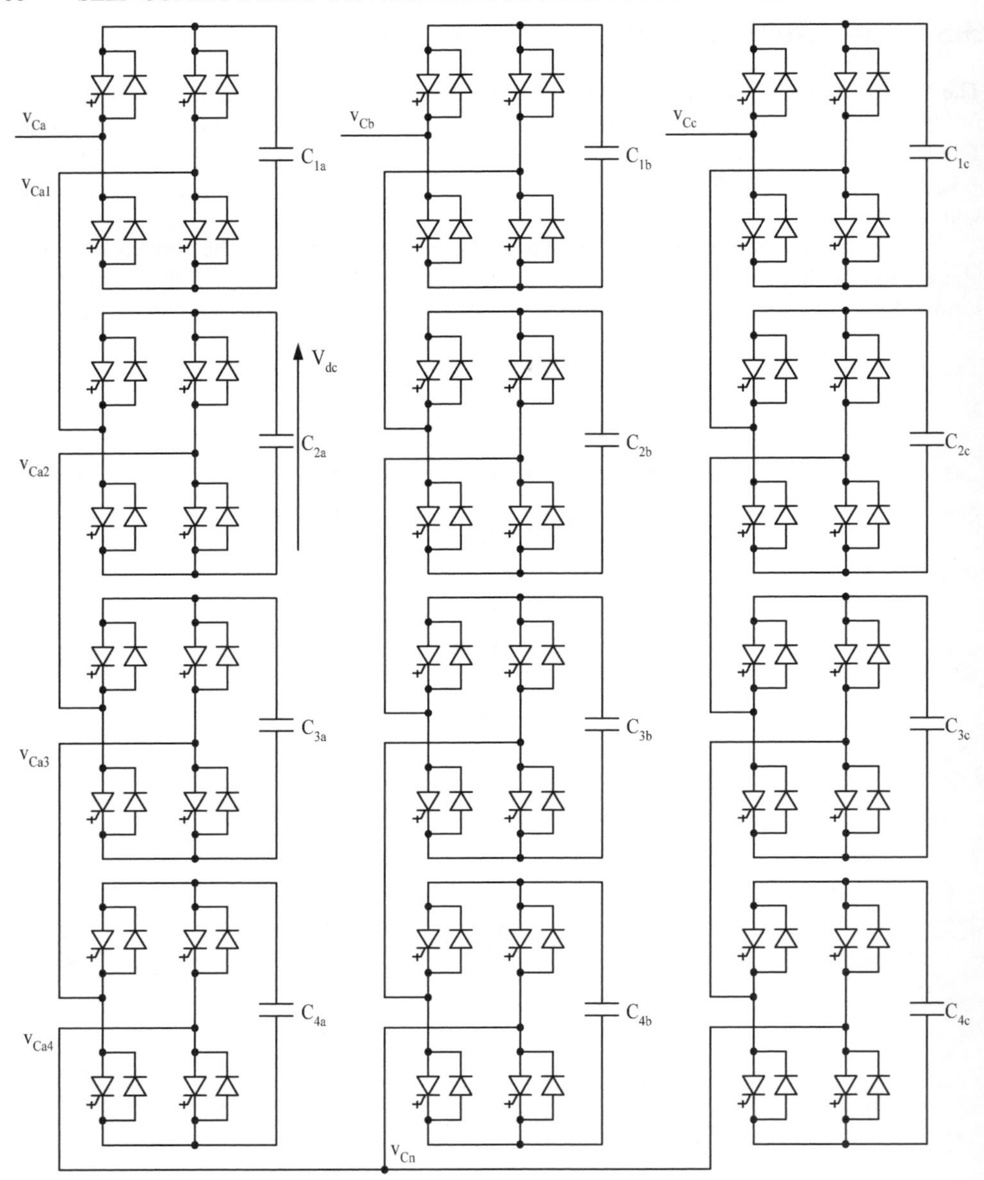

Figure 3.17 A three-phase star-connected H-bridge converter.

The latter is synthesized by the sum of the individual link voltages ($v_{an} = v_1 + v_2 + v_3 + v_4$), with an appropriate selection of their conducting periods and contains nine steps [11]. Thus in general an m-level converter will require the use of $\frac{m-1}{2}$ H links per phase.

The complete structure of the conventional three-phase configuration is shown in Figure 3.17. The stepped waveform appears to be similar to that of the current waveform in a conventional multibridge current source converter. However, while the latter consists of a series of pulses of varying height but equidistant in time, in the H-bridge voltage source converter the pulses are of equal height and the intervals between switching instants vary.

The successive switching angles of the H-circuit may be controlled to minimize the total harmonic content or to eliminate several specific harmonics; in the latter case the H chain can be controlled to eliminate N harmonics per phase, which for the three-phase configuration provides approximately 6N-pulse operation.

When used for AC–DC or DC–AC power conversion, the H-bridge configuration needs separate DC sources. This requirement suits some applications, like the system connection from renewable energy sources such as fuel cells and photo-voltaics.

The H-bridge configuration has considerable merit for applications where no real power transfer is involved, such as active filtering and reactive power compensation. In this case, as the phase current I_{Ca} is leading or lagging the phase voltage v_{Can} by 90°, the average charge to each DC capacitor during each half-cycle is zero, i.e.

$$\int_{\theta_i}^{\pi-\theta_i} \sqrt{2}I \cos\theta d\theta = 0 \tag{3.7}$$

where $i = 1, 2, 3, 4$; $[(\theta_i, (\pi-\theta_i)]$ represents the time interval during which the DC capacitor connects to the AC side and I is the *rms* value of the line current. As a result of the symmetrical charge flow, the voltages in all the DC capacitors remain balanced.

The H-configuration provides a well defined operating environment for the power semiconductors, within a substantially isolated single-phase bridge circuit. Its single-phase converter structure can provide single-phase compensation, an important requirement in applications like arc furnace reactive power compensation. If a module fails (as a result of a short circuit) or is removed from service, the converter can continue to operate at a correspondingly reduced voltage rating.

Under various system conditions the DC capacitor voltages can become unbalanced. This problem is solved by adaptive control action. Provision is made to exchange energy between the capacitors by means of dual IGBT inverters at each link [12]; these are connected to an auxiliary coupling bus via a power isolating transformer (APIT). The inverter/AIPT combination operates at a frequency of 7.5 kHz.

During start up, shut down and system disturbances the links require an independent supply of energy. A ground level power supply (GLPS) is used for this purpose. The GLPS also provides energy for pre-charging the DC capacitors.

3.6 Modular multilevel conversion (MMC)

MMC [13] is a recent concept in multilevel conversion, which, in common with the other VSC multilevel alternatives, provides a fine gradation of the output voltage, thus reducing the harmonic content, the emitted high-frequency radiation and the switching losses. Moreover, unlike previous multilevel proposals, when used as part of a DC link, the flexibility of MMC is comparable to that of PWM, in that it is able to control the reactive power at each converter terminal independently. Since this property is only relevant to HVDC transmission, a detailed description of its operation is given in Chapter 7.

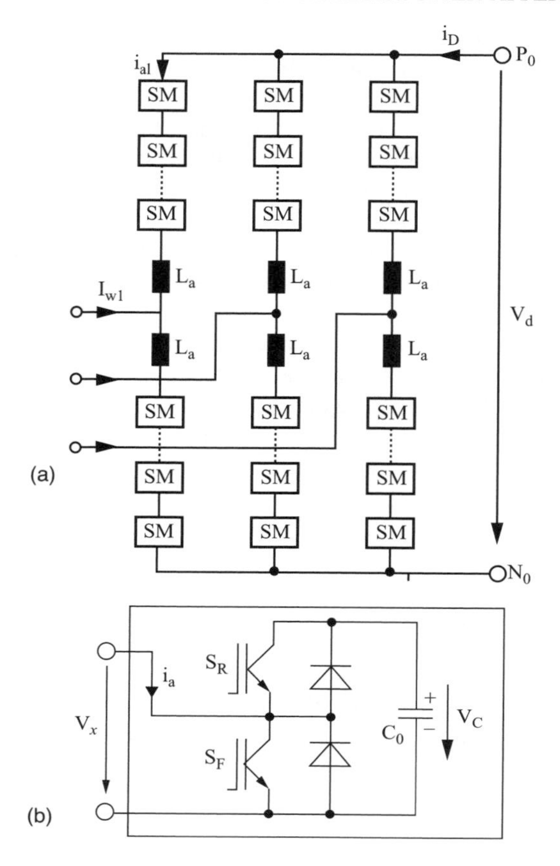

Figure 3.18 Modular multilevel converter: (a) three-phase modular multilevel converter; (b) submodule circuit.

Figure 3.18 shows the modular multilevel converter, consisting of n submodules in each arm. The submodule can be considered as a controlled voltage source with two output states, i.e. 0 V or V_C. The output voltage Vx of each submodule can be switched to either 0 V or to the capacitor voltage Vc. Under balanced conditions all capacitor voltages in every submodule are assumed to be equal, i.e. $V_C = V_O$. By switching a number of submodules in the upper and lower arm to active state, the DC terminal voltage V_d can be adjusted. In a similar manner, the AC output voltage V_{AC} can be adjusted to a desired value.

For each phase leg if the n_1 modules in the upper arm are active $(Vx = V_C = V_O)$, and n_2 modules in the lower arm are active, the DC bus voltage $V_{dc} = (n_1 + n_2)Vo$. By maintaining $n_1 + n_2$ always equal to n, the DC bus voltage, $V_{dc} = nVo$, *is* constant. Moreover, by maintaining $n_1 + n_2 = n$ and adjusting n_1 and n_2 permits generating the frequency, magnitude and phase controllable AC output. For a specific combination of n_1 and n_2 the AC terminal voltage referenced to $V_{dc}/2$ is $V_{ac} = n_2Vo - (n_1 + n_2)Vo/2 = (n_2 - n_1)Vo/2$. The value of $|n_2 - n_1|_{\max}$ determines the peak value of the AC output voltage.

For high-power and high-voltage application there will be over 100 submodules in each arm; the high number of submodules provides the freedom to generate an AC output waveform of high quality for a large range of amplitude adjustment, and without need of PWM switching. For instance for $n = 100$, under the fundamental switching, the 10% AC output voltage waveform is equivalent to 21-level conversion and, therefore, the waveform quality is still guaranteed.

The identical structure and great number of the submodules provide the freedom to control the voltage across the capacitor in every submodule. When active power is exchanged between the AC and DC system interconnected by the converter, the capacitors in the active submodules are charged or discharged. If the active power flow is towards the arm, the capacitor with lower voltage is selected to be active; and if the active power is out of the arm, the capacitor with higher voltage is selected to be active; the capacitor voltages in all the submodules are kept balanced. This balancing control strategy needs the sampling measurements of all of the capacitor voltages in the converter submodules, as well as measurement of the active power direction of six arms; it also needs a quick determination device to select the individual submodules to be active among a large number of submodules. Further processes are required to construct the firing logic for the generation of the gate control sequences for different applications in order to maintain the capacitor voltages balanced.

The simplified electrical equivalent of the MMC topology, shown in Figure 3.19, consists of six converter arms acting as a controllable voltage source. The series combination of many submodules provides a high number of discrete voltage steps. In the steady state the voltages are controlled to achieve one third of the total DC current in each phase unit and to ensure equal AC current in the upper and lower parts of each phase unit.

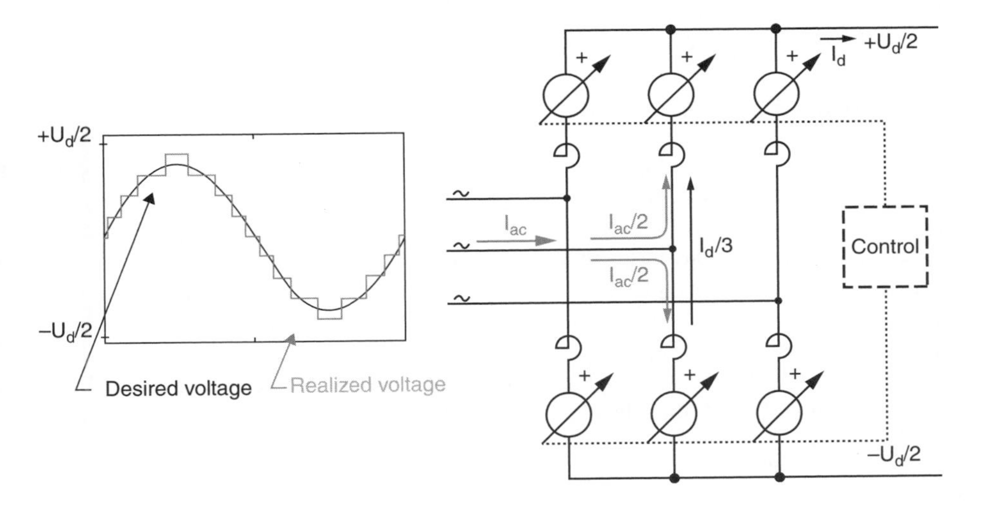

Figure 3.19 Overview of the MMC topology and voltage waveform.

Typical waveforms for a 400 MW converter with 200 submodules per phase arm, without any filter equipment, are shown in Figure 3.20. Figure 3.20(a) shows the DC voltages (±200 kV) and the AC terminal voltages with respect to a virtual reference point. Figure 3.20(b) illustrates the AC terminal converter current and Figure 3.20(c) shows the six phase arm currents in the converter. Clearly, the harmonic content is insignificant.

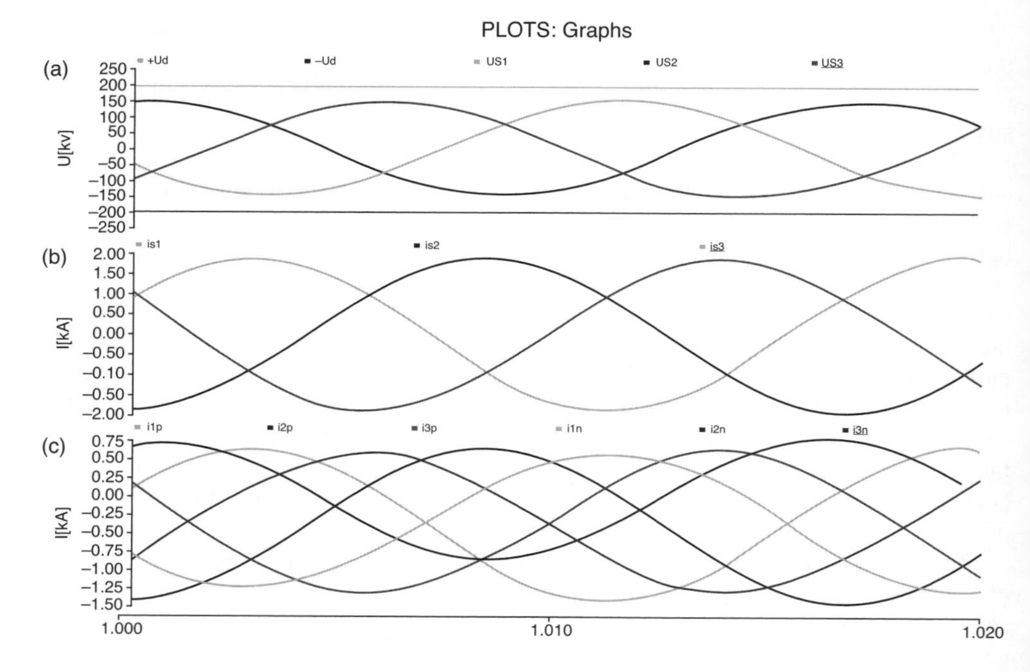

Figure 3.20 Typical waveforms of a 400 MW MMC: (a) AC line-to-line converter voltages; (b) AC converter currents; (c) converter arm currents.

3.7 Summary

Although multilevel converters have been given considerable coverage in the technical press for over a decade, there has been little interest in their commercial realization so far. Despite their more robust and higher rating capability (due to the use of thyristor-type switches) the multilevel configurations have been unable to compete with the transistor-based technology in the mid-power range. Their place in large power conversion is also unclear, due to their complex converter structures and restricted reactive power control flexibility. So the new large HVDC schemes are still opting for the conventional line-commutated technology.

More flexible and structurally simple multilevel concepts are discussed in the following chapters that should make multilevel conversion more attractive to the industry.

References

1. Walker, G. and Ledwich, G. (1994) Bandwidth considerations for multilevel converters. *Proceedings of the Australasian Universities Power Engineering Conference, AUPEC'94, Adelaide, 1994*, pp. 514–19.
2. Nabae, A., Takahashi, I. and Akagi, H. (1981) A new neutral point clamped PWM inverter. *IEEE Transactions on Industry Applications*, **17** (5), 518–23.
3. Choi, N.S., Cho, J.G. and Cho, G.H. (1991) A general circuit topology of multilevel inverter. *Proceedings of Power Electronics Specialists Conference,* 1991, pp. 96–103.

4. Liu, Y.H. *Multi-Level Voltage and Current Reinjection AC–DC conversion*. PhD thesis, University of Canterbury, 2003, New Zealand.
5. Scheidecker, D. and Tennakoon, S.B. Optimisation of the number of levels in a multi-level advanced static VAR compensator. *Proceedings 6th European Conference on Power Electonics and Applications (EPE'95)*, Sevilla, Spain, 1995, pp. 2494–8.
6. Choi, N.S., Jung, Y.C., Liu, H.L. and Cho, G.H. (1992) A high voltage large capacity dynamic VAR compensator using multi-level voltage source inverter. *Proceedings of Power Electronics Specialists Conference*, 1992, pp. 538–45.
7. Hochgraf, C. and Lasseter, R.H. (1997) A transformer-less static synchronous compensator employing a multi-level inverter. *IEEE Transactions on Power Delivery*, **12** (2), 881–7.
8. Meynard, T. and Foch, H. (1992) Multi-level conversion: high voltage choppers and voltage source inverters. *Proceedings of Power Electronics Specialists Conference,* 1992, 397–403.
9. Meynard, T. and Foch, H. (1993) Imbricated cells multi-level voltage source inverters for high voltage applications. *European Power Electronics Journal*, **3** (2), 99–106.
10. Meynard, T. and Foch, H. (1991) *Dispositif electronique de conversion d'energie electrique*. French patent no. 09582. International Patent P.C.T. no. 92.00652: Europe, USA, Japan, Canada.
11. Peng, F.Z., Lai, J.S., McKeever, J.W. and Van Coevering, J. (1996) A multilevel voltage source inverter with separate dc sources for static VAR generation. *IEEE Transactions on Industry Applications*, **32** (5), 1130–7.
12. Hanson, D.J., Woodhouse, M.L., Horwill, C., Monkhouse, D.R. and Osborne, M.M. (2002) STATCOM: a new era of reactive compensation. *IEE Power Engineering Journal*, **16** (3), 151–160.
13. Dorn, J., Huang, H. and Retzmann, D. (2007) Novel voltage source converters for HVDC and FACTS applications. *CIGRE Annual Conference*, Osaka (Japan), November 2007.

4

Multilevel Reinjection

4.1 Introduction

The large harmonic content of the basic three-phase current source converter requires considerable passive filtering investment at the converter terminals. This problem has encouraged the development of alternative configurations aiming at the elimination of the lower harmonic orders.

An early concept applicable to the three-phase line commutated converter, was the addition of a third harmonic current to the quasi-rectangular phase current waveform [1, 2]. This idea, however, presented the following problems for practical implementation:

- The need to provide a controllable source of the triple harmonic current.
- The need to adjust the amplitude and phase of the injected harmonic current under varying operating conditions, and keep it in synchrony with the supply frequency.
- The extra power loss of the injection circuit.

These problems were overcome by a contribution referred to as 'DC ripple reinjection' [3], that used a single-phase bridge transformer connected across the converter DC side with its DC output connected in series with the main converter DC output. The DC ripple was used as the commutating voltage for the reinjection bridge, which, as well as avoiding the need for an independent controllable current source, provided perfect synchronization with the AC system.

However, line-commutated conversion absorbs reactive power for its operation, both during rectification and inversion, much of which is provided by the passive filters and, therefore, the injection and reinjection proposals were not considered cost competitive with the conventional converter technology.

The development of power semiconductors with turn-off capability has eliminated the need to provide reactive power for the conversion process and this has made the reinjection concept a more attractive proposition.

Self-Commutating Converters for High Power Applications J. Arrillaga, Y. H. Liu, N. R. Watson and N. J. Murray

The multilevel clamping configuration has been shown in Chapter 3 to provide sufficient harmonic reduction without the assistance of passive filters. By combining the multilevel and reinjection concepts a much simpler converter structure is developed in this chapter.

4.2 The reinjection concept in line-commutated current source conversion

The single bridge three-phase converter produces a ripple voltage at the DC output, with a period of $(1/6)T$, where $T = 1/f$. However, with respect to the star point of the converter side transformer windings, each DC pole has a non-sinusoidal ripple voltage with a period of $(1/3)T$, i.e. a triple frequency voltage. This voltage has the same phase relationship on each DC pole and is referred to as the common mode DC ripple voltage.

The principle of DC ripple reinjection is applicable to the line-commutated three-phase current source converter bridge and the circuit configuration is shown in Figure 4.1. The converter transformer is star-connected on the converter side and must have either a delta primary or delta tertiary winding. It requires an auxiliary single-phase transformer with two primary windings connected to the common mode DC ripple voltage via blocking capacitors. This transformer provides the commutating voltage for a single-phase full-wave rectifier (or feedback converter) connected to the secondary windings. The output of the feedback converter is connected in series with the DC output of the six-pulse converter bridge.

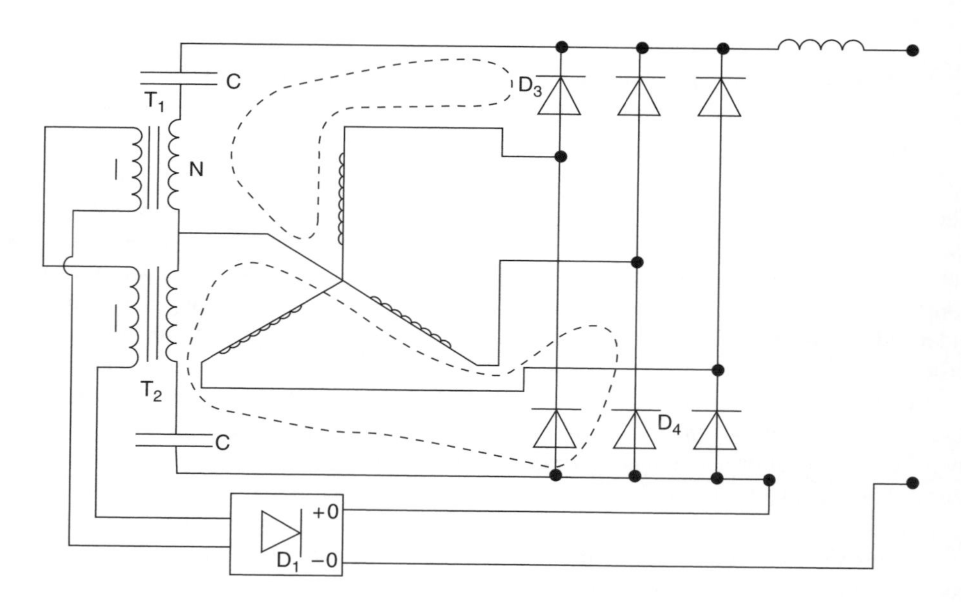

Figure 4.1 Bridge rectifier with ripple reinjection. T_1, T_2, feedback transformers; N, turns ratio; C, blocking capacitor; D_1, feedback rectifier. (The broken lines indicate the paths of injected current, while D_3 and D_4 are on.)

The frequency of the harmonic injection is determined by the supply frequency and, therefore, the harmonic source is always synchronized with the mains frequency.

The reinjection transformer secondary current consists of quasi-rectangular 60° DC current pulses, the magnitude of which is fixed by the main converter DC current, and their phase position is always in synchrony with that of the main converter current. These current pulses, after being appropriately altered by the injection transformer ratio, are added to the otherwise conventional DC current output of the main converter and channelled to the appropriate phases of the main converter current by the conducting thyristors.

The required phase adjustment of the injected current is achieved by thyristor control in the feedback converter, and the firing angle control of the feedback converter is locked to that of the main converter. When the thyristors of the feedback converter are fired 30° after the firings of the main converter valves, the waveforms in Figure 4.2 result. A Fourier analysis of waveform (E) shows that, for a particular ratio of the injected current to rectifier current ratio, all harmonics of orders $6n \pm 1$ (where $n = 1,3,5\ldots.$) are zero, while those of the remaining orders (i.e. for $n = 2,4,6\ldots$) retain the same relationship with the fundamental as before.

The triple frequency commutations of the reinjection bridge also combine with the DC output voltage to produce a 12-pulse voltage waveform. The result is that the original six-pulse converter configuration has been converted to a 12-pulse converter system from the point of view of AC and DC system harmonics.

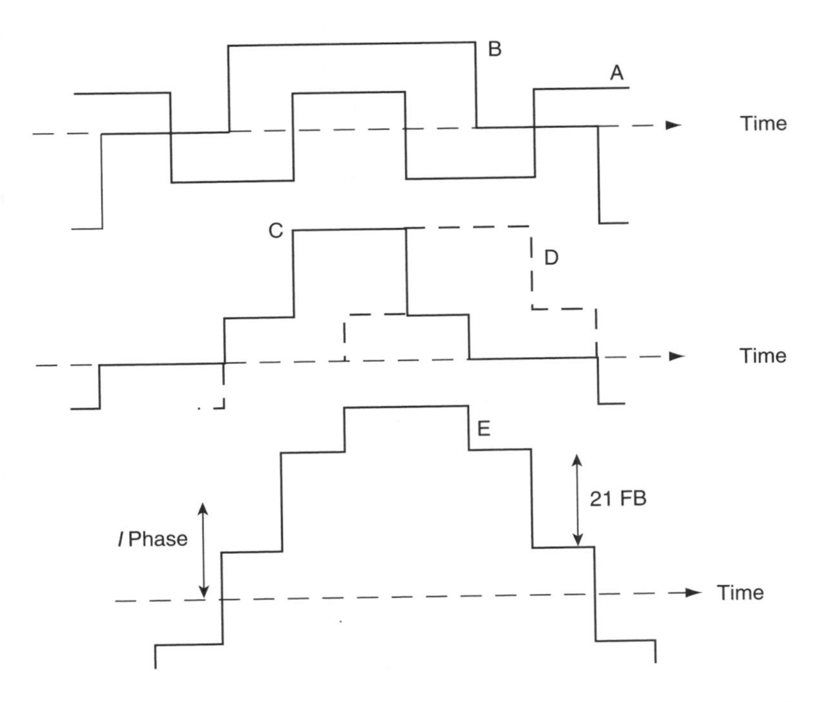

Figure 4.2 Current waveforms for the converter of Figure 4.1. A, *reinjection injected current;* B, *rectifier current before modification;* C, *modified phase current;* D, *second phase current displaced 120°;* E, *resultant phase current on delta primary.*

4.2.1 The reinjection concept in the double-bridge configuration [4]

The reinjection principle is even more effective when the 12-pulse (instead of the six-pulse) converter is used as a basis for the pulse transformation. In this case the centre point between the two bridges is used, instead of the transformer neutral, as the reinjection point and the DC ripple across the bridge (which in this case is $1/6T$) provides the commutating voltage source for the reinjection bridge.

For the series configuration, Figure 4.3 illustrates the modified circuit required to implement the reinjection concept; the additional components are two capacitors C, needed to block the DC component of voltages v_1 and v_2, two transformers T_A (operating at the ripple frequency which is six times the fundamental frequency) and a single-phase converter bridge. The commutating voltage of the ripple reinjection thyristors is the common mode ripple frequency voltage of the 12-pulse converter group.

Figure 4.4 shows the AC current waveforms associated with the circuit of Figure 4.3. Figure 4.4(c), (f) and (h) are the conventional current waveforms in the absence of the

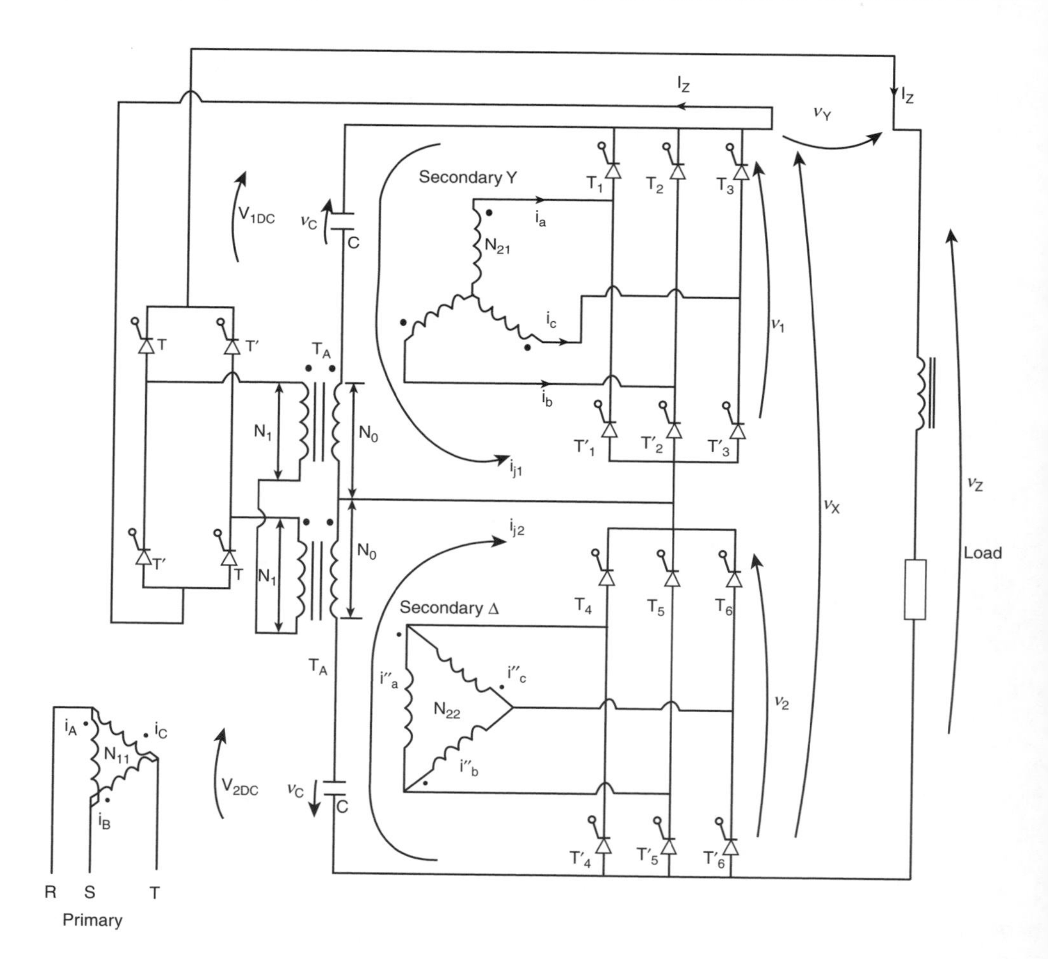

Figure 4.3 24-pulse series bridge configurations.

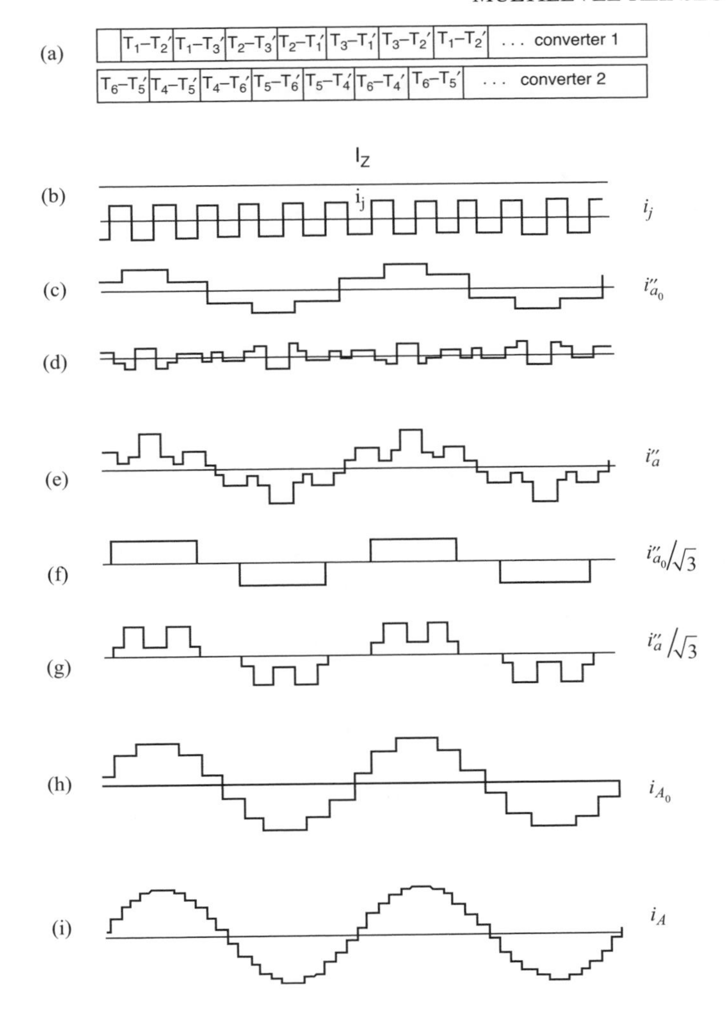

Figure 4.4 Current waveforms for the 24-pulse reinjection configurations.

reinjection circuit. Figure 4.4(a) shows the valve conducting sequences of the main bridges. Using the 12-pulse ripple voltage as commutating voltage for the reinjection bridge, a six-pulse reinjection current (i_j) is produced on the secondary side of the reinjection transformers (shown in Figure 4.4(b)). On the phases of the delta-connected bridge the reinjection current ($i''_{\Delta 0}$) is shown in Figure 4.4(d) and the addition of (d) and (c) produces the delta winding current (Figure 4.4(e)). Similarly, in the star-connected bridge the reinjection current (i_j) added to the basic waveform (f) to produce $i''_a/\sqrt{3}$ (Figure 4.4(g)). Finally the addition of (e) and (g) results in the 24-pulse output current waveform (i_A) shown in Figure 4.4(j).

However, the ripple reinjection concept has not been found competitive with the conventional converter configuration supplemented by filters. This is mainly due to the part played by the filters in the provision of reactive power requirements of the line-commutated conversion process. It is shown next that the use of self-commutating reinjection switches can make this concept far more attractive.

4.3 Application of the reinjection concept to self-commutating conversion

An ideal generalization of the injection concept is made in this section as a first step towards the development of a more practical implementation to be described in later sections. Although the reinjection concept is analysed here with reference to double-bridge (12-pulse) VSC, taking into account the duality principles, the results are equally applicable to current source conversion. However, two main differences apply to these two alternatives. One relates to the converter transformer primary connections to the network (that must be series-connected for the VSC and parallel-connected for the CSC configurations); the second difference relates to the valves' conducting period (which is 180° in the VSC and 120° in the CSC cases respectively).

4.3.1 Ideal injection signal required to produce a sinusoidal output waveform

The bare 12-pulse configuration shown in Figure 4.5 consists of two three-phase full bridges and two interface transformers. The two transformers are arranged in Y/Y and Y/Δ connections, and their primary windings (on the AC side) are connected in series to produce an AC output voltage which is the sum of the two transformers' primary side voltages. The turns ratios are $k_n : 1$ and $k_n : \sqrt{3}$ for the Y/Y and Y/Δ connections respectively. The switches, assumed ideal, have bidirectional blocking and unidirectional current capability.

In Figure 4.5 the S_{Y1} arm of the star-connected bridge is switched on at $\omega t = \ldots, 0, 2\pi, 4\pi, \ldots$ and arm S_{Y4} at $\omega t = \ldots \pi, 3\pi, 5\pi$. Similarly arm $S_{\Delta 1}$ of the delta-connected bridge is switched on at $\omega t = \ldots, \pi/6, 13\pi/6, 25\pi/6, \ldots$ and arm $S_{\Delta 4}$ at $\omega t = \ldots 7\pi/6, 19\pi/6, 31\pi/6, \ldots$.

The secondary winding voltages of the two interface transformers on the bridge side are given by

$$V_{Ya}(\omega t) = \begin{cases} V_{YY}(\omega t)/3 & 0 < \omega t < \pi/3 \\ 2V_{YY}(\omega t)/3 & \pi/3 < \omega t < 2\pi/3 \\ V_{YY}(\omega t)/3 & 2\pi/3 < \omega t < \pi \\ -V_{YY}(\omega t)/3 & \pi < \omega t < 4\pi/3 \\ -2V_{YY}(\omega t)/3 & 4\pi/3 < \omega t < 5\pi/3 \\ -V_{YY}(\omega t)/3 & 5\pi/3 < \omega t < 2\pi \end{cases} \tag{4.1}$$

$$V_{\Delta a}(\omega t) = \begin{cases} 0 & 0 < \omega t < \pi/6 \\ V_{Y\Delta}(\omega t) & \pi/6 < \omega t < 5\pi/6 \\ 0 & 5\pi/6 < \omega t < 7\pi/6 \\ V_{Y\Delta}(\omega t) & 7\pi/6 < \omega t < 11\pi/6 \\ 0 & 11\pi/6 < \omega t < 2\pi \end{cases} \tag{4.2}$$

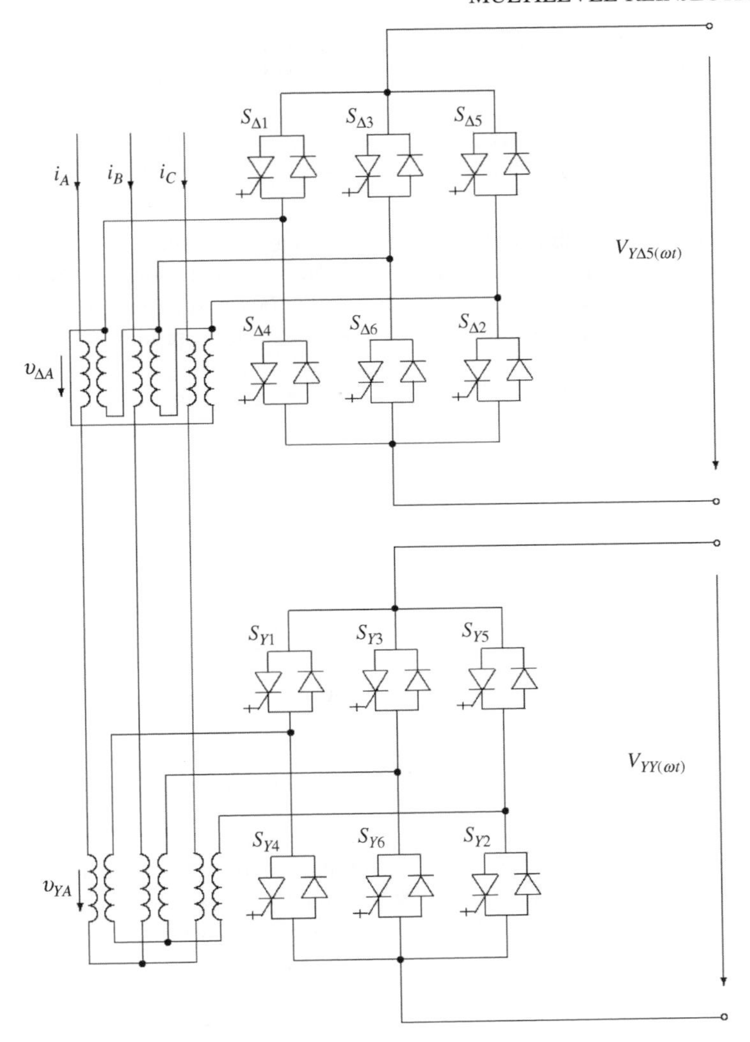

Figure 4.5 The two bridges of the 12-pulse VSC configuration.

If the voltages across the bridges connected to the Y/Y and Y/Δ transformers are made equal to

$$V_{YY}(\omega t) = V_{dc} + \sum_{k=1}^{\infty} A_{Yk}\cos(6k\omega t) \tag{4.3}$$

$$V_{Y\Delta}(\omega t) = V_{dc} + \sum_{k=1}^{\infty} A_{\Delta k}\cos(6k\omega t) \tag{4.4}$$

and considering that the harmonics of orders $n = 6(2l-1) \pm 1$ $(l = 1,2,\ldots)$ cancel in a balanced 12-pulse converter, it can be shown [8] that the addition of the winding voltages on the primary side will produce a perfectly sinusoidal waveform if the following relation exists between the DC and AC components of the bridge voltages:

$$\sum_{k=1}^{\infty} \frac{A_{Yk}}{(12l \pm 1)^2 - 36k^2} = \frac{V_{dc}}{(12l \pm 1)^2} \qquad 1 = 1, 2 \ldots \tag{4.5}$$

which consists of a set of linear algebraic equations and variables.

In practice the harmonics of high orders, i.e. $(12m + 1)$ and above, can be ignored and the number of variables and equations reduces to $2m$. For a sufficiently large value of the parameter m the values of V_k can be approximated by the expressions

$$A_k = \frac{A_{Yk}}{M_{dc}} = \frac{2[2(-1)^k - \sqrt{3}]}{(2-\sqrt{3})(36k^2-1)} \approx \frac{14.9282(-1)^k - 12.9282}{36k^2-1} \qquad \mathrm{k} = 1, 2, \ldots \tag{4.6}$$

$$\frac{A_{\Delta k}}{M_{dc}} = (-1)^k \frac{A_{Yk}}{M_{dc}} \approx \frac{14.9282 - 12.9282(-1)^k}{36k^2-1} \qquad \mathrm{k} = 1, 2, \ldots \tag{4.7}$$

Using the results derived from equations (4.6) and (4.7), the normalized reinjection voltage waveforms to be applied to the star- and delta-connected bridges become

$$V_Y(x) = 1 + \sum_{k=1}^{\infty} A_k \cos(6kx) \tag{4.8}$$

$$V_\Delta(x) = 1 + \sum_{k=1}^{\infty} (-1)^k A_k \cos(6kx) \tag{4.9}$$

The above analysis has been developed in terms of voltages, as required by the VSC configuration of Figure 4.5. However, they are equally applicable to the generic 12-pulse CSC configuration shown in Figure 4.6. In this case the winding currents on the bridge side of the interface transformers are given by:

$$I_{\Delta a}(\omega t) = \begin{cases} I_{B\Delta}(\omega t)/3 & 0 < \omega t < \pi/3 \\ 2I_{B\Delta}(\omega t)/3 & \pi/3 < \omega t < 2\pi/3 \\ I_{B\Delta}(\omega t)/3 & 2\pi/3 < \omega t < \pi \\ -I_{B\Delta}(\omega t)/3 & \pi < \omega t < 4\pi/3 \\ -2I_{B\Delta}(\omega t)/3 & 4\pi/3 < \omega t < 5\pi/3 \\ -I_{B\Delta}(\omega t)/3 & 5\pi/3 < \omega t < 2\pi \end{cases} \tag{4.10}$$

$$I_{Ya}(\omega t) = \begin{cases} 0 & 0 < \omega t) < \pi/6 \\ I_{BY}(\omega t) & \pi/6 < \omega t) < 5\pi/6 \\ 0 & 5\pi/6 < \omega t) < 7\pi/6 \\ -I_{BY}(\omega t) & 7\pi/6 < \omega t) < 11\pi/6 \\ 0 & 11\pi/6 < \omega t) < 2\pi \end{cases} \tag{4.11}$$

By comparing equations (4.1) and (4.2) with (4.10) and (4.11), and using the duality relation between the VSC and CSC configurations, the results obtained for VSC can be easily extended

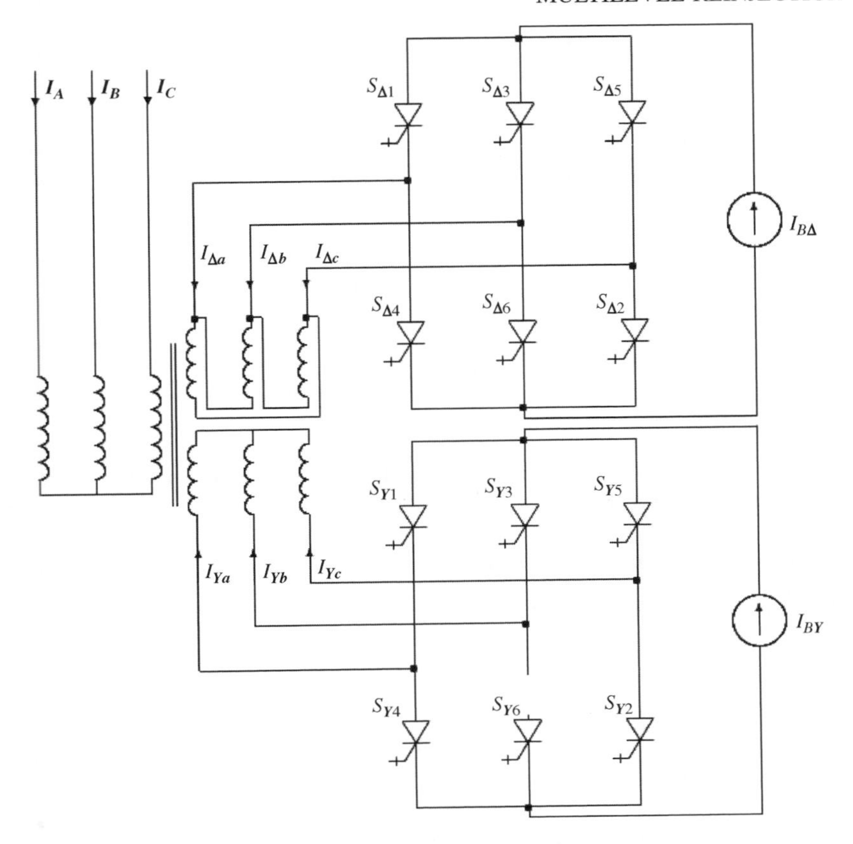

Figure 4.6 The two bridges of the 12-pulse CSC configuration.

to CSC, i.e. the currents to be supplied to the star- and delta-connected bridges are

$$I_{BY}(\omega t) = I_{dc} + \sum_{k=1}^{\infty} A_{Yk}\cos(6k\omega t) \tag{4.12}$$

$$I_{B\Delta}(\omega t) = I_{dc} + \sum_{k=1}^{\infty} A_{\Delta k}\cos(6k\omega t) \tag{4.13}$$

The corresponding currents in the secondary windings $I_{Ya}(\omega t)$ and $I_{\Delta a}(\omega t)$ only include harmonics of orders $n = 6(2l - 1) \pm 1$ ($l = 1,2,\ldots$). Thus, as explained for the VSC configuration, the following relations must exist between the DC and AC components of the bridge currents for complete harmonic cancellation:

$$\sum_{k=1}^{\infty} \frac{(-1)^k A_{Yk}}{(12l \pm 1)^2 - 36k^2} = \frac{I_{dc}}{(12l \pm 1)^2} \qquad l = 1, 2, \ldots \tag{4.14}$$

$$\sum_{k=1}^{\infty} \frac{A_{\Delta k}}{(12l \pm 1)^2 - 36k^2} = \frac{I_{dc}}{(12l \pm 1)^2} \qquad l = 1, 2, \ldots \tag{4.15}$$

Therefore, instead of using the VSC or CSC terminology, the signals to be injected to the star- and delta-connected bridges can be expressed in terms of the variable X, to make them applicable to both, i.e.:

$$X_Y(x) = 1 + \sum_{k=1}^{\infty} A_k \cos(6kx) \tag{4.16}$$

$$X_\Delta(x) = 1 + \sum_{k=1}^{\infty} (-1)^k A_k \cos(6kx) \tag{4.17}$$

The ideal reinjection waveforms for the star and delta bridges are shown in Figures 4.7(a) and (b) respectively. These two waveforms possess the following important characteristics:

1. Zero values appear at the points where the switches in the bridge are turned on and off.
2. The derivatives of the waveforms are limited, particularly around the zero values, where the power switches change their states.
3. The two waveforms add to a DC level with some ripple as shown in Figure 4.7(c).

The first characteristic (1) indicates the possibility of achieving ZVS (zero voltage switching) or ZCS (zero current switching). The second characteristic (2), ensures operation at low dv/dt or di/dt conditions. The third characteristic (3), implies that the two bridges must be supplied by a DC source with some controllable ripple. In practice, however, it is impractical to generate such ripple and some approximations to avoid it are considered next.

4.3.2 Symmetrical approximation to the ideal injection

The reinjection waveforms in Figures 4.7(a) and (b) are symmetrical about the vertical axis but not about any of the cross points of the waveform with its DC average. (That is why their sum is not a constant DC). To avoid the use of complex DC power conditioning to derive the required DC waveform (with controllable ripple), the reinjection signal is approximated by a fully symmetrical waveform. There are two possible solutions for this purpose, (i) using a waveform that minimizes the integration of the error square and the error derivative square (ESEDS) and (ii) using a linearly rising and linearly falling waveform (of constant derivative).

Using the VSC terminology, the ESEDS symmetrical waveform $V_{Ys}(x)$ is obtained by solving the expression

$$\min\left\{\int_0^{\pi/6}\left[[V_Y(x)-V_{Ys}(x)]^2+\left(\frac{d[V_Y(x)-V_{Ys}(x)]}{dx}\right)^2\right]dx+\int_0^{\pi/6}\left[[V_\Delta(x)-V_{\Delta s}(x)]^2\right.\right.$$
$$\left.\left.+\left(\frac{d[V_\Delta(x)-V_{\Delta s}(x)]}{dx}\right)^2\right]dx\right\} \tag{4.18}$$

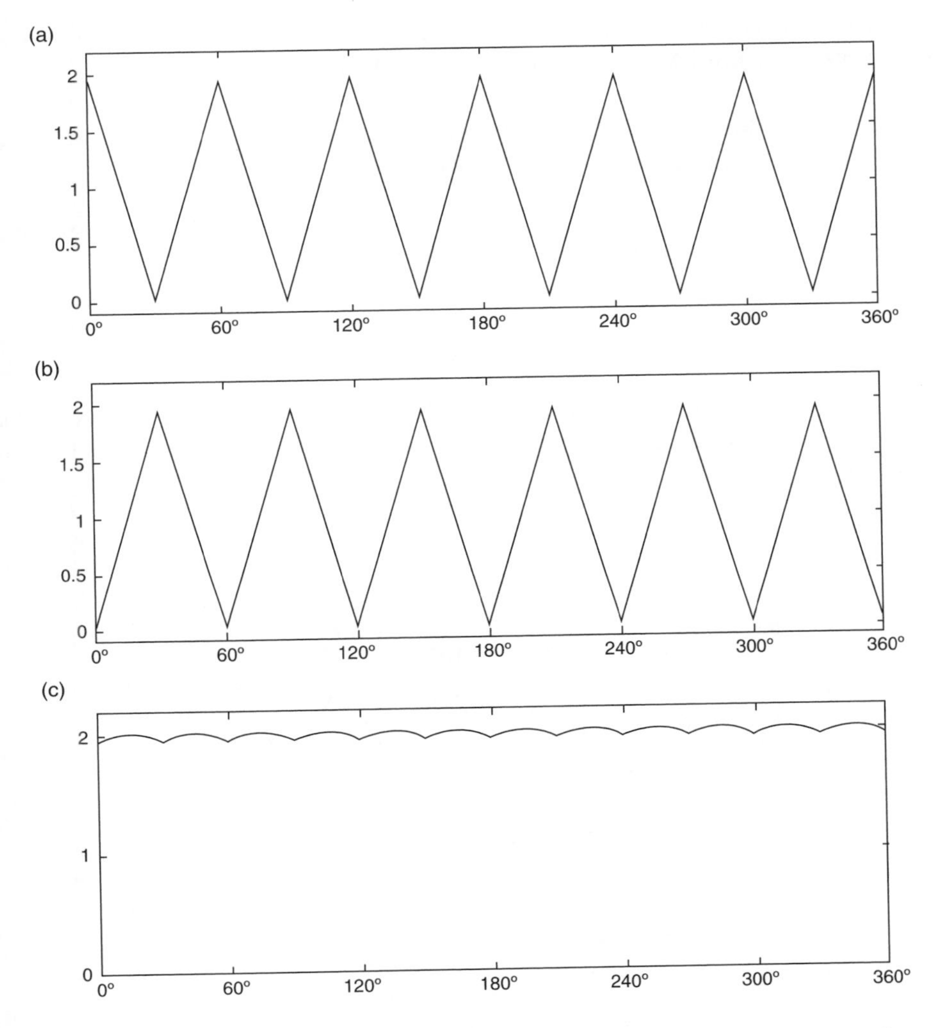

Figure 4.7 The ideal reinjection waveforms. (a) *$X_Y(x)$ The ideal reinjection waveform for Y-bridge.* (b) *$X_\Delta(x)$ The ideal reinjection waveform for Δ-bridge.* (c) *$X_Y(x) + X_\Delta(x)$ The required DC source for the ideal reinjection.*

under the conditions of symmetry and area equality between the two groups of curves, i.e.

$$V_{Ys}(x) + V_{\Delta s}(x) = 2 \qquad \text{for } 0 < x < \pi/6$$

and

$$\int_0^{\pi/12} V_{Ys}(x)dx = \int_0^{\pi/12} V_Y(x)dx \qquad \int_{\pi/12}^{\pi/6} V_{Ys}(x)dx = \int_{\pi/12}^{\pi/6} V_Y(x)dx$$
$$\int_0^{\pi/12} V_{\Delta s}(x)dx = \int_0^{\pi/12} V_\Delta(x)dx \qquad \int_{\pi/12}^{\pi/6} V_{\Delta s}(x)dx = \int_{\pi/12}^{\pi/6} V_\Delta(x)dx \tag{4.19}$$

Based on the numerical results of the symmetrical waveforms V_{Ys} and $V_{\Delta s}$, the Fourier components of the ESEDS waveform $V_{Ys}(x)$ are approximately given by

$$A_k = \frac{(7+4\sqrt{3})[1-(-1)^k]}{(36k^2-1)} \approx \frac{13.9282[1-(-1)^k]}{36k^2-1} \qquad \mathrm{k} = 1,2,\ldots \tag{4.20}$$

Similarly $V_{\Delta s}$ can be obtained by the application of a 30° phase displacement between them.

The spectrum of the linear symmetrical waveform is:

$$A_k = \frac{4[1-(-1)^k]}{k^2\pi^2} \approx \frac{0.4053[1-(-1)^k]}{k^2} \qquad \mathrm{k} = 1,2,\ldots \tag{4.21}$$

The maximum difference between the ESEDS and linear symmetrical waveforms is less than 2.5%. Therefore, considering its simplicity of implementation, the linear waveform is the preferred approximation.

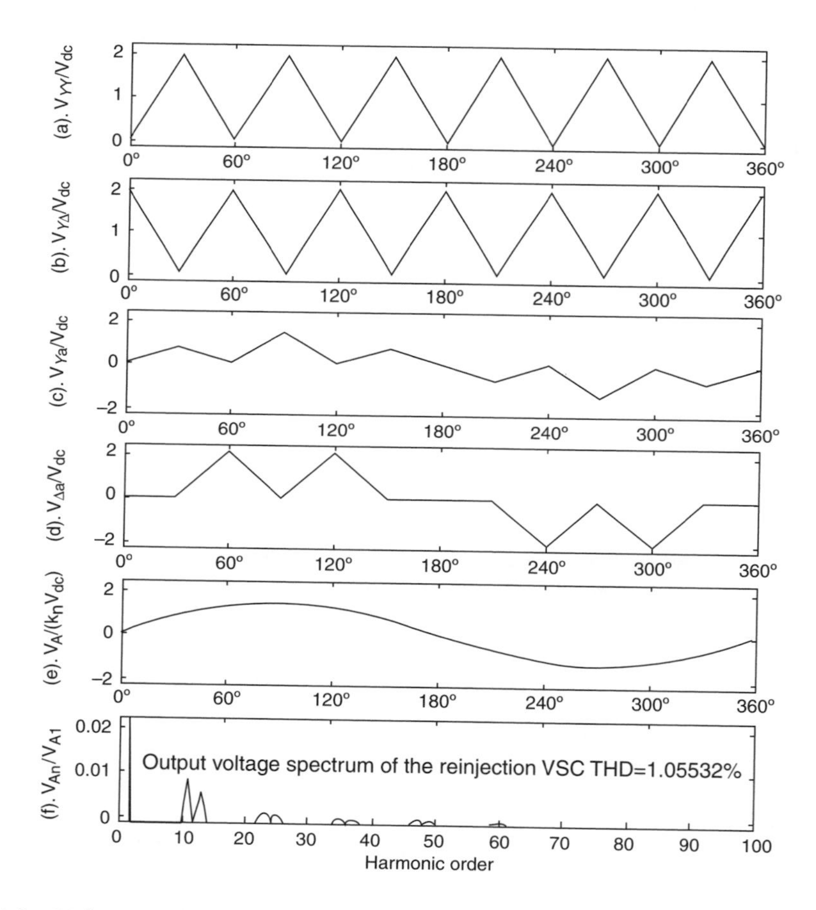

Figure 4.8 Voltage waveforms of a 12-pulse voltage source converter with multilevel linear reinjection.

The voltage waveforms of the linear approximation are shown in Figure 4.8, and the total harmonic distortion of the output voltage is:

$$THD_V = \sqrt{\sum_{k=1}^{\infty} \frac{1}{(12k-1)^4} + \sum_{k=1}^{\infty} \frac{1}{(12k+1)^4}} = \sqrt{\frac{\pi^4(40+23\sqrt{3})}{8x12x81} - 1} \approx 1.05532\% \tag{4.22}$$

The use of the ESEDS or linear symmetrical reinjection waveforms produces output waveforms with under 1% total harmonic distortion (THD), without the need to modify the DC power source. However, to derive these waveforms from the DC voltage (or current) is still impractical and a further approximation of the triangular waveform in the form of a multilevel reinjection [8] is described next.

4.4 Multilevel reinjection (MLR) – the waveforms

Figure 4.9 illustrates an eight-step approximation to the triangular reinjection concept developed in Section 4.3.2 and shown in Figure 4.8. It must be understood that these waveforms (shown for the case of VSC) are equally applicable to CSC, in which case the waveforms would be currents instead of voltages. Figure 4.9(a) and (b) are the voltage waveforms on the converter side of the star- and delta-connected bridges; they result from the addition of the DC voltage and the reinjected waveform. Figure 4.9(c) and (d) show the corresponding AC voltage waveforms on the primary sides of the converter transformers and Figure 4.9(e) is the converter output voltage, which is the addition of (c) and (d). The frequency spectrum shown in Figure 4.9(f) indicates that the harmonic content of the output voltage is very small. The fully symmetrical multilevel reinjection waveforms of the two bridges always add to a perfect DC.

The waveforms are analysed here under ideal conditions, i.e. all the DC capacitances are of infinite size, the on state switches volt-drop is zero and their off state impedance infinite, the free-wheeling diodes are ideal and the transformer leakage reactance is ignored. Also the voltages across the capacitors are constant and of the same value, i.e. $V_L = \frac{V_{dc}}{m-1}$.

The voltage waveforms across the main bridges, shown in Figures 4.9(a) and (b) for the eight-level configuration, illustrate that these change every 30/$(m-1)$ degrees in the sequence $0, V_L, 2V_L, \ldots (m-1)V_L, (m-2)V_L, \ldots, V_L, 0$ and that the zero voltages (coinciding with the valve commutations) occur around the points (0°,60°,...360°) and (30°,90°,...330°) for the Y/Y and Y/Δ connected bridges respectively.

In Figures 4.9(a) and (b) the voltages across the bridges connected to the Y/Y and Y/Δ interface transformers vary between zero and full DC voltage every 30° in $(m-1)$ steps. The voltage across a secondary phase of the star-connected transformer (V_{Ya}), shown in Figure 4.9(c), is one third (between 0 and 60°), two thirds (between 60 and 120°) and again one third (between 120 and 180°).

A secondary winding of the delta-connected bridge is either directly connected to the bridge terminal through the on state valves or short-circuited by them, thus $V_{\Delta a}$ (shown in Figure 4.9(d)) is zero (between 0 and 30°), the bridge voltage (between 30 and 150°), and zero again (between 150 and 180°). The last two voltages add together on the series-connected primary sides of the transformers to constitute the output voltage, shown in Figure 4.9(e).

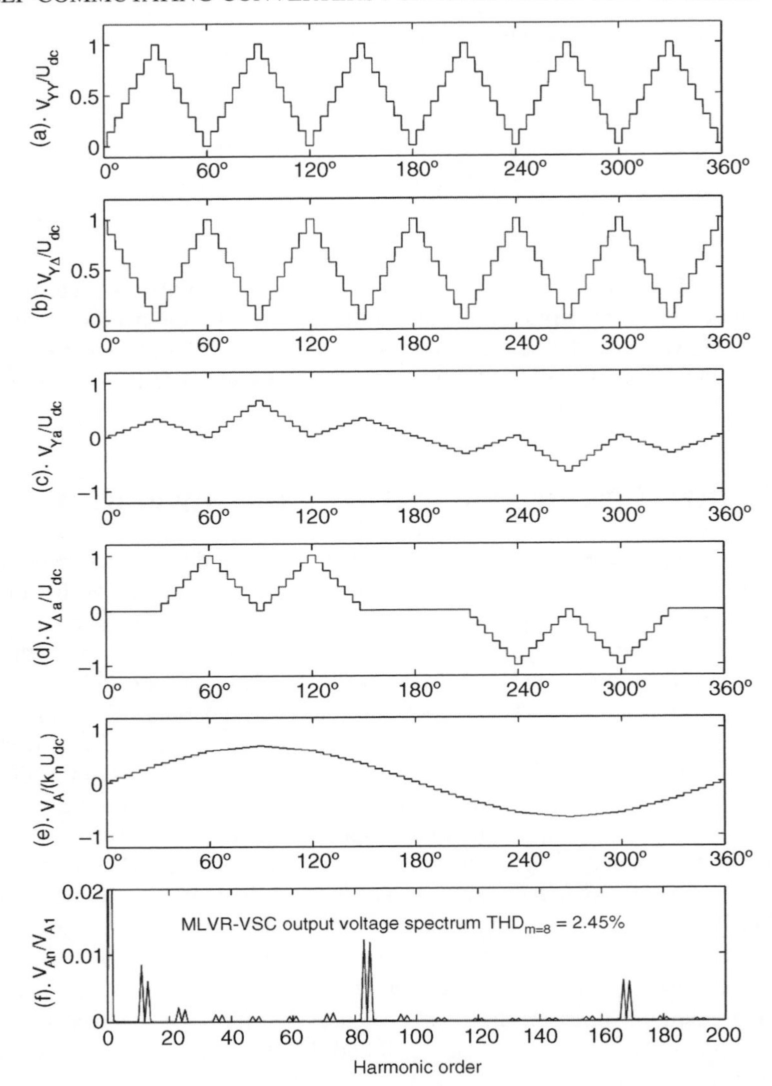

Figure 4.9 Voltage waveforms of the eight-level MLVR voltage source converter.

The Fourier components of the output voltage (V_A) are given by the expression:

$$V_{An} = \frac{8[1-(-1)^n]k_n V_{dc}}{3n(m-1)} \sin\left(\frac{n\pi}{12(m-1)}\right)\cos\left(\frac{n\pi}{6}\right)\left[\cos\left(\frac{n\pi}{6}\right)+\frac{\sqrt{3}}{2}\right]\left[(m-1)\sin\left(\frac{n\pi}{6}\right)\right.$$
$$\left. +\sqrt{3}\sum_{i=1}^{m-2} i\sin\left(\frac{n\pi}{6}+\frac{in\pi}{6(m-1)}\right) - \sum_{i=1}^{m-2} i\sin\left(\frac{in\pi}{6(m-1)}\right)\right] \quad \text{for } m \geq 3, n = 1,2,3,\ldots \tag{4.23}$$

From the above expression, the fundamental peak value of the output voltage is:

$$V_{A1} = \frac{16k_n V_{dc}}{\pi(m-1)} \sin\left(\frac{\pi}{12(m-1)}\right) \left[\frac{(m-1)}{2} + \sum_{i=1}^{m-2} i\cos\left(\frac{\pi}{6}\right) - \frac{i\pi}{6(m-1)}\right)\right] \tag{4.24}$$

and the total harmonic distortion:

$$THDv_A = \sqrt{\frac{\pi^2(4+\sqrt{3})(m-1)^2\left[1+\frac{11-6\sqrt{3}}{13(m-1)^2}\right]}{27x128\sin^2\left(\frac{\pi}{12(m-1)}\right)\left[\frac{(m-1)}{2}+\sum_{i=1}^{m-2} i\cos\left(\frac{\pi}{6}-\frac{i\pi}{6(m-1)}\right)\right]^2} - 1} \tag{4.25}$$

The approximate output voltage total harmonic distortions for different reinjection levels numbers are:

MLVR level	3	4	5	6	7	8
THD_{MLVR} %	7.8	5.3	4	3.3	2.77	2.5

4.5 MLR implementation – the combination concept

As the AC component parts of the reinjection waveform of the two bridges always have the same level but they are in opposite directions, a common AC component 'generator' can be used by the two bridges. The common AC component combines with the DC voltage (or current) of the two main bridges by adding to or substracting from them. To maintain coordination between the DC source and the generated AC component, the latter must be derived from (or follow) the DC source. Moreover, the derived AC component has to be isolated from the DC supply.

4.5.1 CSC configuration [5]

Figure 4.10 shows the reinjection scheme for the series-connected 12-pulse current source converter. The AC components of the reinjection waveform are step levels of the DC current determined by the turn-ratios of a multi-tapped reinjection transformer. The self-commutating switches of the reinjection circuit are switched on and off to approximate the required waveform by selecting the appropriate turns ratio for specified time intervals.

4.5.1.1 Zero current switching [6]

An important property of the multilevel current reinjection scheme is its capability to control the position, magnitude and duration of the reinjection steps. If these parameters are optimized to achieve maximum harmonic cancellation, for every pair of taps symmetrically placed with respect to the two reinjection transformer secondaries, the pulse number is doubled, with the midpoint tap and short-circuiting switch pair ($S_{pj0}-S_{nj0}$) adding an extra multiplication factor. Thus the five-level configuration can achieve 60-pulse conversion [5(reinjection level number) × 6(reinjection frequency ratio) × 2(number of bridges)]. The converter valve currents are forced

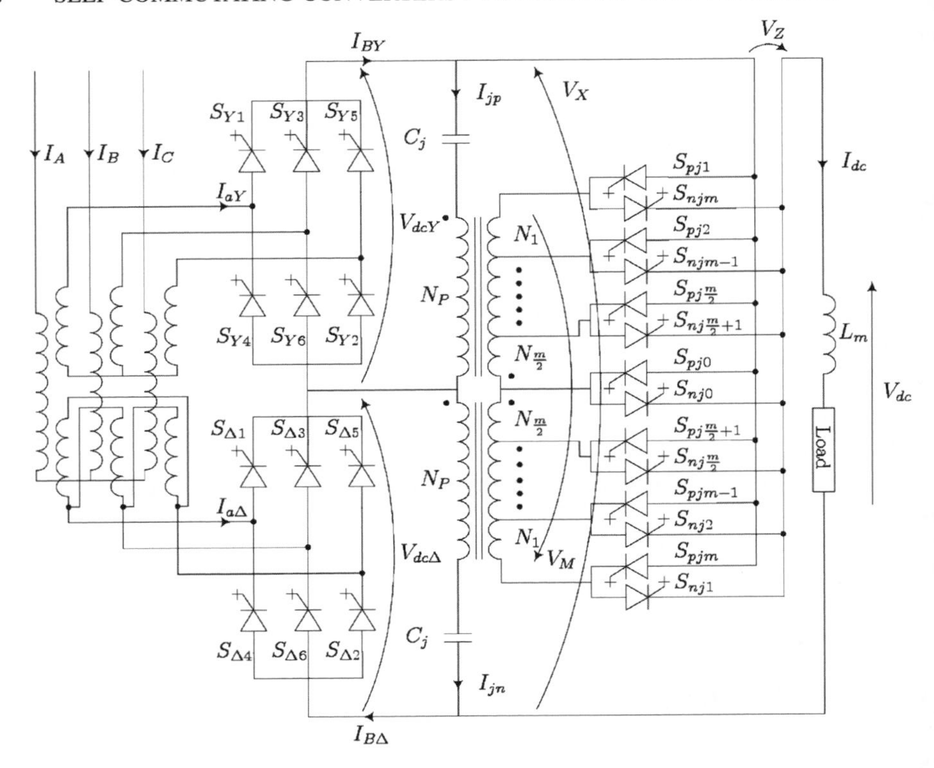

Figure 4.10 Structure of the series MLCR current source converter.

to a very low value (i.e. to an almost zero current switching condition (ZCS)) during the commutations, which simplifies the design of the snubber circuits.

Moreover, for the five-level reinjection configuration, the use of a non-optimal reinjection waveform (from the harmonics viewpoint) can force a ZCS condition for an interval of 6° (or 333 μs at 50 Hz) during the commutation. This should permit the outgoing thyristor to recover its blocking capability and thus make the conventional thyristor converter self-commutating. To achieve the ZCS condition the quality of the AC current and DC voltage waveforms is somewhat reduced (from the optimum 60 to 48-pulse for the five-level reinjection scheme). If necessary, larger zero current regions can be achieved at the expense of further reduction in the harmonic reduction capability.

Let us consider the circuit of Figure 4.10 (with five levels) operating in the steady state, with valve (S_{Y1}) conducting. When the reinjection current forces the current of valve S_{Y1} to zero, none of the valves connected to the common cathode (CC) conduct and the DC current will continue to flow via the reinjection path. However, the next level of the multistep reinjection current should force a change in the DC current. This is prevented by the large DC reactor, that develops the necessary transient *emf* (with negative polarity on the bridge CC bus) to ensure that the anode of S_{Y3} becomes positive with respect to its cathode, irrespective of the potential of the AC voltage. Therefore, provided that valve S_{Y1} has by then recovered its blocking capability, it is possible to turn on valve S_{Y3} to provide a new path for the converter current. Thus the converter can commutate without the assistance of a turn-off pulse or a line-commutating voltage, i.e. it can be of the conventional thyristor type.

It is therefore possible to achieve self-commutation, as well as pulse multiplication, using conventional thyristors for the converters and GTOs (or IGCTs) for the reinjection switches. This alternative gives the thyristor converter flexibility similar to that of a forced-commutated voltage source converter, i.e. the ability to control both the DC voltage and the DC current, with leading or lagging power factor, as well as reducing the harmonic content. In other words, it combines the benefits of the robust and efficient conventional converter and the controllability of the advanced self-commutated technology. This is an important breakthrough that should give greater flexibility to thyristor-based HVDC transmission.

The following theoretical waveforms are shown in Figure 4.11 for the five-level reinjection scheme for the case when the converter bridges operate with a firing angle of -45° (i.e. supplying reactive power, even though the switches are thyristors):

- Figure 4.11(a) and (b) – $I_{B\Delta}$ and I_{BY} are the DC currents of the bridge converters modified by their respective reinjection currents I_{jn} and I_{jp}.
- Figure 4.11(c) and (d) – $I_{ca\Delta}$ and I_{aY} are the phase 'a' currents in the secondary windings of the delta- and star-connected windings respectively.

The total harmonic distortion (THD) of the output current waveform is 4%. On the DC side, the reinjection circuit increases the pulse number of the voltage waveform by a factor of four, i.e. to 48 pulses per cycle (for the five-level reinjection configuration).

It follows that the combination of a conventional thyristor converter and a self-commutated multilevel reinjection circuit can provide reactive power, as well as active power controllability. It also transforms the conventional CSC waveforms into multistep AC current and multipulse DC voltage waveforms. This applies equally to rectifier and inverter operation and to variable frequency supplies. The power conversion efficiency is high, because the rectified harmonic power is reinjected into the DC system. Moreover, the larger number of taps of the multilevel arrangement does not increase the total current rating of the reinjection switches very much, as the average currents of the individual reinjection switches reduce in inverse proportion to their number; of course the *rms* currents of the individual switches (and therefore the total *rms* current) will be higher than their average values.

4.5.2 VSC configuration

The multilevel configuration based on the series connection of H-bridge units, described in Section 3.5, is already used extensively in STATCOM controllers. However, it is not suitable for applications requiring a constant DC voltage, such as HVDC transmission. A new concept [7] of more general applicability is multilevel H-bridge reinjection, where the standard double-bridge converter constitutes the main converter and the series-connected H-bridge units are used as a level-controlling circuit separated from and common to the three phases of the main converter bridges. In this configuration, shown in Figure 4.12, the AC components of the reinjection waveform are generated by a chain of $\frac{m-1}{2}$ H-bridge cells. Each cell can output three different levels and their combination forms a bipolar m-level waveform. The voltage of each capacitor in an H-bridge cell is forced to be equal to $1/(m-1)$ of the total DC voltage (the voltage across the two main capacitors) during the maximum and minimum levels of the reinjection voltage.

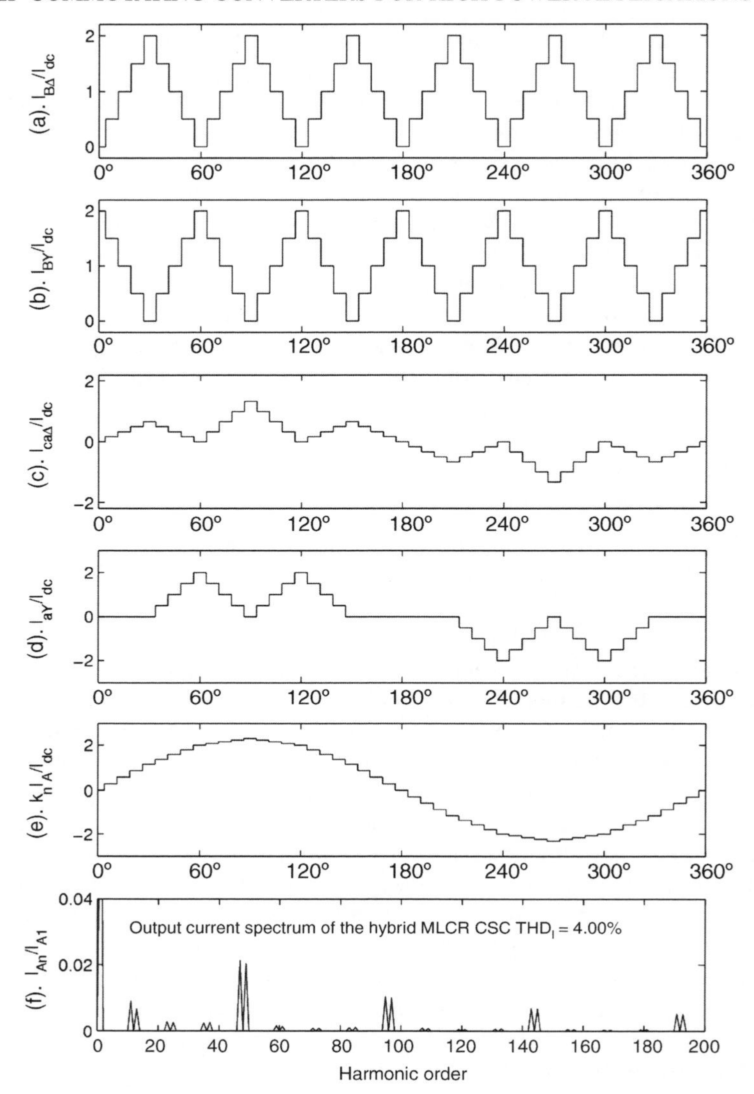

Figure 4.11 Theoretical current waveforms of the five-level reinjection current source converter.

Since each H-bridge generates three different output states, i.e. $-\frac{V_{dc}}{m-1}$ (state S_{-1}), 0 (state S_0) and $+\frac{V_{dc}}{m-1}$ (state S_{+1}), an appropriate combination of the $\frac{m-1}{2}$ bridge outputs can generate any one of the possible m voltage levels, i.e.:

$$-\frac{V_{dc}}{2}, -\frac{(m-3)V_{dc}}{2(m-1)}, \ldots, -\frac{V_{dc}}{m-1} \cdot \frac{2V_{dc}}{m-1}, \ldots, \frac{V_{dc}}{2}.$$

To achieve the linear reinjection waveform, the H-bridges are fired periodically at six times the fundamental frequency and the firings are synchronized with those of the two main bridges.

The multilevel reinjection voltage is added to the two DC capacitor voltages to shape the voltages across the two main bridges into equal linear staircase waveforms, but with a 30° phase difference. Finally, the series combination of the two main bridge voltages produces a high-quality output voltage waveform at the converter terminals.

The theoretical reinjection waveforms produced by the cascaded H-bridge configuration are also those illustrated in Figure 4.9; again, all the valve switchings of the main bridge valves take place under a zero voltage condition, both during leading and lagging voltage operation. This implies that four-quadrant power controllability can be achieved with only a few of the series-connected switches being of the self-commutating type, the remaining being thyristors to sustain the gradually increased voltage. The voltage rating of the chain of H-bridge cells is only half of the DC voltage, the number of extra switches required is proportional to the voltage rating and the current rating of the chain is only that of the AC component of the converter current. The main structural advantage of this configuration is that the number of main and reinjection switches is proportional to the converter voltage rating, as compared with the diode clamping alternative, where the number of diodes increases with the square of the level number; therefore this configuration is perfectly suited for high-voltage applications.

4.5.2.1 Switching pattern of the reinjection circuit

As explained earlier, following a change of switching state in either of the two main bridges, the reinjection waveform is increased or decreased in m step levels of equal width with a period of $\pi/6$. In the configuration of Figure 4.12, following a switching state change in the Y/Y bridge, the reinjection voltage is increased step by step; similarly, after a change of switching state in the Y/Δ bridge, the reinjection waveform is decreased step by step. In each case the reinjection voltage reference is the voltage from the neutral point of the two main bridges to the common node of the two DC capacitors.

To generate the required multilevel waveform, while achieving the lowest switching frequency (i.e. the reinjection voltage frequency), the $\frac{m-1}{2}$ H-bridges can be triggered in a variety of patterns. The preferred pattern for high-voltage application should use the lowest switching frequency (i.e. six times the fundamental) as well as provide self-balancing capacitor voltage capability.

Figure 4.13 illustrates a firing pattern where the output states of the H-bridges are all symmetrical about the vertical line at $\pi/6$, while the widths of the states (S_{+1}, S_0 and S_{-1}) of the individual H-bridges are not equal. If the $\frac{m-1}{2}$ H-bridges are numbered H_1, $H_2, \ldots, H_{(m-1)/2}$, the four state changing points in the $\pi/3$ period of bridge H_k are

$$\begin{aligned} a_{H_k} &= \frac{k\pi}{6m} \\ b_{H_k} &= \frac{\pi}{6} - \frac{k\pi}{6m} \\ c_{H_k} &= \frac{\pi}{6} + \frac{k\pi}{6m} \\ d_{H_k} &= \frac{\pi}{3} - \frac{k\pi}{6m} \end{aligned} \tag{4.26}$$

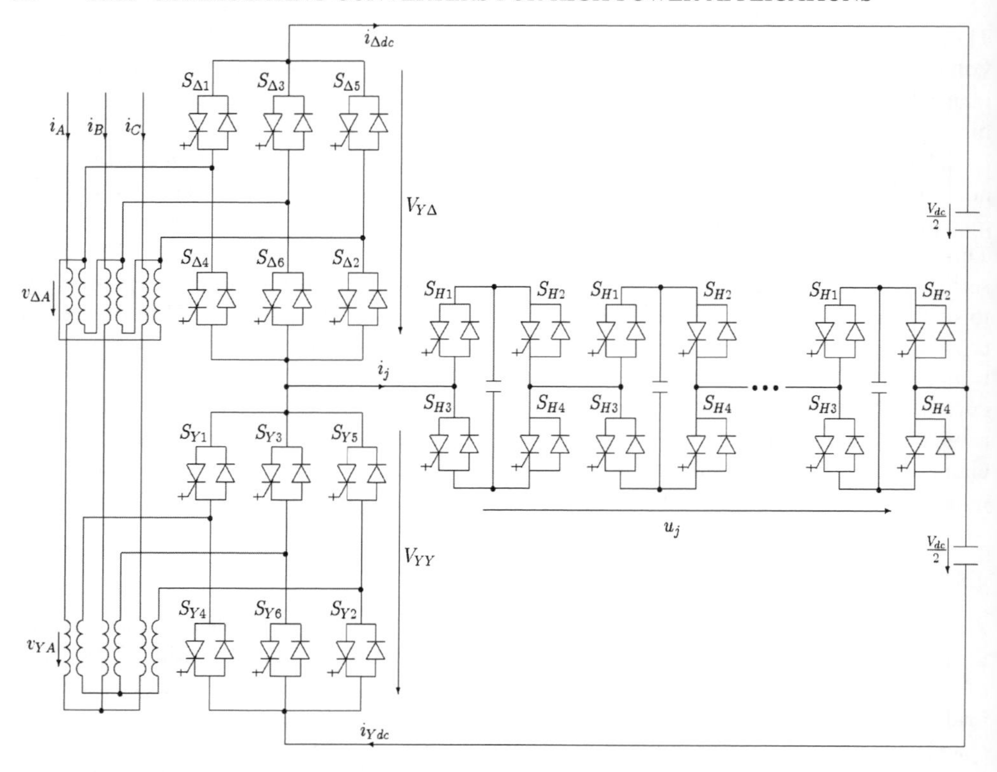

Figure 4.12 MLVR-HB voltage source converter configuration.

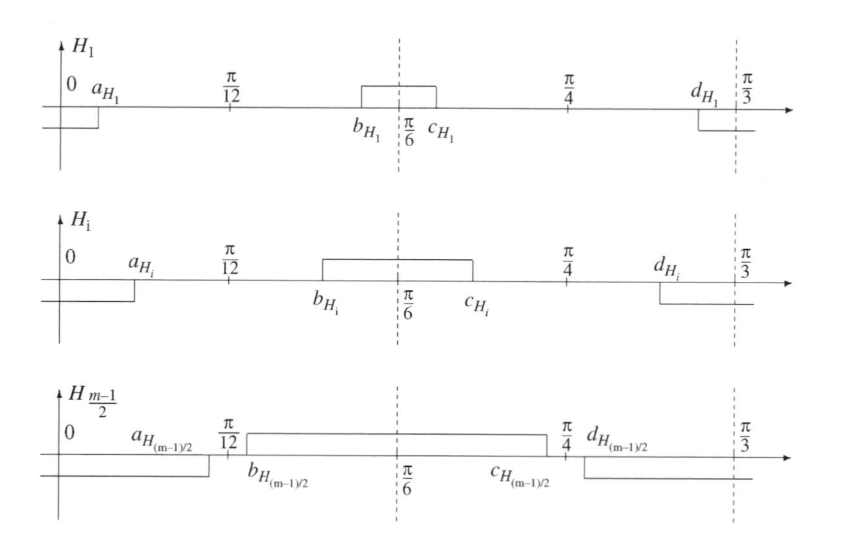

Figure 4.13 Switching pattern of the reinjection bridges.

The firing sequence of the four switches in the kth bridge can be arranged as shown in Figure 4.14. $S_{H_{k1}}$ is changed from off-state to on-state at point b_{H_k}, and from on to off-state at d_{H_k};$S_{H_{k2}}$ is changed from on-state to off-state at point a_{H_k}, and from off to on-state at c_{H_k};$S_{H_{k3}}$ is changed from on-state to off-state at point b_{H_k}, and from off to on-state at d_{Hi};$S_{H_{k4}}$ is changed from off-state to on-state at point a_{H_k}, and from on to off-state at c_{H_k}. With this arrangement there are always two of the four switches in on-state and the kth H-bridge output states are determined by two switches in on-state, i.e.:

$$H_k = S_{-1} \text{ is determined by } S_{H_{k2}} \text{ and } S_{H_{k3}} \text{ in on-state}$$

$$H_k = S_0 \text{ is determined by } S_{H_{k3}} \text{ and } S_{H_{k4}} \text{ in on-state}$$

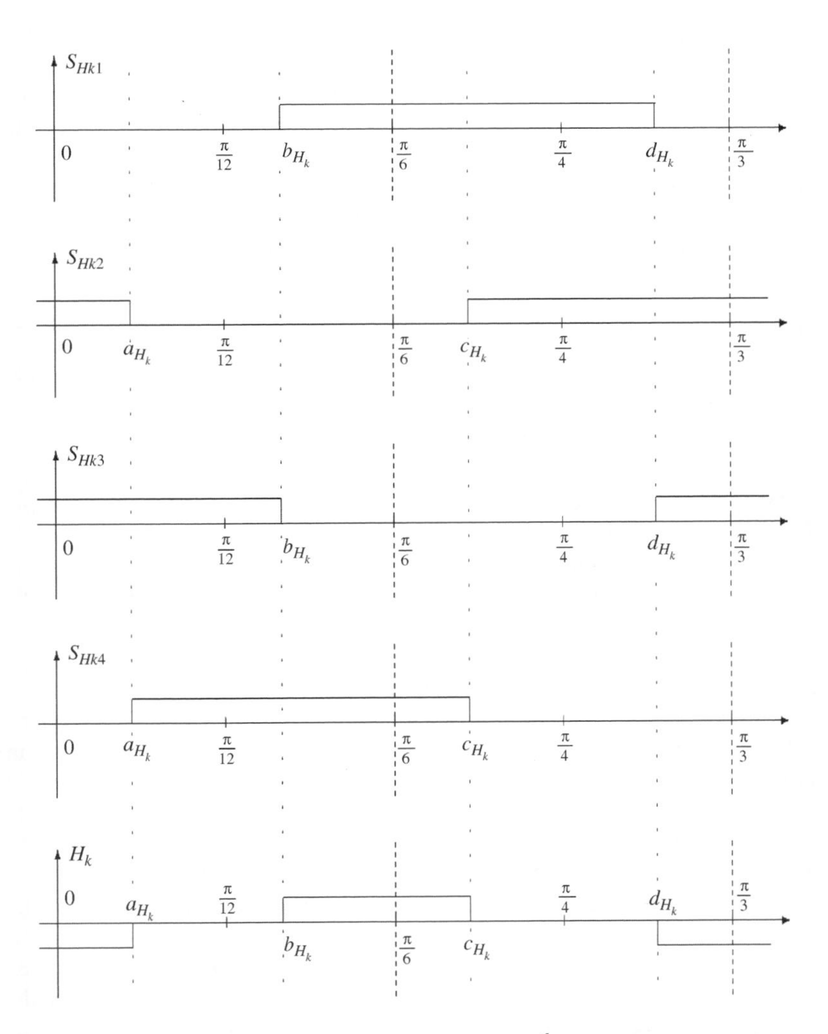

Figure 4.14 Switching sequence for the kth reinjection bridge.

$$H_k = S_{+1} \text{ is determined by } S_{H_{k1}} \text{ and } S_{H_{k4}} \text{ in on-state}$$

$$H_k = S_0 \text{ is determined by } S_{H_{k1}} \text{ and } S_{H_{k2}} \text{ in on-state}$$

$$H_k = S_{-1} \text{ is determined by } S_{H_{k2}} \text{ and } S_{H_{k3}} \text{ in on-state}$$

It is clear that the state $H_k = S_0$ can be generated either by $S_{H_{k1}}$ and $S_{H_{k2}}$ or $S_{H_{k3}}$ and $S_{H_{k4}}$ in on-state. With the firing sequences of Figure 4.14 for the four switches of the H-bridge, the firing frequency is six times the fundamental and the two zero output stages are used equally. These two properties, low-frequency switching and equal rating of the switches, make this arrangement attractive for HVDC application.

4.6 MLR implementation – the distribution concept

The linear reinjection waveforms supplied to the two main bridges add up to a constant DC at all instants and they can, thus, be generated by distributing appropriately the DC voltage (in the VSC case) or current (in the CSC case). The reinjection circuit operates like a controllable voltage or current divider to distribute the DC voltage or current to the main bridges. If the DC source is appropriately distributed to the two main bridges, the required relation between the AC and DC components of the reinjection waveforms is automatically achieved.

4.6.1 CSC configuration [8]

Figure 4.15 shows the $(m + 1)$ level MLCR configuration based on the 12-pulse parallel-connected converter. The multitapped reactor assisted by the switching action of the reinjection switches distributes the DC current I_{dc} to the two bridges in $(m + 1)$ level waveforms.

The main property of the multilevel current reinjection scheme is its capability to control the position, magnitude and duration of the reinjection steps. If these parameters are optimized to achieve maximum harmonic cancellation, for every pair of taps symmetrically placed with respect to the two reinjection transformer secondaries, the pulse number is doubled, with the midpoint tap and short-circuiting switch pair (S_{pj0}–S_{nj0}) adding an extra multiplication factor. Thus, as in the reinjection by combination case described earlier (for the series connection), a five-level reinjection scheme can achieve 60-pulse conversion (when the reinjection pulses are optimized to produce minimal harmonic content) or 48 pulses (when the reinjection pulses are controlled to ensure zero current during the commutations). The ZSC condition does not apply to the reinjection switches, but, due to the unidirectional nature of the current, the snubbers required can be of the simple resistor capacitor diode (RCD) type.

The commutations in one of the main bridges always occur in the regions where the reinjection circuit distributes the full current to the second bridge. That means that the valves in the two main bridges commutate after their currents decay to zero (i.e. under a ZCS condition) and this implies that the valves in the main bridges can be conventional thyristor chains. Moreover, the forced commutating switches in the reinjection circuit only sustain the AC component of the voltage difference between the two main bridge valves. Therefore the

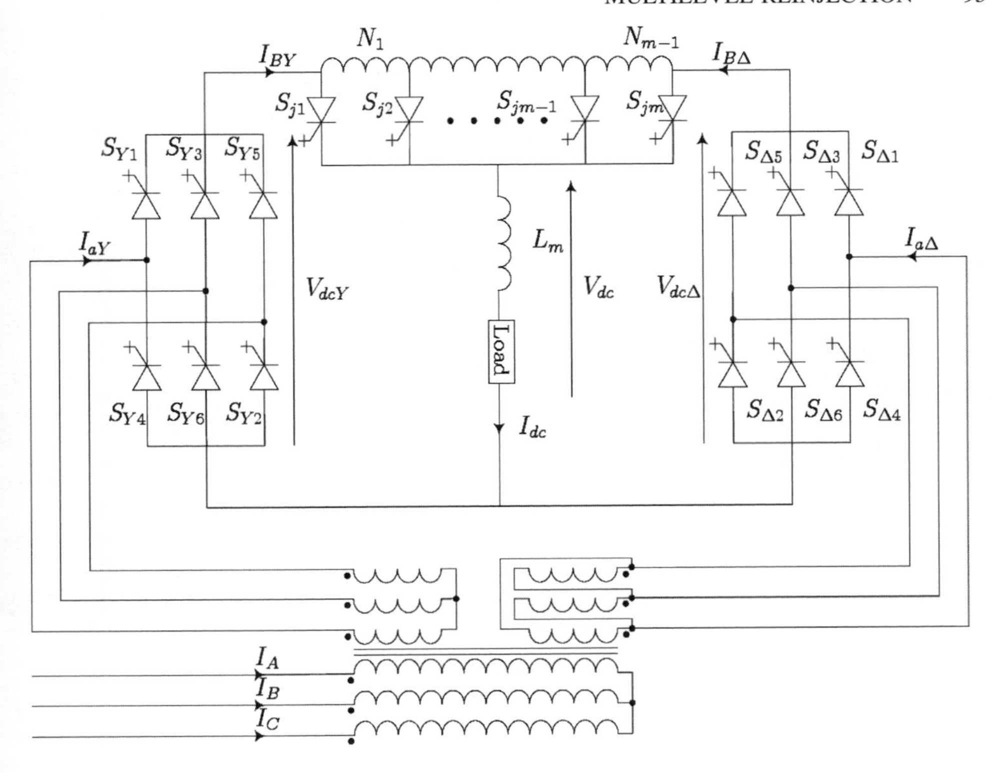

Figure 4.15 Structure of the parallel MLCR current source converter.

voltage rating of the reinjection circuit valves is lower than that of the main bridge valves. Also, since during the commutation of the reinjection valves, the inductive energy involved is restricted to that of the multitapped reactor, the rating of the reinjection switches snubber is relatively low.

4.6.2 VSC configuration [9]

Figure 4.16 shows a simplified clamping structure for the series-connected VSC configuration. In this case the series-connected DC capacitors and the forced commutated switches function as a controllable voltage divider to distribute the DC voltage to the main bridges. Again, the commutations in one of the main bridges occur in the regions where the reinjection circuit distributes all the DC voltage to the second bridge. Therefore the valves in the two main bridges commutate after their voltage is clamped to zero. This effect, combined with the gradually increasing voltage across the main bridge valves, should permit the use of only a few forced commutating switches (needed to switch the valve current off), while the remaining switches can be conventional thyristors (needed to sustain the voltage). As the DC component of the output current of the main bridges does not pass through the reinjection switches, the current rating of the reinjection circuit is relatively low.

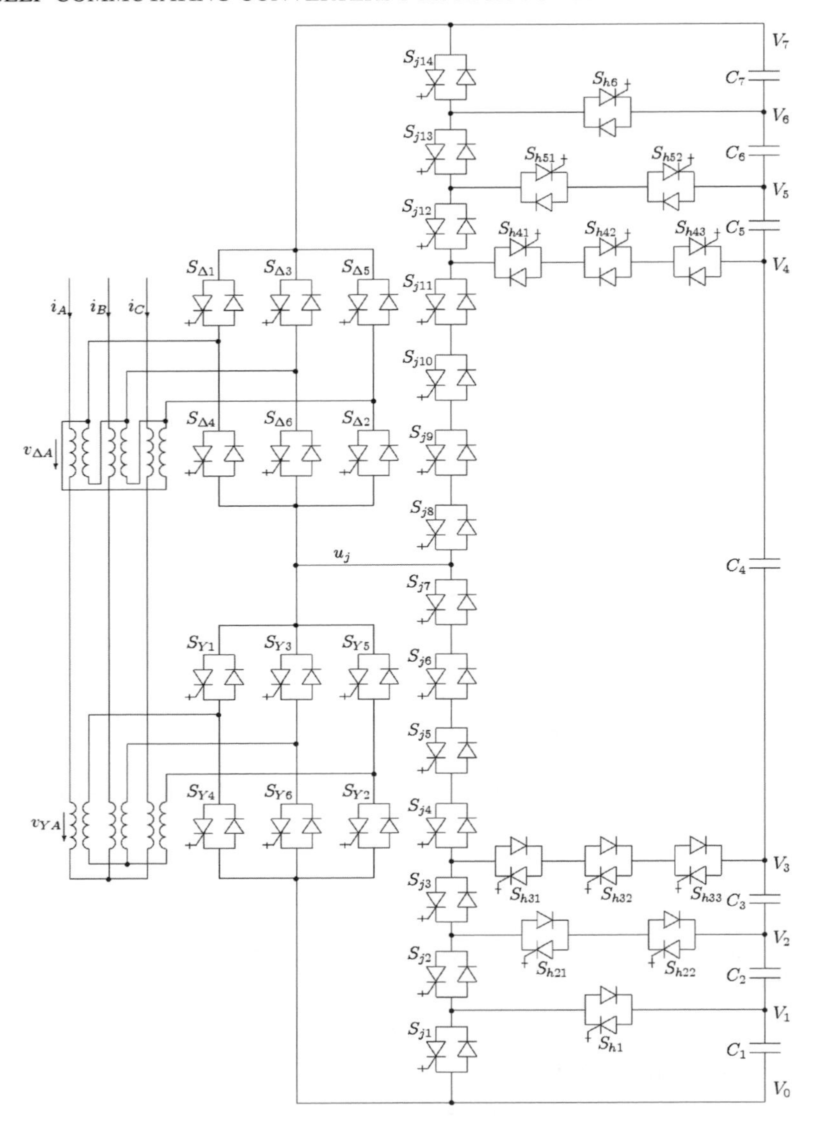

Figure 4.16 Structure of an eight-level MLVR voltage source converter.

4.7 Summary

It has been shown theoretically that a perfectly sinusoidal conversion can be achieved by injecting an ideal waveform at six times fundamental frequency to the two main bridges of the twelve-pulse VSC and CSC configurations. However, the power conditioning required to derive such a waveform is impractical. A symmetrical triangular approximation has then been developed that achieves a high-quality AC waveform (with a THD of 1%). Although the latter is still impractical to implement, it provides the basis for an interesting solution in which the triangular reinjection signal is further approximated by a multilevel waveform. This is an effective alternative, as the waveform can be derived from the DC side voltage (in the case of VSC) or current (in the case of CSC) of the converter.

The main features of the reinjection conversion are:

- Quality voltage and current waveforms and low switching frequency (the main bridges at fundamental and the reinjection switches at six times fundamental frequency).
- High voltage rating, as the zero switching condition achieved by the reinjection system enables an effective direct series connection of the main bridge switching devices with reduced dv/dt stress without dynamic voltage balancing problems.
- High efficiency (due to the use of thyristors as the main bridge switches, low switching frequency and soft switching of the main valves).
- Reduced switching complexity as the number of switching devices only increases in direct proportion to the level number.
- Flexible reactive power controllability.

References

1. Bird, B.M., Marsh, J.F. and McLellan, P.R. (1969) Harmonic reduction in multiplex convertors by triplen frequency current injection. *IEE Proceedings*, **116** (10), 1730–34.
2. Ametani, A. (1972) Generalised method of harmonic reduction in a.c.-d. c. converter by harmonic current injection. *IEE Proceedings*, **119**, 857–64.
3. Baird, J.F. and Arrillaga, J. (1980) Harmonic reduction in dc ripple reinjection. *IEE Proceedings*, **127**, 294–303.
4. Arrillaga, J. and Villablanca, M. (1991) 24-pulse HVDC conversion. *IEE Proceedings*, **138** (1), 57–64.
5. Liu, Y.H., Perera, L.B., Arrillaga, J. and Watson, N.R. (2007) Application of the multi-level current reinjection concept to HVDC Transmission, *IET Generation, Transmission & Distribution*, **1** (3), 399–404.
6. Arrillaga, J., Liu, Y.H., Perera, L.B. and Watson, N.R. (2006) A current reinjection scheme that adds self-commutation and pulse multiplication to the thyristor converter. *IEEE Transactions on Power Delivery*, **21** (3), 1593–9.
7. Liu, Y.H., Arrillaga, J. and Watson, N.R. (2008) Cascaded H-bridge voltage reinjection – Part I: a new concept in multi-level voltage source conversion. *IEEE Transactions on Power Delivery*, **23** (2), 1175–82.
8. Perera, L.B., Liu, Y.H., Watson, N.R. and Arrillaga, J. (2005) Multi-level current reinjection in double-bridge self-commutated current source conversion. *IEEE Transactions on Power Delivery*, **20** (2), 984–91
9. Liu, Y.H., Arrillaga, J. and Watson, N.R. (2004) Multi-level voltage reinjection – a new concept in high power voltage source conversion. *IEE Proceedings on Generation, Transmission & Distribution*, **151** (3), 290–8.

5

Modelling and Control of Converter Dynamics

5.1 Introduction

A common feature of self-commutated VSC and CSC power converters is the control of the electrical power conversion by the generation of either a controllable fundamental frequency AC voltage or current derived from a DC source, or the generation of controllable DC voltage and current averages derived from an AC source. The AC fundamental component and/or DC average of the converter output voltage and current are controlled by the conducting state combination of the converter switching devices driven by their gate signals.

The control of the magnitude and phase angle of the fundamental frequency voltage and current determines the active and reactive power output of a converter on its AC side, while the control of the DC voltage and current averages determine the active power on its DC side. The active power transfer can be either from the AC system to the DC source or vice versa and that implies that the power converter can act as a rectifier or an inverter. Moreover, by controlling the converter such that the AC voltage leads or lags the current, the converter can either absorb or generate reactive power. The four-quadrant active and reactive controllability of the self-commutated power converters makes them ideally suited for a variety of applications.

The main differences between the self-commutated VSC and CSC configurations are:

- The voltage source converter controls its AC terminal fundamental voltage directly and its AC current indirectly. It provides a fully rated range of DC current control, but only a relatively small range of DC voltage control.
- The current source converter controls its AC fundamental output current directly and its AC terminal voltage indirectly. It provides a fully rated range of DC voltage and current control.

High-power self-commutated VSC and CSC controllers have now become important components of power transmission systems in the form of FACTS and HVDC.

To achieve the expected control objectives a suitable control strategy must be implemented based on the available devices and technologies. The control strategies for the power converter systems are closely related to their applications and the objectives to be achieved.

The converter operation states are controlled by the gate driving signals of the switching devices and this chapter derives the dynamic models required to assess the relations between the gate orders and the operating states for both the VSC and CSC converter alternatives.

5.2 Control system levels

The converter system control is conveniently divided into three levels, namely firing, state and system control. These three levels are related as shown in the block diagram of Figure 5.1.

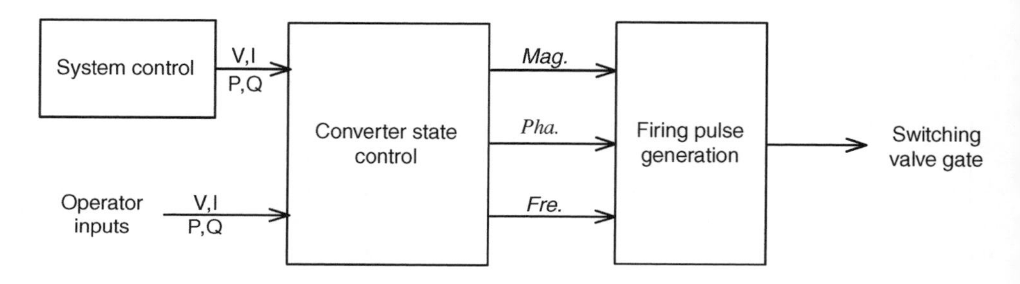

Figure 5.1 Control levels of a power converter system.

5.2.1 Firing control

The converter firing control depends on the conducting state combinations of the switching devices and the converter configuration. The valve firing signal sequence is related to the firing patterns applied. A well constructed firing pattern can generate frequency, magnitude and phase angle controllable fundamental voltage or current. If the converter is connected to an active AC power system, the fundamental frequency must be synchronized to the AC source. If the converter is connected to a passive load the frequency has to be controlled independently by the firing unit. The AC voltage magnitude can be controlled either within a small range or a wide range depending on the specific application requirement. The phase angle is commonly referenced to the AC source phase angle; therefore, the firing signal unit has to monitor the AC source voltage signal in real time for the frequency synchronization and phase displacement control.

The frequency, magnitude and phase angle parameters of a firing pattern are the control input variables to the gate firing signal unit. The gate firing signal unit generates the firing signals for the converter switches to produce the required fundamental component defined by these input variables. An important factor in determining the converter system dynamic response is the time delay taken by the gate firing signal unit to respond to the control input variables in real time; for quick dynamic response, this delay must therefore be minimized.

5.2.2 Converter state control

The converter state control level performs the AC and DC side terminal voltage and current regulation, as well as the active and reactive power control, to achieve the objectives required at the system level. The converter state control also maintains other required converter operating conditions, such as the DC voltage level (in a voltage source converter), the DC current (in a current source converter) and the capacitor voltage balance (in a multilevel voltage source converter).

The main states to be controlled are:

- *The AC voltage.* The voltage source converter AC terminal voltage can be directly controlled by regulating the magnitude of the fundamental frequency component of its AC voltage; while the current source converter AC terminal voltage is indirectly controlled by regulating the magnitude and the phase angle of the fundamental frequency component of the AC current. The directly controlled voltage source converter terminal voltage is regulated either by altering the DC capacitor voltage (in fundamental switching patterns) or by varying the modulation index (in PWM switching patterns).
- *The AC current.* The control of the converter AC terminal current is probably the most critical state, as it ensures that the converter valves are not overloaded. The current source converter AC terminal current is directly controlled, either by altering the DC current (in fundamental switching patterns) or by varying the modulation index (in PWM switching patterns). The voltage source converter AC terminal current is indirectly controlled by regulating the magnitude and the phase angle of the fundamental frequency component of its AC output voltage.
- *The DC voltage and current.* When the voltage source converter is connected to an active AC system, the voltage source converter DC side voltage must be maintained around a specific level in order to ensure operating normally without serious over-current. For some applications, however, the DC voltage is required to vary within a large range and in such cases a chopper needs to be inserted between the voltage source converter and the receiving circuit in order to provide the large range of voltage variation while the voltage source converter DC voltage is kept nearly constant. The current source converter can be controlled to vary its DC voltage polarity and magnitude within its rated range; this DC voltage profile is required by many DC loads. The current source converter DC current can also be adjusted from zero to its rated level.
- *The active power.* The control of active power transfer is achieved by regulating the phase displacement angle between the AC source voltage and the fundamental frequency component of the converter-generated AC voltage. Power is taken from or delivered to the AC system depending on the sign of this angle. A controllable active power transfer between two asynchronous AC systems can be achieved by two converters with their AC terminals connected to the AC systems and their DC terminals connected together via a DC link. This type of converter configuration is typical of HVDC transmission, back-to-back interconnections and motor drives; in such cases the control of the two converters needs to be coordinated to achieve the active power balance.

- *The reactive power.* The reactive power generated or absorbed by a VSC power converter is controlled by the fundamental magnitude of the converter AC voltage, which in PWM conversion is determined by the modulation index, and in fundamental switching conversion by DC voltage magnitude control. In CSC power conversion, both the phase displacement between the AC source voltage and the converter AC output current at its AC terminal and its AC output current fundamental magnitude can be controlled directly; thus, in this case the reactive power generated or absorbed by the current source converter can be controlled by both the AC current magnitude and phase angle, under normal AC source voltage conditions.

5.2.3 System control level

Based on the objectives to be achieved in each particular application, the system control level provides the appropriate state orders to the converter state controller, and the power converter is controlled to follow these orders to achieve the required system operation. The power converter system can achieve a variety of important functions for a wide range of applications. In power transmission systems the power converters are used to control the real and reactive power flow, the bus voltage magnitude and phase angle, thus enhancing the system transient stability, helping to damp system oscillations, providing frequency control, etc.

The firing control patterns have already been introduced in previous chapters. The rest of this chapter describes the converter system modelling and the converter state control at the converter control level. The system level involved in the specific applications will be described in later chapters.

5.3 Non-linearity of the power converter system

While the firing control level determines the power conversion performance, the converter state control level determines the converter system dynamic and steady performance. A good understanding of the converter system is thus essential.

The modelling and operation state control of self-commutating conversion has been studied for decades [1–3]. In the following sections the state equations are derived describing the relations between the gate orders (the control inputs) and the converter operation states (such as AC and DC voltage and current, active and reactive power). These derivations are based on the generalized state–space averaging method and the principle of AC and DC power balance. For a good description of the converter system dynamics, the power balance components include the active power transfer between the AC and DC sides, the power losses on both sides, the DC side energy storage increments in the DC capacitance and inductance, as well as the AC side energy storage increments in the AC capacitance and inductance (most of the models proposed in the literature do not include the AC side energy storage increments stored in the AC capacitance and inductance) [4–6].

The non-linear nature of self-commutated conversion under the four-quadrant operation requirement presents the main difficulty to the control system design. In line-commutated conversion (LCC), the line commutation restriction forces the firing delay angle (α) to be within the restricted operation region, where the DC voltage, being proportional to $\cos(\alpha)$, can be linearized around $\alpha = 90\circ$ without significant difference. This linearization has been proved reliable in practical thyristor LCC system control of active power.

The self-commutated converter system models discussed here using the averaging method and the power balance principle are of the so-called 'affine' non-linear type. An affine non-linear system can be linearized by feedback linearization [7, 8], and a non-linear controller can then be designed based on linear control theory. The main difficulty for transforming the affine non-linear equations into linear form is in synthesizing the non-linear transformations, which for the self-commutated converter systems can be achieved based on the electrical technology.

The affine system control permits the design of a non-linear control system for the self-commutated VSC and CSC systems that achieve high performance globally. The non-linear transformations presented in the following sections should provide a clear understanding without relying on advanced mathematical techniques based on differential geometry and differential algebra.

5.4 Modelling the voltage source converter system

A model for the control of the AC side fundamental and DC side components of the converter is derived by using the generalized state--space averaging method together with the principle of power balance. In the case of VSC, the accuracy of the model is assured by the stiffness of the DC voltage provided by the voltage source converter and the quality of the waveforms achieved by the PWM or multilevel configurations.

5.4.1 Conversion under pulse width modulation

The circuit shown in Figure 5.2 is a general equivalent suitable for the derivation of the average model of the voltage source converter for different applications. This general model can then be adapted for specific applications by setting the parameters of the appropriate components to zero or infinite.

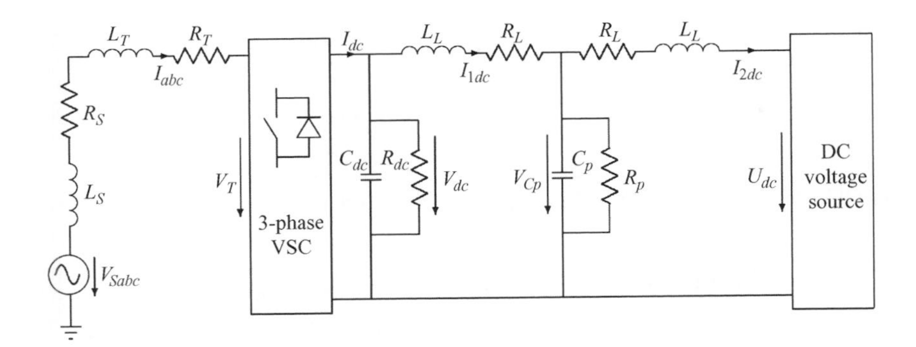

Figure 5.2 The equivalent circuit of the voltage source converter system.

In Figure 5.2 the converter AC terminals are connected to the AC system represented by an equivalent Thevenin circuit (V_S, L_S, R_S) via the inductance and resistance (L_T, R_T) of the interface transformer (or interface reactor). The converter DC terminal is connected to a shunt capacitance (C_{dc}) and resistance (R_{dc}) representing the losses of the switching and affiliated components of the voltage source converter.

The three-phase AC side voltage balancing equations of the voltage source converter are expressed as:

$$L_S \frac{d\mathbf{I_{abc}}}{dt} + R_S\mathbf{I_{abc}} + L_T \frac{d\mathbf{I_{abc}}}{dt} + R_T\mathbf{I_{abc}} = \mathbf{V_{Sabc}} - \mathbf{V_T}$$

where

$\mathbf{I_{abc}} = [I_a I_b I_c]^T$ is the three-phase current vector,

$\mathbf{V_{Sabc}} = [V_{Sa} V_{Sb} V_{Sc}]^T$ is the three-phase AC source voltage vector,

$\mathbf{V_T} = [V_{Ta} V_{Tb} V_{Tc}]^T = \sqrt{2} V_{Tm} \left[\sin(\omega t + \phi_T) \sin\left(\omega t - \frac{2\pi}{3} + \phi_T\right) \sin\left(\omega t + \frac{2\pi}{3} + \phi_T\right)\right]^T$ is the voltage source converter AC terminal three-phase voltage vector.

The *rms* amplitude $V_{Tm} = m_o V_{dc}$ ($0 < m_o < 1$, related to the PWM modulation index), the frequency ω and the phase angle ϕ_T are controllable variables of the PWM voltage source converter. When connected to a constant frequency AC system only $V_{Tm} = m_o V_{dc} (0 < m_o < 1)$ and ϕ_T need to be used for the control of the VSC operation.

The DC current to the T-network and the DC current to the DC source are:

$$I_{1dc} = I_{dc} - \frac{V_{dc}}{R_{dc}} - C_{dc} \frac{dV_{dc}}{dt}$$

$$I_{2dc} = I_{1dc} - \frac{V_{cp}}{R_p} - C_p \frac{dV_{cp}}{dt}$$

The DC side voltage balance equations are:

$$V_{dc} - V_{cp} = R_L I_{1dc} + L_L \frac{dI_{1dc}}{dt}$$

$$V_{cp} - U_{dc} = R_L I_{2dc} + L_L \frac{dI_{2dc}}{dt}$$

and the DC and AC side power balancing:

$$\begin{aligned} V_{dc} I_{dc} = \mathbf{I}^{\mathbf{T}}_{\mathbf{abc}} \mathbf{V_T} &= \mathbf{I}^{\mathbf{T}}_{\mathbf{abc}} \mathbf{V_{Sabc}} - R_S \mathbf{I}^{\mathbf{T}}_{\mathbf{abc}} \mathbf{I_{abc}} - R_T \mathbf{I}^{\mathbf{T}}_{\mathbf{abc}} \mathbf{I_{abc}} - \frac{dW_{L_S}}{2dt} - \frac{dW_{L_T}}{dt} \\ &= \mathbf{I}^{\mathbf{T}}_{\mathbf{abc}} \mathbf{V_{Sabc}} - R_S \mathbf{I}^{\mathbf{T}}_{\mathbf{abc}} \mathbf{I_{abc}} - R_T \mathbf{I}^{\mathbf{T}}_{\mathbf{abc}} \mathbf{I_{abc}} - \frac{L_S}{2} \frac{d(\mathbf{I}^{\mathbf{T}}_{\mathbf{abc}} \mathbf{I_{abc}})}{dt} - \frac{L_T}{2} \frac{d(\mathbf{I}^{\mathbf{T}}_{\mathbf{abc}} \mathbf{I_{abc}})}{dt} \end{aligned}$$

where W_{LS} and W_{LT} are the energies stored in all of the three-phase inductances for L_S and L_T respectively.

By substituting $I_{dc} = \mathbf{I}^{\mathbf{T}}_{\mathbf{abc}} V_T / V_{dc}$ into the expression $I_{1dc} = I_{dc} - \frac{V_{dc}}{R_{dc}} - C_{dc} \frac{dV_{dc}}{dt}$, and rearranging all the other equations, the following system equations result:

$$\frac{d\mathbf{I_{abc}}}{dt} = \frac{1}{L_S + L_T} [-(R_S + R_T)\mathbf{I_{abc}} + \mathbf{V_{Sabc}} - \mathbf{V_T}] \tag{5.1}$$

$$\frac{dV_{dc}}{dt} = \frac{1}{C_{dc}}\left[-I_{1dc} - \frac{V_{dc}}{R_{dc}} + \mathbf{I}_{\mathbf{abc}}^{\mathbf{T}} \frac{1}{V_{dc}} \mathbf{V}_{\mathbf{T}}\right] \tag{5.2}$$

$$\frac{dV_{cp}}{dt} = \frac{1}{C_p}\left(I_{1dc} - I_{2dc} - \frac{V_{cp}}{R_p}\right) \tag{5.3}$$

$$\frac{dI_{1dc}}{dt} = \frac{1}{L_L}(V_{dc} - V_{cp} - R_L I_{1dc}) \tag{5.4}$$

$$\frac{dI_{2dc}}{dt} = \frac{1}{L_L}(V_{cp} - U_{dc} - R_L I_{2dc}) \tag{5.5}$$

Using the orthogonal transformation

$$\mathbf{M} = (\mathbf{M}^{-1})^{\mathbf{T}} = \sqrt{\frac{2}{3}} \begin{bmatrix} \sin(\omega t) & \sin(\omega t - 120°) & \sin(\omega t + 120°) \\ \cos(\omega t) & \cos(\omega t - 120°) & \cos(\omega t + 120°) \\ \frac{1}{\sqrt{2}} & \frac{1}{\sqrt{2}} & 1\sqrt{2} \end{bmatrix} \tag{5.6}$$

the three-phase vectors in equations 5.1 and 5.2 are next converted to the dqo-frame.

Thus by substituting

$$\frac{d\mathbf{I}_{\mathbf{abc}}}{dt} = \frac{d[\mathbf{M}^{-1}\mathbf{I}_{\mathbf{dqo}}]}{dt} = \frac{d\mathbf{M}^{-1}}{dt}\mathbf{I}_{\mathbf{dqo}} + \mathbf{M}^{-1}\frac{d\mathbf{I}_{\mathbf{dqo}}}{dt}$$

and

$$\mathbf{I}_{\mathbf{abc}} = \mathbf{M}^{-1}\mathbf{I}_{\mathbf{dqo}},\ \mathbf{V}_{\mathbf{Sabc}} = \mathbf{M}^{-1}\mathbf{V}_{\mathbf{Sdqo}},\ \mathbf{V}_{\mathbf{T}} = \mathbf{M}^{-1}\mathbf{V}_{\mathbf{Tdqo}}$$

into equations 5.1 and 5.2, the following equations result in the dqo-frame of reference:

$$\frac{d\mathbf{I}_{\mathbf{dqo}}}{dt} = -\mathbf{M}\frac{d\mathbf{M}^{-1}}{dt}\mathbf{I}_{\mathbf{dqo}} - \left(\frac{R_S + R_T}{L_S + L_T}\right)\mathbf{I}_{\mathbf{dqo}} + \frac{1}{L_S + L_T}\left[\mathbf{V}_{\mathbf{Sdqo}} - \mathbf{V}_{\mathbf{Tdqo}}\right]$$

and

$$V_{dc}\frac{dV_{dc}}{dt} = \frac{1}{C_{dc}}\left[-V_{dc}I_{1dc} - V_{dc}\frac{V_{dc}}{R_{dc}} + \mathbf{I}_{\mathbf{dqo}}^{\mathbf{T}}\mathbf{V}_{\mathbf{Tdqo}}\right]$$

Since

$$\mathbf{M}\frac{d\mathbf{M}^{-1}}{dt} = \begin{bmatrix} 0 & -\omega & 0 \\ \omega & 0 & 0 \\ 0 & 0 & 0 \end{bmatrix}$$

if the three-phase currents and voltages are symmetrical (i.e. the three-phase components add to a zero at any instant), all the last components of

$$\mathbf{I_{dqo}},\ \mathbf{V_{Sdqo}}\ \text{and}\ \mathbf{V_{Tdqo}} = [V_{Td} V_{Tq} 0]^T = [\sqrt{3} V_{dc} m_o \cos(\phi_T) \sqrt{3} V_{dc} m_o \sin(\phi_T) 0]^T$$

are zero, and the system state equations become:

$$\frac{dI_d}{dt} = -aI_d + \omega I_q + bV_{Sd} - bV_{Td} \tag{5.7}$$

$$\frac{dI_q}{dt} = -\omega I_d - aI_q + bV_{Sq} - bV_{Tq} \tag{5.8}$$

$$\frac{dV_{dc}}{dt} = \frac{1}{C_{dc}}\left[-I_{1dc} - \frac{V_{dc}}{R_{dc}} + \frac{1}{V_{dc}}(I_d V_{Td} + I_q V_{Tq})\right] \tag{5.9}$$

$$\frac{dV_{cp}}{dt} = \frac{1}{C_p}\left(I_{1dc} - I_{2dc} - \frac{V_{cp}}{R_p}\right) \tag{5.10}$$

$$\frac{dI_{1dc}}{dt} = \frac{1}{L_L}(V_{dc} - V_{cp} - R_L I_{1dc}) \tag{5.11}$$

$$\frac{dI_{2dc}}{dt} = \frac{1}{L_L}(V_{cp} - U_{dc} - R_L I_{2dc}) \tag{5.12}$$

where $a = \frac{R_S + R_T}{L_S + L_T}$, $b = \frac{1}{L_S + L_T}$, $R_{ST} = R_S + R_T$ and $L_{ST} = L_S + L_T$.

The generalized equations 5.7 to 5.12 can be directly used for the control of an HVDC terminal and they can be modified as follows for other applications.

For STATCOM application, as $C_p = 0$, $I_{1dc} = I_{2dc} = 0$ and $L_L = \infty$, the state equations become

$$\frac{dI_d}{dt} = -aI_d + \omega I_q + bV_{Sd} - bV_{Td} \tag{5.13}$$

$$\frac{dI_q}{dt} = -\omega I_d - aI_q + bV_{Sq} - bV_{Tq} \tag{5.14}$$

$$\frac{dV_{dc}}{dt} = \frac{1}{C_{dc}}\left[-\frac{V_{dc}}{R_{dc}} + \frac{1}{V_{dc}}(I_d V_{Td} + I_q V_{Tq})\right] \tag{5.15}$$

For a back-to-back interconnection, the DC voltage source block in Figure 5.2 is an equivalent mirror system of the voltage source converter and AC system of the left-hand side. As $C_p = 0$, $R_p = \infty$, $I_{1dc} = I_{2dc}$, $V_{Cp} = V_{dc} = U_{dc}$, $L_L = 0$ and $R_L = 0$, all the terms in equations 5.10 to 5.12 are zero; the subscript '1' is added to the left-side voltage source converter and AC system variables and the subscript '2' is added to the right-side voltage source converter and AC system variables in equations 5.7 and 5.8. The term $-V_{dc}I_{1dc}$ in equation 5.9, used to represent the power transferred from the right-side system, can be simplified by referring to the other terms in equation 5.9 with subscript '2': $-V_{dc}I_{1dc} = (I_{d2}V_{Td2} + I_{q2}V_{Tq2})$.

The state equations for the back-to-back application are thus:

$$\frac{dI_{d1}}{dt} = -a_1 I_{d1} + \omega_1 I_{q1} + b_1 V_{Sd1} - b_1 V_{Td1} \tag{5.16}$$

$$\frac{dI_{q1}}{dt} = -\omega_1 I_{d1} - a_1 I_{q1} + b_1 V_{Sq1} - b_1 V_{Tq1} \tag{5.17}$$

$$\frac{dI_{d2}}{dt} = -a_2 I_{d2} + \omega_2 I_{q2} + b_2 V_{Sd2} - b_2 V_{Td2} \tag{5.18}$$

$$\frac{dI_{q2}}{dt} = -\omega_2 I_{d2} - a_1 I_{q2} + b_2 V_{Sq2} - b_2 V_{Tq2} \tag{5.19}$$

$$\frac{dV_{dc}}{dt} = \frac{1}{C_{dc}}\left[-\frac{V_{dc}}{R_{dc}} + \frac{1}{V_{dc}}(I_{d1}V_{Td1} + I_{q1}V_{Tq1}) + \frac{1}{V_{dc}}(I_{d2}V_{Td2} + I_{q2}V_{Tq2})\right] \tag{5.20}$$

These equations can also be used for the UPFC (unified power flow controller) system by making $\omega_1 = \omega_2 = \omega$.

5.5 Modelling grouped voltage source converters operating with fundamental frequency switching

The losses involved in high-frequency switching, as required in PWM, can be minimized by using only fundamental frequency switching, but in this case the direct AC voltage amplitude control freedom is lost.

However, several converter groups are likely to be needed for high-power applications and in that case an appropriate combination of the individual fundamental switching voltage source converter groups can provide the AC voltage amplitude with control freedom. This can be achieved by the series connection of the individual voltage source converters on the AC side and parallel connection of the groups on the DC side. The series connection on the AC side provides the opportunity of adding the AC voltages with different phase shifts and, thus, forming an amplitude-controllable AC output. The parallel connection on the DC side provides the freedom of adding different level DC currents at the common stiff DC voltage terminal. The resulting configuration for the case of two voltage source converters is shown in Figure 5.3.

In the diagram of Figure 5.4 the AC voltage phasors of the two voltage source converters are denoted (V_{T1}, V_{T2}) and the combined AC line voltage phasor (V_t) is the vectorial sum of these two voltages ($V_t = V_{T1} + V_{T2}$). Phasor V_X represents the voltage across the impedance between the summed voltage source converters AC terminal and the AC source.

As the amplitudes of $|V_{T1}|$ and $|V_{T2}|$ are exactly the same, and determined by the converter DC side voltage, the following relations can be written for the amplitude and phase angle

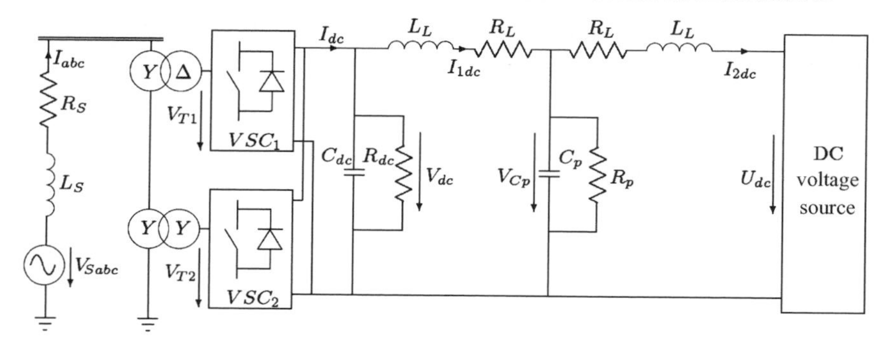

Figure 5.3 The equivalent circuit with two voltage source converters.

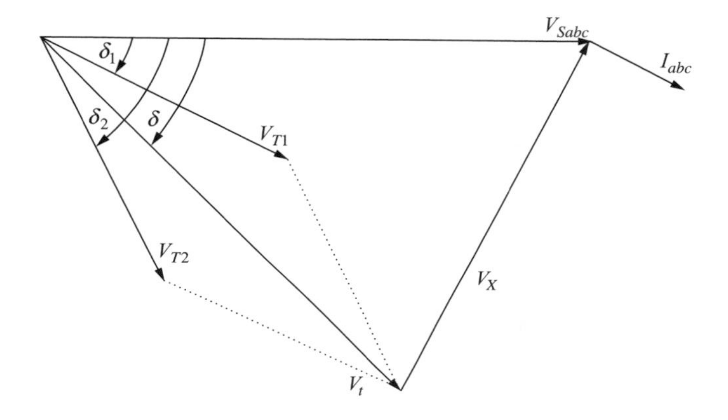

Figure 5.4 The phasor diagram of the two voltage source converters system.

(related to the source voltage) of the converter terminal voltage:

$$|V_t| = |V_{T1} + V_{T2}| = 2|V_{T1}|\cos\left(\frac{\delta_1 - \delta_2}{2}\right) \tag{5.21}$$

$$\delta = \frac{\delta_1 + \delta_2}{2} \tag{5.22}$$

It is clear that the amplitude of V_t can be controlled by $\cos\left(\frac{\delta_1-\delta_2}{2}\right)$, and its phase angle by $\left(\frac{\delta_1+\delta_2}{2}\right)$. Thus the combined double voltage source converter group provides phase and amplitude controllability independently from the DC voltage.

To derive the state equations of the grouped voltage source converters the interfacing transformers are modelled by an ideal transformer (k_n : 1) and a series-connected impedance consisting of the resistance and inductance (R_{T1}, L_{T1} and R_{T2}, L_{T2} for the interfacing transformers connected to the VSC_1 and VSC_2 respectively). The AC side voltage balancing equations written in vector form are:

$$\mathbf{V_{Sabc}} - k_n(\mathbf{V_{T1}} + \mathbf{V_{T2}}) = R_S\mathbf{I_{abc}} + L_S\frac{d\mathbf{I_{abc}}}{dt} + k_n^2(R_{T1} + R_{T2})\mathbf{I_{abc}} + k_n^2(L_{T1} + L_{T2})\frac{d\mathbf{I_{abc}}}{dt}$$

and by substituting $L_T = k_n^2(L_{T1} + L_{T2})$, $R_T = k_n^2(R_{T1} + R_{T2})$ and $\mathbf{V_t} = k_n(\mathbf{V_{T1}} + \mathbf{V_{T2}})$ a more compact form results, i.e.

$$L_S\frac{d\mathbf{I_{abc}}}{dt} + R_S\mathbf{I_{abc}} + L_T\frac{d\mathbf{I_{abc}}}{dt} + R_T\mathbf{I_{abc}} = \mathbf{V_{Sabc}} - \mathbf{V_t} \tag{5.23}$$

Under the fundamental switching control strategy the AC output voltages of the two voltage source converters have a strict relation to the common DC voltage:

$$\mathbf{V_{T1}} = \sqrt{2}V_{tm}\left[\sin(\omega t + \phi_{T_1})\sin\left(\omega t - \frac{2\pi}{3} + \phi_{T_1}\right)\sin\left(\omega t + \frac{2\pi}{3} + \phi_{T_1}\right)\right]^T$$

$$\mathbf{V_{T2}} = \sqrt{2}V_{tm}\left[\sin(\omega t + \phi_{T_2})\sin\left(\omega t - \frac{2\pi}{3} + \phi_{T_2}\right)\sin\left(\omega t + \frac{2\pi}{3} + \phi_{T_2}\right)\right]^T$$

where the *rms* voltage $V_{tm} = k_o V_{dc}$ (k_o is a constant related to the voltage source converter topology).

The combined AC voltage of the two voltage source converters is

$$\mathbf{V_t} = k_n(\mathbf{V_{T1}} + \mathbf{V_{T2}}) = \sqrt{2}V_{tm}\left[\sin(\omega t + \phi_T)\sin\left(\omega t - \frac{2\pi}{3} + \phi_T\right)\sin\left(\omega t + \frac{2\pi}{3} + \phi_T\right)\right]^T$$

where $V_{tm} = 2k_n k_o V_{dc}\cos\left(\frac{\phi_{T_1} - \phi_{T_2}}{2}\right)$ and $\phi_t = \frac{\phi_{T_1} + \phi_{T_2}}{2}$.

The power balance between the AC and DC sides can be expressed as:

$$V_{dc}I_{dc} = \mathbf{I_{abc}^T}\mathbf{V_t} = \mathbf{I_{abc}^T}\mathbf{V_{Sabc}} - R_S\mathbf{I_{abc}^T}\mathbf{I_{abc}} - R_T\mathbf{I_{abc}^T}\mathbf{I_{abc}} - \frac{L_S}{2}\frac{d(\mathbf{I_{abc}^T}\mathbf{I_{abc}})}{dt} - \frac{L_T}{2}\frac{d(\mathbf{I_{abc}^T}\mathbf{I_{abc}})}{dt}$$

The state equations in the three-phase frame of reference are:

$$\frac{d\mathbf{I_{abc}}}{dt} = \frac{1}{L_S + L_T}[-(R_S + R_T)\mathbf{I_{abc}} + \mathbf{V_{Sabc}} - \mathbf{V_t}] \tag{5.24}$$

$$\frac{dV_{dc}}{dt} = \frac{1}{C_{dc}}\left[-I_{1dc} - \frac{V_{dc}}{R_{dc}} + \mathbf{I_{abc}^T}\frac{1}{V_{dc}}\mathbf{V_t}\right] \tag{5.25}$$

$$\frac{dV_{cp}}{dt} = \frac{1}{C_p}\left(I_{1dc} - I_{2dc} - \frac{V_{cp}}{R_p}\right) \tag{5.26}$$

$$\frac{dI_{1dc}}{dt} = \frac{1}{L_L}(V_{dc} - V_{cp} - R_L I_{1dc}) \tag{5.27}$$

$$\frac{dI_{2dc}}{dt} = \frac{1}{L_L}(V_{cp} - U_{dc} - R_L I_{2dc}) \tag{5.28}$$

In the dq-frame of reference the state equations are

$$\frac{dI_d}{dt} = -aI_d + \omega I_q + bV_{Sd} - bV_{td} \tag{5.29}$$

$$\frac{dI_q}{dt} = -\omega I_d - aI_q + bV_{Sq} - bV_{tq} \tag{5.30}$$

$$\frac{dV_{dc}}{dt} = \frac{1}{C_{dc}}\left[-I_{1dc} - \frac{V_{dc}}{R_{dc}} + \frac{1}{V_{dc}}(I_d V_{td} + I_q V_{tq})\right] \tag{5.31}$$

$$\frac{dV_{cp}}{dt} = \frac{1}{C_p}\left(I_{1dc} - I_{2dc} - \frac{V_{cp}}{R_p}\right) \tag{5.32}$$

$$\frac{dI_{1dc}}{dt} = \frac{1}{L_L}(V_{dc} - V_{cp} - R_L I_{1dc}) \tag{5.33}$$

$$\frac{dI_{2dc}}{dt} = \frac{1}{L_L}(V_{cp} - U_{dc} - R_L I_{2dc}) \tag{5.34}$$

where, $a = \frac{R_S + R_T}{L_S + L_T}$, $b = \frac{1}{L_S + L_T}$, $R_{ST} = R_S + R_T$ and $L_{ST} = L_S + L_T$.

The state equations for the PWM voltage source converter and grouped voltage source converters are nearly in the same form except for the substitution of V_{td} and V_{tq} for V_{Td} and V_{Tq}.

If V_{Tdq} and V_{tdq} are expressed as

$$V_{Tdq} = MV_T = [V_{Td} V_{Tq}]^T = V_{dc}[\sqrt{3}m_o][\cos(\phi_T)\sin(\phi_T)]^T = V_{dc}[u_1 u_2]^T$$

$$V_{tdq} = MV_t = [V_{td} V_{tq}]^T = V_{dc}\left[2\sqrt{3}k_n k_o \cos\left(\frac{\phi_{T_1} - \phi_{T_2}}{2}\right)\right][\cos(\phi_t)\sin(\phi_t)]^T = V_{dc}[v_1 v_2]^T$$

the state equations for all the applications can be written in the affine form:

$$\frac{\mathbf{dX}}{\mathbf{dt}} = \mathbf{F}(\mathbf{X}) + \mathbf{G}(\mathbf{X})\mathbf{U} \tag{5.35}$$

where, $\mathbf{X} = [x_1 x_2 \ldots x_n]^T$, $\mathbf{U} = [u_1 u_2 \ldots u_m]^T$, $\mathbf{F}(\mathbf{X}) = [f_1(\mathbf{X}) f_2(\mathbf{X}) \ldots f_n(\mathbf{X})]^T$,

$$\mathbf{G}(\mathbf{X}) = \begin{pmatrix} g_{11}(\mathbf{X}) & g_{12}(\mathbf{X}) & \cdots & g_{1m}(\mathbf{X}) \\ g_{21}(\mathbf{X}) & g_{22}(\mathbf{X}) & \cdots & g_{2m}(\mathbf{X}) \\ \vdots & \vdots & \vdots & \vdots \\ g_{n1}(\mathbf{X}) & g_{n2}(\mathbf{X}) & \cdots & g_{nm}(\mathbf{X}) \end{pmatrix}.$$

For HVDC terminal control:

$$\mathbf{X} = [I_d I_q V_{dc} V_{cp} I_{1dc} I_{2dc}]^T, \quad \mathbf{U} = [u_1 u_2]^T \quad \text{or} \quad \mathbf{U} = [v_1 v_2]^T,$$

$$\mathbf{F}(\mathbf{X}) = \begin{pmatrix} -aI_d + \omega I_q + bV_{Sd} \\ -\omega I_d - aI_q + bV_{Sq} \\ \frac{1}{C_{dc}}\left(-I_{1dc} - \frac{V_{dc}}{R_{dc}}\right) \\ \frac{1}{C_p}\left(I_{1dc} - I_{2dc} - \frac{V_{cp}}{R_p}\right) \\ \frac{1}{L_L}(V_{dc} - V_{cp} - R_L I_{1dc}) \\ \frac{1}{L_L}(V_{cp} - U_{dc} - R_L I_{2dc}) \end{pmatrix}, \quad \mathbf{G}(\mathbf{X}) = \begin{pmatrix} -bV_{dc} & 0 \\ 0 & -bV_{dc} \\ \frac{I_d}{C_{dc}} & \frac{I_q}{C_{dc}} \\ 0 & 0 \\ 0 & 0 \\ 0 & 0 \end{pmatrix}.$$

For STATCOM control:

$$\mathbf{X} = [I_d I_q V_{dc}]^T, \quad \mathbf{U} = [u_1 u_2]^T \quad \text{or} \quad \mathbf{U} = [v_1 v_2]^T \quad \mathbf{F}(\mathbf{X}) = \begin{pmatrix} -aI_d + \omega I_q + bV_{Sd} \\ -\omega I_d - aI_q + bV_{Sq} \\ -\frac{V_{dc}}{R_{dc} C_{dc}} \end{pmatrix},$$

$$\mathbf{G}(\mathbf{X}) = \begin{pmatrix} -bV_{dc} & 0 \\ 0 & -bV_{dc} \\ \frac{I_d}{C_{dc}} & \frac{I_q}{C_{dc}} \end{pmatrix}.$$

For back-to-back control:

$$\mathbf{X} = [I_{d1} I_{q1} I_{d2} I_{q2} V_{dc}]^T, \quad \mathbf{U} = [u_{11} u_{12} u_{21} u_{22}]^T \quad \text{or} \quad \mathbf{U} = [v_{11} v_{12} v_{21} v_{22}]^T,$$

$$\mathbf{F}(\mathbf{X}) = \begin{pmatrix} -a_1 I_{d1} + \omega_1 I_{q1} + b_1 V_{Sd1} \\ -\omega_1 I_{d1} - a_1 I_{q1} + b1 V_{Sq1} \\ -a_2 I_{d2} + \omega_2 I_{q2} + b_2 V_{Sd2} \\ -\omega_2 I_{d2} - a_2 I_{q2} + b_2 V_{Sq2} \\ -\frac{V_{dc}}{R_{dc} C_{dc}} \end{pmatrix}, \quad \mathbf{G}(\mathbf{X}) = \begin{pmatrix} -b_1 V_{dc} & 0 & 0 & 0 \\ 0 & -b_1 V_{dc} & 0 & 0 \\ 0 & 0 & -b_2 V_{dc} & 0 \\ 0 & 0 & 0 & -b_2 V_{dc} \\ \frac{I_{d1}}{C_{dc}} & \frac{I_{q1}}{C_{dc}} & \frac{I_{d2}}{C_{dc}} & \frac{I_{q2}}{C_{dc}} \end{pmatrix}.$$

These equations present the affine non-linearity, as there are state variables (V_{dc}, I_d and I_q) in the input matrix $G(X)$. They could be linearized, and the controller designed, around a specific operating state, which would only be valid in a small region around that specific operation state. In practical voltage source converter operation, although the DC side voltage V_{dc} is controlled within a small margin, the AC side current (determined by I_d and I_q) needs to be varied gradually or sharply between zero and the rated level. The challenge is, thus, to find a design method which is valid globally and independently of the operation states.

To convert the affine non-linear system into a linear system [7] the controller is designed based on the non-linear transformation of the system state variables and the input variables. The transformed state variables can then be controlled by the transformed input variables, to vary within desired paths by using well developed linear control techniques. Then performing the reversed transformation from the transformed input variables to the original input variables enables the control of the VSC system, through the practical control variables, to achieve the desired performance.

The main difficulty in the process of transforming the affine non-linear equations into linear form is synthesizing the non-linear transformations; however, for a VSC system this can be achieved based on the electrical technology.

To simplify understanding, the following derivations are presented without the use of differential geometry and differential algebra. The simpler and lower order STATCOM equations are transformed first. The STATCOM state equations are rewritten in the form:

$$\begin{pmatrix} \dfrac{dI_{Sd}}{dt} \\ \dfrac{dI_{Sq}}{dt} \\ \dfrac{dV_{dc}}{dt} \end{pmatrix} = \begin{pmatrix} -aI_{Sd} + \omega I_{Sq} + bV_{Sd} \\ -\omega I_{Sd} - aI_{Sq} + bV_{Sq} \\ -\dfrac{V_{dc}}{R_{dc}C_{dc}} \end{pmatrix} + \begin{pmatrix} -bV_{dc} & 0 \\ 0 & -bV_{dc} \\ \dfrac{I_{Sd}}{C_{dc}} & \dfrac{I_{Sq}}{C_{dc}} \end{pmatrix} \begin{pmatrix} u_1 \\ u_2 \end{pmatrix} \tag{5.36}$$

The energies stored in the AC side inductors and the DC side capacitor are related to the currents in the inductors and the voltage across the capacitor. They are independent of the input or control variables and their instantaneous power is the derivative of the stored energy.

If the energy stored in the AC side inductors and the DC side capacitor and their derivatives, as well as the DC voltage, are chosen as the transformed state variables the following expressions result:

$$z_3 = 0.5\left[L_{ST}(I_{Sd}^2 + I_{Sq}^2) + C_{dc}V_{dc}^2\right]$$

$$z_2 = \frac{dz_3}{dt} = L_{ST}\left(I_{Sd}\frac{dI_{Sd}}{dt} + I_{Sq}\frac{dI_{Sq}}{dt}\right) + C_{dc}V_{dc}\frac{dV_{dc}}{dt} = V_{Sd}I_{Sd} + V_{Sq}I_{Sq} - R_{ST}\left(I_{Sd}^2 + I_{Sq}^2\right) - \frac{V_{dc}^2}{R_{dc}}$$

$$z_1 = V_{dc}$$

Thus the non-linear transformation $Z=\Phi(X)$ is:

$$\begin{pmatrix} z_1 \\ z_2 \\ z_3 \end{pmatrix} = \begin{pmatrix} V_{dc} \\ V_{Sd}I_{Sd}+V_{Sq}I_{Sq}-R_{ST}\left(I_{Sd}^2+I_{Sq}^2\right)-\dfrac{V_{dc}^2}{R_{dc}} \\ 0.5\left[L_{ST}\left(I_{Sd}^2+I_{Sq}^2\right)+C_{dc}V_{dc}^2\right] \end{pmatrix} \tag{5.37}$$

The transformed state equations for the STATCOM application are then

$$\begin{pmatrix} \dfrac{dz_1}{dt} \\ \dfrac{dz_2}{dt} \\ \dfrac{dz_3}{dt} \end{pmatrix} = \begin{pmatrix} w_1 \\ w_2 \\ z_2 \end{pmatrix} = \begin{pmatrix} -\dfrac{V_{dc}}{R_{dc}C_{dc}} \\ h_{sta}(X) \\ 0.5\dfrac{d}{dt}\left[L_{ST}\left(I_d^2+I_q^2\right)+C_{dc}V_{dc}^2\right] \end{pmatrix} + \begin{pmatrix} \dfrac{I_d}{C_{dc}} & \dfrac{I_q}{C_{dc}} \\ g_{1sta}(X) & g_{2stat}(X) \\ 0 & 0 \end{pmatrix} \begin{pmatrix} u_1 \\ u_2 \end{pmatrix} \tag{5.38}$$

where

$$h_{sta}(X) = (V_{Sd}-2R_{ST}I_d)(-aI_d+\omega I_q+bV_{Sd})+(V_{Sq}-2R_{ST}I_q)(-\omega I_d-aI_q+bV_{Sq})+\frac{2V_{dc}^2}{R_{dc}^2C_{dc}}$$

$$g_{1sta}(X) = -b(V_{Sd}-2R_{ST}I_d)V_{dc}-\frac{2}{R_{dc}C_{dc}}I_dV_{dc}$$

$$g_{2sta}(X) = -b(V_{Sq}-2R_{ST}I_q)V_{dc}-\frac{2}{R_{dc}C_{dc}}I_qV_{dc}$$

They can be derived from

$$\frac{dz_2}{dt}=\frac{d}{dt}\left[V_{Sd}I_d+V_{Sq}I_q-R_{ST}\left(I_d^2+I_q^2\right)-\frac{V_{dc}^2}{R_{dc}}\right].$$

The left-hand side terms of equation 5.38 are the linearized Z state equations and the top two rows of the right-hand side equations define the relationship between the transformed input variables $[w_1 w_2]$ and the original input variables $[u_1 u_2]$.

If

$$\Delta_{sta} = \frac{I_d}{C_{dc}} g_{2sta}(X) - \frac{I_q}{C_{dc}} g_{1sta}(X) = \frac{b}{C_{dc}} V_{dc}(V_{Sd}I_q - V_{Sq}I_d) \neq 0,$$
$$\text{i.e.}\quad V_{dc} \neq 0 \quad \text{and} \quad V_{Sq}/V_{Sd} \neq I_q/I_d,$$

$$\begin{pmatrix} u_1 \\ u_2 \end{pmatrix} = \begin{pmatrix} \frac{I_d}{C_{dc}} & \frac{I_q}{C_{dc}} \\ g_{1sta}(X) & g_{2sta}(X) \end{pmatrix}^{-1} \begin{pmatrix} w_1 + \frac{V_{dc}}{R_{dc}C_{dc}} \\ w_2 - h_{sta}(X) \end{pmatrix} \tag{5.39}$$

The state equations for HVDC terminal control are rewritten in the form:

$$\begin{pmatrix} \frac{dI_d}{dt} \\ \frac{dI_q}{dt} \\ \frac{dV_{dc}}{dt} \\ \frac{dV_{cp}}{dt} \\ \frac{dI_{1dc}}{dt} \\ \frac{dI_{2dc}}{dt} \end{pmatrix} = \begin{pmatrix} -aI_d + \omega I_q + bV_{Sd} \\ -\omega I_d - aI_q + bV_{Sq} \\ \frac{1}{C_{dc}}\left(-I_{1dc} - \frac{V_{dc}}{R_{dc}}\right) \\ \frac{1}{C_p}\left(I_{1dc} - I_{2dc} - \frac{V_{cp}}{R_p}\right) \\ \frac{1}{L_L}(V_{dc} - V_{cp} - R_L I_{1dc}) \\ \frac{1}{L_L}(V_{cp} - U_{dc} - R_L I_{2dc}) \end{pmatrix} + \begin{pmatrix} -bV_{dc} & 0 \\ 0 & -bV_{dc} \\ \frac{I_d}{C_{dc}} & \frac{I_q}{C_{dc}} \\ 0 & 0 \\ 0 & 0 \\ 0 & 0 \end{pmatrix} \begin{pmatrix} u_1 \\ u_2 \end{pmatrix} \tag{5.40}$$

As in the STATCOM model, the energies stored in the AC side inductors and the DC side capacitor are related to the currents in the inductors and the voltage across the capacitor. They are independent of the input or control variables and their instantaneous power is the derivative of the stored energy. If the energy stored in the AC side inductors and the DC side capacitor and their derivatives, as well as $V_{dc}, V_{cp}, I_{1dc}, I_{2dc}$, are chosen as the transformed state variables:

$$z_1 = V_{dc}$$

$$z_2 = \frac{dz_3}{dt} = L_{ST}\left(I_d \frac{dI_d}{dt} + I_q \frac{dI_q}{dt}\right) + C_{dc}V_{dc}\frac{dV_{dc}}{dt}$$
$$= V_{Sd}I_d + V_{Sq}I_q - R_{ST}\left(I_d^2 + I_q^2\right) - \frac{V_{dc}^2}{R_{dc}} - I_{1dc}V_{dc}$$

$$z_3 = 0.5\left[L_{ST}\left(I_d^2 + I_q^2\right) + C_{dc}V_{dc}^2\right]$$

$$z_4 = V_{cp}$$

$$z_5 = I_{1dc}$$

$$z_6 = I_{2dc}$$

Thus the transformed state equations for HVDC terminal control application are

$$\begin{pmatrix} \frac{dz_1}{dt} \\ \frac{dz_2}{dt} \\ \frac{dz_3}{dt} \\ \frac{dz_4}{dt} \\ \frac{dz_5}{dt} \\ \frac{dz_6}{dt} \end{pmatrix} = \begin{pmatrix} w_1 \\ w_2 \\ z_2 \\ -\frac{1}{R_pC_p}z_4 + \frac{1}{C_p}z_5 - \frac{1}{C_p}z_6 \\ \frac{1}{L_L}z_1 - \frac{1}{L_L}z_4 - \frac{R_L}{L_L}z_5 \\ \frac{1}{L_L}z_4 - \frac{R_L}{L_L}z_6 - \frac{1}{L_L}U_{dc} \end{pmatrix} = \begin{pmatrix} -\frac{I_{1dc}}{C_{dc}} - \frac{V_{dc}}{R_{dc}C_{dc}} \\ h_{ter}(X) \\ 0.5\frac{d}{dt}\left[L_{ST}\left(I_d^2 + I_q^2\right) + C_{dc}V_{dc}^2\right] \\ \frac{1}{C_p}\left(I_{1dc} - I_{2dc} - \frac{V_{cp}}{R_p}\right) \\ \frac{1}{L_L}(V_{dc} - V_{cp} - R_LI_{1dc}) \\ \frac{1}{L_L}(V_{cp} - U_{dc} - R_LI_{2dc}) \end{pmatrix}$$

$$+ \begin{pmatrix} \frac{I_d}{C_{dc}} & \frac{I_q}{C_{dc}} \\ g_{1ter}(X) & g_{2ter}(X) \\ 0 & 0 \\ 0 & 0 \\ 0 & 0 \\ 0 & 0 \end{pmatrix} \begin{pmatrix} u_1 \\ u_2 \end{pmatrix} \tag{5.41}$$

where

$$h_{ter}(X) = (V_{Sd} - 2R_{ST}I_d)(-aI_d - \omega I_q + bV_{Sd}) + (V_{Sq} - 2R_{ST}I_q)(-\omega I_d - aI_q + bV_{Sq}) + \left(I_{1dc} + \frac{2V_{dc}}{R_{dc}}\right)\left(\frac{I_{1dc}}{C_{dc}} + \frac{V_{dc}}{R_{dc}C_{dc}}\right) - \frac{V_{dc}}{L_L}(V_{dc} - V_{cp} - R_LI_{1dc})$$

$$g_{1ter}(X) = -b(V_{Sd} - 2R_{ST}I_d)V_{dc} - \left(I_{1dc} + \frac{2V_{dc}}{R_{dc}}\right)\frac{I_d}{C_{dc}}$$

$$g_{2ter}(X) = -b(V_{Sq} - 2R_{ST}I_q)V_{dc} - \left(I_{1dc} + \frac{2V_{dc}}{R_{dc}}\right)\frac{I_q}{C_{dc}}$$

They can be derived from $\frac{dz_2}{dt} = \frac{d}{dt}\left[V_{Sd}I_d + V_{Sq}I_q - R_{ST}\left(I_d^2 + I_q^2\right) - \frac{V_{dc}^2}{R_{dc}} - I_{1dc}V_{dc}\right]$.

The transformed linear Z state equations can be written in the standard linear form of:

$$\begin{pmatrix} \frac{dz_1}{dt} \\ \frac{dz_2}{dt} \\ \frac{dz_3}{dt} \\ \frac{dz_4}{dt} \\ \frac{dz_5}{dt} \\ \frac{dz_6}{dt} \end{pmatrix} = \begin{pmatrix} 0 & 0 & 0 & 0 & 0 & 0 \\ 0 & 0 & 0 & 0 & 0 & 0 \\ 0 & 1 & 0 & 0 & 0 & 0 \\ 0 & 0 & 0 & -\frac{1}{R_p C_p} & \frac{1}{C_p} & -\frac{1}{C_p} \\ \frac{1}{L_L} & 0 & 0 & -\frac{1}{L_L} & -\frac{R_L}{L_L} & 0 \\ 0 & 0 & 0 & \frac{1}{L_L} & 0 & -\frac{R_L}{L_L} \end{pmatrix} \begin{pmatrix} z_1 \\ z_2 \\ z_3 \\ z_4 \\ z_5 \\ z_6 \end{pmatrix} + \begin{pmatrix} 1 & 0 & 0 \\ 0 & 1 & 0 \\ 0 & 0 & 0 \\ 0 & 0 & 0 \\ 0 & 0 & 0 \\ 0 & 0 & -\frac{1}{L_L} \end{pmatrix} \begin{pmatrix} w_1 \\ w_2 \\ U_{dc} \end{pmatrix} \tag{5.42}$$

and the relationship between the transformed input variables $[w_1 w_2]$ and the original input variables $[u_1 u_2]$ is given by the top two rows of equation 5.41:

$$\begin{pmatrix} w_1 \\ w_2 \end{pmatrix} = \begin{pmatrix} -\frac{I_{1dc}}{C_{dc}} - \frac{V_{dc}}{R_{dc} C_{dc}} \\ h_{ter}(X) \end{pmatrix} + \begin{pmatrix} \frac{I_d}{C_{dc}} & \frac{I_q}{C_{dc}} \\ g_{1ter}(X) & g_{2ter}(X) \end{pmatrix} \begin{pmatrix} u_1 \\ u_2 \end{pmatrix}$$

If

$$\Delta_{ter} = \frac{I_d}{C_{dc}} g_{2ter}(X) - \frac{I_q}{C_{dc}} g_{1ter}(X) = \frac{b}{C_{dc}} V_{dc}(V_{Sd} I_q - V_{Sq} I_d) \neq 0,$$

$$\text{i.e. } V_{dc} \neq 0 \text{ and } V_{Sq}/V_{Sd} \neq I_q/I_d,$$

$$\begin{pmatrix} u_1 \\ u_2 \end{pmatrix} = \begin{pmatrix} \frac{I_d}{C_{dc}} & \frac{I_q}{C_{dc}} \\ g_{1ter}(X) & g_{2ter}(X) \end{pmatrix}^{-1} \begin{pmatrix} w_1 + \frac{V_{dc}}{R_{dc} C_{dc}} \\ w_2 - h_{ter}(X) \end{pmatrix} \tag{5.43}$$

The state equations for back-to-back application are rewritten in the form:

$$\begin{pmatrix} \dfrac{dI_{d1}}{dt} \\ \dfrac{dI_{q1}}{dt} \\ \dfrac{dI_{d2}}{dt} \\ \dfrac{dI_{q2}}{dt} \\ \dfrac{dV_{dc}}{dt} \end{pmatrix} = \begin{pmatrix} -a_1 I_{d1} + \omega_1 I_{q1} + b_1 V_{Sd1} \\ -\omega_1 I_{d1} - a_1 I_{q1} + b_1 V_{Sq1} \\ -a_2 I_{d2} + \omega_2 I_{q2} + b_2 V_{Sd2} \\ -\omega_2 I_{d2} - a_2 I_{q2} + b_2 V_{Sq2} \\ -\dfrac{V_{dc}}{R_{dc} C_{dc}} \end{pmatrix} + \begin{pmatrix} -b_1 V_{dc} & 0 & 0 & 0 \\ 0 & -b_1 V_{dc} & 0 & 0 \\ 0 & 0 & -b_2 V_{dc} & 0 \\ 0 & 0 & 0 & -b_2 V_{dc} \\ \dfrac{I_{d1}}{C_{dc}} & \dfrac{I_{q1}}{C_{dc}} & \dfrac{I_{d2}}{C_{dc}} & \dfrac{I_{q2}}{C_{dc}} \end{pmatrix} \begin{pmatrix} u_{11} \\ u_{12} \\ u_{21} \\ u_{22} \end{pmatrix} \tag{5.44}$$

Also the energy stored in the inductors on both AC sides and the DC side capacitor are only related to the currents in the inductors and the voltage across the capacitor. They are independent of the input or control variables and their total instantaneous power is the derivative of the stored energy. If the energy stored in the AC side inductors and DC side capacitor and their derivatives as well as I_{d1}, I_{q1}, I_{d2} are chosen as the transformed state variables, i.e.

$$z_1 = I_{d1}$$

$$z_2 = I_{q1}$$

$$z_3 = I_{d2}$$

$$\begin{aligned} z_4 &= \frac{dz_5}{dt} = L_{ST1}\left(I_{d1}\frac{dI_{d1}}{dt} + I_{q1}\frac{dI_{q1}}{dt}\right) + L_{ST2}\left(I_{d2}\frac{dI_{d2}}{dt} + I_{q2}\frac{dI_{q2}}{dt}\right) + C_{dc}V_{dc}\frac{dV_{dc}}{dt} \\ &= V_{Sd1}I_{d1} + V_{Sq1}I_{q1} - R_{ST1}\left(I_{d1}^2 + I_{q1}^2\right) + V_{Sd2}I_{d2} + V_{Sq2}I_{q2} - R_{ST2}\left(I_{d2}^2 + I_{q2}^2\right) - \frac{V_{dc}^2}{R_{dc}} \end{aligned}$$

$$z_5 = 0.5\left[L_{ST1}\left(I_{d1}^2 + I_{q1}^2\right) + L_{ST2}\left(I_{d2}^2 + I_{q2}^2\right) + C_{dc}V_{dc}^2\right]$$

The non-linear transformation $Z = \Phi(X)$ is:

$$\begin{pmatrix} z_1 \\ z_2 \\ z_3 \\ z_4 \\ z_5 \end{pmatrix} = \begin{pmatrix} I_{d1} \\ I_{q1} \\ I_{d2} \\ V_{Sd1}I_{d1} + V_{Sq1}I_{q1} - R_{ST1}\left(I_{d1}^2 + I_{q1}^2\right) + V_{Sd2}I_{d2} + V_{Sq2}I_{q2} - R_{ST2}\left(I_{d2}^2 + I_{q2}^2\right) - \dfrac{V_{dc}^2}{R_{dc}} \\ 0.5\left[L_{ST1}\left(I_{d1}^2 + I_{q1}^2\right) + L_{ST2}\left(I_{d2}^2 + I_{q2}^2\right) + C_{dc}V_{dc}^2\right] \end{pmatrix} \tag{5.45}$$

The transformed state equations for back-to-back application are then

$$\begin{pmatrix} \frac{dz_1}{dt} \\ \frac{dz_2}{dt} \\ \frac{dz_3}{dt} \\ \frac{dz_4}{dt} \\ \frac{dz_5}{dt} \end{pmatrix} = \begin{pmatrix} w_{11} \\ w_{12} \\ w_{21} \\ w_{22} \\ z_4 \end{pmatrix} \begin{pmatrix} -a_1 I_{d1} + \omega_1 I_{q1} + b_1 V_{Sd1} \\ -\omega_1 I_{d1} - a_1 I_{q1} + b_1 V_{Sq1} \\ -a_2 I_{d2} + \omega_2 I_{q2} + b_2 V_{Sd2} \\ h_{btb}(X) \\ z_4 \end{pmatrix}$$

$$+ \begin{pmatrix} -b_1 V_{dc} & 0 & 0 & 0 \\ 0 & -b_1 V_{dc} & 0 & 0 \\ 0 & 0 & -b_2 V_{dc} & 0 \\ g_{1btb}(X) & g_{2btb}(X) & g_{3btb}(X) & g_{4btb}(X) \\ 0 & 0 & 0 & 0 \end{pmatrix} \begin{pmatrix} u_{11} \\ u_{12} \\ u_{21} \\ u_{22} \end{pmatrix} \tag{5.46}$$

where

$$\begin{aligned} h_{btb}(X) = &(V_{Sd1} - 2R_{ST1} I_{d1})(-a_1 I_{d1} - \omega_1 I_{q1} + b_1 V_{Sd1}) \\ &+ (V_{Sq1} - 2R_{ST1} I_{q1})(-\omega_1 I_{d1} - a_1 I_{q1} + b_1 V_{Sq1}) \\ &+ (V_{Sd2} - 2R_{ST2} I_{d2})(-a_2 I_{d2} - \omega_2 I_{q2} + b_2 V_{Sd2}) \\ &+ (V_{Sq2} - 2R_{ST2} I_{q2})(-\omega_2 I_{d2} - a_2 I_{q2} + b_2 V_{Sq2}) + \frac{V_{dc}^2}{R_{dc}^2 C_{dc}} \end{aligned}$$

$$g_{1btb}(X) = -b_1 (V_{Sd1} - 2R_{ST1} I_{d1}) V_{dc} - \frac{2V_{dc} I_{d1}}{R_{dc} C_{dc}}$$

$$g_{2btb}(X) = -b_1 (V_{Sq1} - 2R_{ST1} I_{q1}) V_{dc} - \frac{2V_{dc} I_{q1}}{R_{dc} C_{dc}}$$

$$g_{3btb}(X) = -b_2 (V_{Sd2} - 2R_{ST2} I_{d2}) V_{dc} - \frac{2V_{dc} I_{d2}}{R_{dc} C_{dc}}$$

$$g_{4btb}(X) = -b_2 (V_{Sq2} - 2R_{ST2} I_{q2}) V_{dc} - \frac{2V_{dc} I_{q2}}{R_{dc} C_{dc}}$$

They can be derived from

$$\frac{dz_4}{dt}=\frac{d}{dt}\left[V_{Sd1}I_{d1}+V_{Sq1}I_{q1}-R_{ST1}(I_{d1}^2+I_{q1}^2)+V_{Sd2}I_{d2}+V_{Sq2}I_{q2}-R_{ST2}(I_{d2}^2+I_{q2}^2)-\frac{V_{dc}^2}{R_{dc}}\right].$$

The state equations of the Z variables can be written in the standard linear form of:

$$\begin{pmatrix}\frac{dz_1}{dt}\\ \frac{dz_2}{dt}\\ \frac{dz_3}{dt}\\ \frac{dz_4}{dt}\\ \frac{dz_5}{dt}\end{pmatrix}=\begin{pmatrix}0&0&0&0&0&0\\0&0&0&0&0&0\\0&0&0&0&0&0\\0&0&0&0&0&0\\0&0&0&0&1&0\end{pmatrix}\begin{pmatrix}z_1\\z_2\\z_3\\z_4\\z_5\end{pmatrix}+\begin{pmatrix}1&0&0&0\\0&1&0&0\\0&0&1&0\\0&0&0&1\\0&0&0&0\end{pmatrix}\begin{pmatrix}w_{11}\\w_{12}\\w_{21}\\w_{22}\end{pmatrix} \tag{5.47}$$

and the relationship between the transformed input variables $[w_1 w_2]$ and the original input variables $[u_1 u_2]$ is given by the top four rows of equation 5.46:

$$\begin{pmatrix}w_{11}\\w_{12}\\w_{21}\\w_{22}\end{pmatrix}=-\begin{pmatrix}-a_1I_{d1}+\omega_1I_{q1}+b_1V_{Sd1}\\-\omega_1I_{d1}-a_1I_{q1}+b_1V_{Sq1}\\-a_2I_{d2}+\omega_2I_{q2}+b_2V_{Sd2}\\h_{btb}(X)\end{pmatrix}+\begin{pmatrix}-b_1V_{dc}&0&0&0\\0&-b_1V_{dc}&0&0\\0&0&-b_2V_{dc}&0\\g_{1btb}(X)&g_{2btb}(X)&g_{3btb}(X)&g_{4btb}(X)\end{pmatrix}\begin{pmatrix}u_{11}\\u_{12}\\u_{21}\\u_{22}\end{pmatrix}$$

If

$$\Delta_{btb}=-b_1^2b_2V_{dc}^3g_{4btb}(X)=-b_1^2b_2V_{dc}^4\left[b_2V_{Sq2}-2\left(a_2+\frac{1}{R_{dc}C_{dc}}\right)I_{q2}\right]\neq 0,$$

$$\text{i.e.}\quad V_{dc}\neq 0\quad\text{and}\quad I_{q2}\neq\frac{b_2R_{dc}C_{dc}}{2(a_2R_{dc}C_{dc}+1)}V_{Sq2},$$

$$\begin{pmatrix}u_{11}\\u_{12}\\u_{21}\\u_{22}\end{pmatrix}=\begin{pmatrix}-b_1V_{dc}&0&0&0\\0&-b_1V_{dc}&0&0\\0&0&-b_2V_{dc}&0\\g_{1btb}(X)&g_{2btb}(X)&g_{3btb}(X)&g_{4btb}(X)\end{pmatrix}^{-1}\begin{pmatrix}w_{11}+a_1I_{d1}-\omega_1I_{q1}-b_1V_{Sd1}\\w_{12}+\omega_1I_{d1}+a_1I_{q1}-b_1V_{Sq1}\\w_{21}+a_2I_{d2}-\omega_2I_{q2}-b_2V_{Sd2}\\w_{22}-h_{btb}(X)\end{pmatrix} \tag{5.48}$$

5.6 Modelling the current source converter system

In theory both self-commutated voltage source converters and self-commutated current source converters can be used for the control of active and reactive powers, but most of the existing schemes use the VSC topology. The main reasons behind the choice of VSC over CSC are:

1. The interfacing capacitors, required for absorbing the energy stored in the AC side inductance when the current through it is switched to zero, not only add to the overall cost but also present the risk of harmonic resonance.
2. The switching devices required must be symmetrical in terms of voltage blocking capability. That rejects the use the asymmetrical switches like the IGBT, unless they are combined with a series diode (which results in extra losses). The symmetrical switches are restricted in switching frequency but they have a lower voltage drop; they are better suited to CSC application.
3. The energy storage inductor on the current source converter DC side causes greater losses than the capacitor of the voltage source converter; something that the superconductive inductor technology is promising to improve.

The first two problems can be solved by progressive CSC topologies and control technology. The MLCR current source converter (described in Chapter 4) uses the multilevel and reinjection concepts, as well as fundamental switching to produce controllable and continuous AC output currents without the need of AC interfacing capacitors to interface with the power grid. The advantages provided by the current source converter with respect to the voltage source converter are mainly due to its direct control of the DC current and the fully rated range of DC voltage.

The CSC models derived in this section are also based on the generalized state–space averaging method and the principle of power balance. The accuracy of the CSC average models is ensured by the high-quality waveforms (achieved by either PWM or multilevel techniques) and the stiff DC current of the high-power current source converter.

5.6.1 Current source converters with pulse width modulation

The circuit shown in Figure 5.5 is used in this section for the derivation of a generalized average model of the current source converter. Then, by setting the parameters of the respective components to zero or infinite, the general model will be made suitable for specific applications.

In Figure 5.5 the current source converter AC terminals are connected to a power system, represented by a Thevenin equivalent in series with the leakage inductance (L_T) and resistance (R_T) of the interfacing transformer. The DC terminal is connected via a smoothing reactor (represented by L_{dc}, R_{dc}) and a T-network equivalent of the transmission line or cable to a DC voltage source or to a load.

The three-phase AC side voltage balancing equations are written in the form:

$$L_S \frac{d\mathbf{I_{abc}}}{dt} + R_S\mathbf{I_{abc}} + L_T \frac{d\mathbf{I_{abc}}}{dt} + R_T\mathbf{I_{abc}} = \mathbf{V_{Sabc}} - \mathbf{V_T}$$

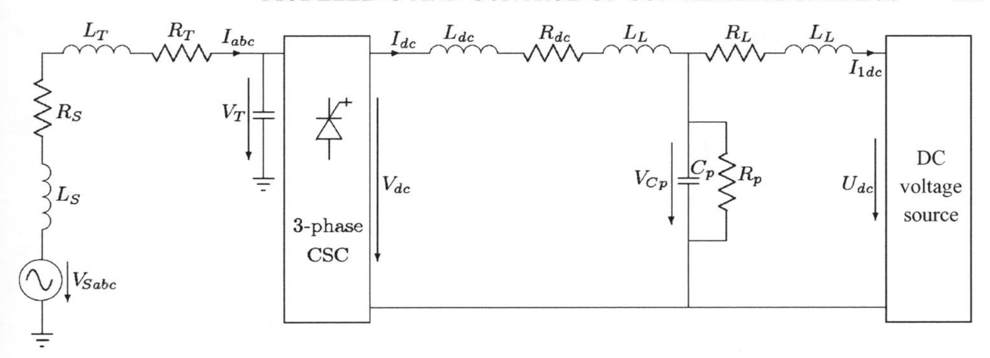

Figure 5.5 The equivalent circuit of the current source converter system.

where

$\mathbf{I_{abc}} = [I_a I_b I_c]^T$ is the three-phase current vector,

$\mathbf{V_{Sabc}} = [V_{Sa} V_{Sb} V_{Sc}]^T$ is the three-phase AC source voltage vector and

$\mathbf{V_T} = [V_{Ta} V_{Tb} V_{Tc}]^T$ is the three-phase voltage vector at the current source converter AC terminal.

The current balance equations at the current source converter AC terminals are expressed as

$$C_T \frac{d\mathbf{V_T}}{dt} = \mathbf{I_{abc}} - \mathbf{I_T}$$

where the control variables are:

$\mathbf{I_T} = [I_{Ta} I_{Tb} I_{Tc}]^T = \sqrt{2} I_{Tm} \left[\sin(\omega t + \phi_T) \sin\left(\omega t - \frac{2\pi}{3} + \phi_T\right) \sin\left(\omega t + \frac{2\pi}{3} + \phi_T\right)\right]^T$ (the current source converter AC output current vector), $I_{Tm} = m_o I_{dc} (0 < m_o < 1)$ (the *rms* amplitude),

ω (the frequency) and

ϕ_T (the phase angle).

When the current source converter is connected to an AC system operating at constant frequency, $I_{Tm} = m_o I_{dc} (0 < m_o < 1)$ and ϕ_T can be used for the control of the current source converter operation.

The current source converter DC output current I_{dc} flowing from the T-network to the DC source is:

$$I_{1dc} = I_{dc} - \frac{V_{cp}}{R_p} - C_p \frac{dV_{cp}}{dt}$$

The DC side voltage balance equations are:

$$V_{dc} - V_{cp} = R_{dc} I_{dc} + (L_{dc} + L_L)\frac{dI_{dc}}{dt}$$

$$V_{cp} - U_{dc} = R_L I_{1dc} + L_L \frac{dI_{1dc}}{dt}$$

and the DC and AC side power balancing:

$$\begin{aligned} V_{dc} I_{dc} &= \mathbf{V_T^T I_T} = \mathbf{I_{abc}^T V_{Sabc}} - (R_S + R_T)\mathbf{I_{abc}^T I_{abc}} - \frac{dW_{L_S}}{2dt} - \frac{dW_{L_T}}{dt} - \frac{dW_{C_T}}{dt} \\ &= \mathbf{I_{abc}^T V_{Sabc}} - (R_S + R_T)\mathbf{I_{abc}^T I_{abc}} - \frac{(L_S + L_T)}{2}\frac{d(\mathbf{I_{abc}^T I_{abc}})}{dt} - C_T \mathbf{V_T}\frac{d\mathbf{V_T}}{dt} \end{aligned}$$

where W_{LS}, W_{LT} and W_{C_T} represent the energies stored in the three phase inductances (L_S and L_T) and the shunt capacitance respectively.

Substituting $V_{dc} = \mathbf{V_T^T I_T}/I_{dc}$ into the expression $V_{dc} - V_{cp} = R_{dc} I_{dc} + (L_{dc} + L_L)\frac{dI_{dc}}{dt}$, and rearranging the other equations, the system equations become:

$$\frac{d\mathbf{I_{abc}}}{dt} = -a\mathbf{I_{abc}} + b\mathbf{V_{Sabc}} - \mathbf{b}\mathbf{V_T} \tag{5.49}$$

$$\frac{d\mathbf{V_T}}{dt} = \frac{1}{C_T}[\mathbf{I_{abc}} - \mathbf{I_T}] \tag{5.50}$$

$$\frac{dI_{dc}}{dt} = \frac{1}{L_{DC}}\left[\frac{\mathbf{V_T^T I_T}}{I_{dc}} - V_{cp} - R_{dc} I_{dc}\right] \tag{5.51}$$

$$\frac{dV_{cp}}{dt} = \frac{1}{C_p}\left(I_{dc} - I_{1dc} - \frac{V_{cp}}{R_p}\right) \tag{5.52}$$

$$\frac{dI_{1dc}}{dt} = \frac{1}{L_L}(V_{cp} - U_{dc} - R_L I_{2dc}) \tag{5.53}$$

where $a = \frac{R_S + R_T}{L_S + L_T}$, $b = \frac{1}{L_S + L_T}$, $R_{ST} = R_S + R_T$ and $L_{ST} = L_S + L_T$, $L_{DC} = L_{dc} + L_L$.

Performing the orthogonal transformation $\mathbf{M}$ defined by equation 5.6 (as in the case of VSC modelling), the three-phase vectors in equations 5.49 to 5.51 are transformed into the dq-frame and the CSC system state equations become:

$$\frac{dI_d}{dt} = -aI_d + \omega I_q + bV_{Sd} - bV_{Td} \tag{5.54}$$

$$\frac{dI_q}{dt} = -\omega I_d - aI_q + bV_{Sq} - bV_{Tq} \tag{5.55}$$

$$\frac{dV_{Td}}{dt} = \omega V_{Tq} + \frac{1}{C_T}(I_d - I_{Td}) \tag{5.56}$$

$$\frac{dV_{Tq}}{dt} = -\omega V_{Td} + \frac{1}{C_T}(I_q - I_{Tq}) \tag{5.57}$$

$$\frac{dI_{dc}}{dt} = \frac{1}{L_{DC}}\left[-V_{cp} - R_{dc}I_{dc} + \frac{1}{I_{dc}}(V_{Td}I_{Td} + V_{Tq}I_{Tq})\right] \tag{5.58}$$

$$\frac{dV_{cp}}{dt} = \frac{1}{C_p}\left(I_{dc} - I_{1dc} - \frac{V_{cp}}{R_p}\right) \tag{5.59}$$

$$\frac{dI_{1dc}}{dt} = \frac{1}{L_L}(V_{cp} - U_{dc} - R_L I_{1dc}) \tag{5.60}$$

The system of generalized equations 5.54 to 5.60 can be used directly for the control of an HVDC terminal or it can be simplified for the STATCOM and back-to-back applications.

For the STATCOM application, as $C_p = \infty, I_{1dc} = 0, R_L = 0, L_L = 0$ and $U_{dc} = V_{cp} = 0$, the state equations become

$$\frac{dI_d}{dt} = -aI_d + \omega I_q + bV_{Sd} - bV_{Td} \tag{5.61}$$

$$\frac{dI_q}{dt} = -\omega I_d - aI_q + bV_{Sq} - bV_{Tq} \tag{5.62}$$

$$\frac{dV_{Td}}{dt} = \omega V_{Tq} + \frac{1}{C_T}(I_d - I_{Td}) \tag{5.63}$$

$$\frac{dV_{Tq}}{dt} = -\omega V_{Td} + \frac{1}{C_T}(I_q - I_{Tq}) \tag{5.64}$$

$$\frac{dI_{dc}}{dt} = \frac{1}{L_{DC}}\left[-R_{dc}I_{dc} + \frac{1}{I_{dc}}(V_{Td}I_{Td} + V_{Tq}I_{Tq})\right] \tag{5.65}$$

For the back-to-back application the DC voltage source block in Figure 5.5 also consists of the AC system and current source converter on the left side of the diagram. However, as $C_p = 0$, $R_p = \infty$, $I_{1dc} = I_{dc}$, $V_{Cp} = U_{dc}$, $L_L = 0$ and $R_L = 0$, equations 5.59 and 5.60 are removed. Equations 5.54 to 5.57 are kept for the two AC systems and current source converters, with subscript 'A' added to the variables of the left-hand side and subscript 'B' to those of the right-hand side.

As $V_{cp} = U_{dc}$ and $-U_{dc}I_{dc} = I_{d2}V_{Td2} + I_{q2}V_{Tq2}$ is the active power transferred from the right-hand side system, substituting $-V_{cp}$ in equation 5.58 by $\frac{1}{I_{dc}}(I_{d2}V_{Td2} + I_{q2}V_{Tq2})$, the state equations for back-to-back application become:

$$\frac{dI_{dA}}{dt} = -a_A I_{dA} + \omega_A I_{qA} + b_A V_{SdA} - b_A V_{TdA} \tag{5.66}$$

$$\frac{dI_{qA}}{dt} = -\omega_A I_{dA} - a_A I_{qA} + b_A V_{SqA} - b_A V_{TqA} \tag{5.67}$$

$$\frac{dI_{dB}}{dt} = -a_B I_{dB} + \omega_B I_{qB} + b_B V_{SdB} - b_B V_{TdB} \tag{5.68}$$

$$\frac{dI_{qB}}{dt} = -\omega_B I_{dB} - a_B I_{qB} + b_B V_{SqB} - b_B V_{TqB} \tag{5.69}$$

$$\frac{dV_{TdA}}{dt} = \omega_A V_{TqA} + \frac{1}{C_{TA}}(I_{dA} - I_{TdA}) \tag{5.70}$$

$$\frac{dV_{TqA}}{dt} = -\omega_A V_{TdA} + \frac{1}{C_{TA}}(I_{qA} - I_{TqA}) \tag{5.71}$$

$$\frac{dV_{TdB}}{dt} = \omega_B V_{TqB} + \frac{1}{C_{TB}}(I_{dB} - I_{TdB}) \tag{5.72}$$

$$\frac{dV_{TqB}}{dt} = -\omega_B V_{TdB} + \frac{1}{C_{TB}}(I_{qB} - I_{TqB}) \tag{5.73}$$

$$\frac{dI_{dc}}{dt} = \frac{1}{L_{dc}}\left[-R_{dc}I_{dc} + \frac{1}{I_{dc}}(I_{TdA}V_{TdA} + I_{TqA}V_{TqA}) + \frac{1}{I_{dc}}(I_{TdB}V_{TdB} + I_{TqB}V_{TqB})\right] \tag{5.74}$$

These equations can also be used for an UPFC system by making the change $\omega_A = \omega_B = \omega$.

By expressing $\mathbf{I_{Tdq}}$ (the control variable vector of the PWM current source converter) as:

$$\mathbf{I_{Tdq}} = \mathbf{MI_T} = [I_{Td} I_{Tq}]^T = I_{dc}\sqrt{3}m_o[\cos(\phi_T)\sin(\phi_T)]^T = I_{dc}[u_1 u_2]^T$$

the state equations of the PWM current source converter for general application can be written in the affine form

$$\dot{\mathbf{X}} = \mathbf{F}(\mathbf{X}) + \mathbf{G}(\mathbf{X})\mathbf{U}$$

For PWM current source converter HVDC terminal control application, the state equations in the affine form are:

$$
\begin{pmatrix} \frac{dI_d}{dt} \\ \frac{dI_q}{dt} \\ \frac{dV_{Td}}{dt} \\ \frac{dV_{Tq}}{dt} \\ \frac{dI_{dc}}{dt} \\ \frac{dV_{cp}}{dt} \\ \frac{dI_{1dc}}{dt} \end{pmatrix} = \begin{pmatrix} -aI_d + \omega I_q + bV_{Sd} - bV_{Td} \\ -\omega I_d - aI_q + bV_{Sq} - bV_{Tq} \\ \omega V_{Tq} + \frac{1}{C_T} I_d \\ -\omega V_{Td} + \frac{1}{C_T} I_q \\ \frac{1}{L_{DC}}(-V_{cp} - R_{dc} I_{dc}) \\ \frac{1}{C_p}(I_{dc} - I_{1dc} - \frac{V_{cp}}{R_p}) \\ \frac{1}{L_L}(V_{cp} - U_{dc} - R_L I_{1dc}) \end{pmatrix} + \begin{pmatrix} 0 & 0 \\ 0 & 0 \\ -\frac{1}{C_T} I_{dc} & 0 \\ 0 & -\frac{1}{C_T} I_{dc} \\ \frac{V_{Td}}{L_{DC}} & \frac{V_{Tq}}{L_{DC}} \\ 0 & 0 \\ 0 & 0 \end{pmatrix} \begin{pmatrix} u_1 \\ u_2 \end{pmatrix} \quad (5.75)
$$

For a PWM current source converter STATCOM application the state equations in the affine form are:

$$
\begin{pmatrix} \frac{dI_d}{dt} \\ \frac{dI_q}{dt} \\ \frac{dV_{Td}}{dt} \\ \frac{dV_{Tq}}{dt} \\ \frac{dV_{dc}}{dt} \end{pmatrix} = \begin{pmatrix} -aI_d + \omega I_q + bV_{Sd} - bV_{Td} \\ -\omega I_d - aI_q + bV_{Sq} - bV_{Tq} \\ \omega V_{Tq} + \frac{1}{C_T} I_d \\ -\omega V_{Td} + \frac{1}{C_T} I_d \\ \frac{-R_{dc} I_{dc}}{L_{dc}} \end{pmatrix} + \begin{pmatrix} 0 & 0 \\ 0 & 0 \\ -\frac{1}{C_T} I_{dc} & 0 \\ 0 & -\frac{1}{C_T} I_{dc} \\ \frac{V_{Td}}{L_{dc}} & \frac{V_{Tq}}{L_{dc}} \end{pmatrix} \begin{pmatrix} u_1 \\ u_2 \end{pmatrix} \quad (5.76)
$$

For the PWM current source converter back-to-back application the state equations in the affine form are:

$$\begin{pmatrix} \dfrac{dI_{dA}}{dt} \\ \dfrac{dI_{qA}}{dt} \\ \dfrac{dI_{dB}}{dt} \\ \dfrac{dI_{qB}}{dt} \\ \dfrac{dV_{TdA}}{dt} \\ \dfrac{dV_{TqA}}{dt} \\ \dfrac{dV_{TdB}}{dt} \\ \dfrac{dV_{TqB}}{dt} \\ \dfrac{dI_{dc}}{dt} \end{pmatrix} = \begin{pmatrix} -a_A I_{dA} + \omega_A I_{qA} + b_A V_{SdA} - b_A V_{TdA} \\ -\omega_A I_{dA} - a_A I_{qA} + b_A V_{SqA} - b_B V_{TqA} \\ -a_B I_{dB} + \omega_B I_{qB} + b_B V_{SdB} - b_B V_{TdB} \\ -\omega_B I_{dB} - a_B I_{qB} + b_B V_{SqB} - b_B V_{TqB} \\ \omega_A V_{TqA} + \dfrac{1}{C_{TA}} I_{dA} \\ -\omega_A V_{TdA} + \dfrac{1}{C_{TA}} I_{qA} \\ \omega_B V_{TqB} + \dfrac{1}{C_{TB}} I_{dB} \\ -\omega_B V_{TdB} + \dfrac{1}{C_{TB}} I_{qB} \\ -\dfrac{R_{dc}}{L_{dc}} I_{dc} \end{pmatrix}$$

$$+ \begin{pmatrix} 0 & 0 & 0 & 0 \\ 0 & 0 & 0 & 0 \\ 0 & 0 & 0 & 0 \\ 0 & 0 & 0 & 0 \\ -\dfrac{I_{dc}}{C_{TA}} & 0 & 0 & 0 \\ 0 & -\dfrac{I_{dc}}{C_{TA}} & 0 & 0 \\ 0 & 0 & -\dfrac{I_{dc}}{C_{TB}} & 0 \\ 0 & 0 & 0 & -\dfrac{I_{dc}}{C_{TB}} \\ \dfrac{V_{TdA}}{L_{dc}} & \dfrac{V_{TqA}}{L_{dc}} & \dfrac{V_{TdB}}{L_{dc}} & \dfrac{V_{TqB}}{L_{dc}} \end{pmatrix} 0 \begin{pmatrix} u_{A1} \\ u_{A2} \\ u_{B1} \\ u_{B2} \end{pmatrix} \tag{5.77}$$

These affine non-linear equations do not satisfy the state-input linearizable condition and, therefore, do not provide the exact linearized state equations. However, by ignoring some terms in the dq-frame state equations, the linearized equations can be obtained at somewhat reduced accuracy. The approximations proposed by various researchers are focused on the last of the state equations of the STATCOM and back-to-back applications. The last equation is derived based on the AC and DC active power balancing; the AC side active power of the current source converter at its AC terminals can be expressed by $\mathbf{V_{Tdq}^T I_{Tdq}}$ or by

$$\begin{aligned}\mathbf{V_{Tdq}^T I_{Tdq}} &= \mathbf{V_{sdq}^T I_{dq}} - \mathbf{I_{dq}^T} L_{ST}\frac{d}{dt}(\mathbf{I_{dq}}) - \mathbf{V_{Tdq}^T} C_T \frac{d}{dt}(\mathbf{V_{Tdq}}) - R_{ST}(\mathbf{I_{dq}^T I_{dq}}) \\ &= \mathbf{V_{sdq}^T I_{dq}} - \frac{d}{dt}(0.5L_{ST}\mathbf{I_{dq}^T I_{dq}}) - \frac{d}{dt}(0.5C_T\mathbf{V_{Tdq}^T V_{Tdq}}) - R_{ST}(\mathbf{I_{dq}^T I_{dq}})\end{aligned}$$

The term $R_{ST}(\mathbf{I_{dq}^T I_{dq}})$, corresponding to AC side resistive losses, is a small percentage of the power controlled by the current source converter, and can be neglected in practice without causing significant difference. The energy terms $\frac{d}{dt}(0.5L_{ST}\mathbf{I_{dq}^T I_{dq}})$ and $\frac{d}{dt}(0.5C_T\mathbf{V_{Tdq}^T V_{Tdq}})$, corresponding to the changing rate of the energy stored in inductance L_{ST} and capacitance C_T respectively, are directly related to the *rms* levels of the inductance current and capacitance voltage ($I_{dq}^T I_{dq} = 3I_{rms}^2$ and $V_{Tdq}^T V_{Tdq} = 3V_{Trms}^2$). If (and only if) the current and current *rms*-levels change slowly or L_{ST} and C_T are relatively small, ignoring them will not cause great difference to the model in practice. However, ignoring the two terms will not be suitable for system analysis and controller design with fast changing AC voltage and current.

After removing the three terms, the linearized state equations for STATCOM and back-to-back applications become:

$$\begin{pmatrix} \frac{dI_d}{dt} \\ \frac{dI_q}{dt} \\ \frac{dV_{Td}}{dt} \\ \frac{dV_{Tq}}{dt} \\ \frac{d(I_{dc}^2)}{dt} \end{pmatrix} = \begin{pmatrix} -aI_d + \omega I_q + bV_{Sd} - bV_{Td} \\ -\omega I_d - aI_q + bV_{Sq} - bV_{Tq} \\ \omega V_{Tq} + \frac{1}{C_T} I_d \\ -\omega V_{Td} + \frac{1}{C_T} I_q \\ \frac{-R_{dc}}{L_{dc}} I_{dc}^2 + \frac{V_{Sd}}{L_{dc}} I_d + \frac{V_{Sq}}{L_{dc}} I_q \end{pmatrix} + \begin{pmatrix} 0 & 0 \\ 0 & 0 \\ -\frac{1}{C_T} & 0 \\ 0 & -\frac{1}{C_T} \\ 0 & 0 \end{pmatrix} \begin{pmatrix} I_{dc}u_1 \\ I_{dc}u_2 \end{pmatrix} \tag{5.78}$$

and

$$\begin{pmatrix} \dfrac{dI_{dA}}{dt} \\ \dfrac{dI_{qA}}{dt} \\ \dfrac{dI_{dB}}{dt} \\ \dfrac{dI_{qB}}{dt} \\ \dfrac{dV_{TdA}}{dt} \\ \dfrac{dV_{TqA}}{dt} \\ \dfrac{dV_{TdB}}{dt} \\ \dfrac{dV_{TqB}}{dt} \\ \dfrac{d(I_{dc}^2)}{dt} \end{pmatrix} = \begin{pmatrix} -a_A I_{dA} + \omega_A I_{qA} + b_A V_{SdA} - b_A V_{TdA} \\ -\omega_A I_{dA} - a_A I_{qA} + b_A V_{SqA} - b_A V_{TqA} \\ -a_B I_{dB} + \omega_B I_{qB} + b_B V_{SdB} - b_B V_{TdB} \\ -\omega_B I_{dB} - a_B I_{qB} + b_B V_{SqB} - b_B V_{TqB} \\ \omega_A V_{TqA} + \dfrac{1}{C_{TA}} I_{dA} \\ -\omega_A V_{TdA} + \dfrac{1}{C_{TA}} I_{qA} \\ \omega_B V_{TqB} + \dfrac{1}{C_{TB}} I_{dB} \\ -\omega_B V_{TdB} + \dfrac{1}{C_{TB}} I_{qB} \\ \dfrac{V_{SdA} I_{dA}}{L_{dc}} + \dfrac{V_{SqA} I_{qA}}{L_{dc}} + \dfrac{V_{SdB} I_{dB}}{L_{dc}} + \dfrac{V_{SqB} I_{qB}}{L_{dc}} - \dfrac{R_{dc} I_{dc}^2}{L_{dc}} \end{pmatrix}$$

$$+ \begin{pmatrix} 0 & 0 & 0 & 0 \\ 0 & 0 & 0 & 0 \\ 0 & 0 & 0 & 0 \\ 0 & 0 & 0 & 0 \\ \dfrac{-1}{C_{TA}} & 0 & 0 & 0 \\ 0 & \dfrac{-1}{C_{TA}} & 0 & 0 \\ 0 & 0 & \dfrac{-1}{C_{TB}} & 0 \\ 0 & 0 & 0 & \dfrac{-1}{C_{TB}} \\ 0 & 0 & 0 & 0 \end{pmatrix} \begin{pmatrix} I_{dc} u_{A1} \\ I_{dc} u_{A2} \\ I_{dc} u_{B1} \\ I_{dc} u_{B2} \end{pmatrix} \tag{5.79}$$

where $[I_{dc}u_1 I_{dc}u_2]^T$ and $[I_{dc}u_{A1} I_{dc}u_{A2} I_{dc}u_{B1} I_{dc}u_{B2}]^T$ are the new input vectors, and I_{dc}^2 is the new state variable (instead of I_{dc}).

5.7 Modelling grouped current source converters with fundamental frequency switching

As in the voltage source converter case, the use of fundamental switching minimizes the CSC switching losses but then the direct AC current amplitude control freedom is lost. However, high-power applications are likely to need more than one CSC group and these grouped configurations, if appropriately combined, can provide the AC current amplitude control freedom even operating at fundamental frequency. This can be achieved by a parallel connection of the CSC groups on the AC side and series connection on the DC side. The parallel connection on the AC side provides the freedom of adding AC currents with different phase shifts to form an amplitude-controllable AC current output. On the DC side the series connection provides the freedom of adding DC voltages with different levels to form the required higher DC voltage output.

The MLCR current source converter (described in Chapter 4) is the preferred topology for the grouped CSC configuration, as it uses the multilevel and reinjection concepts, as well as fundamental switching, to produce phase-controllable and continuous AC output currents without the need of AC capacitors to interface the converter with the power grid.

The grouped configuration is shown in Figure 5.6 for the case of two MLCR current source converters. Two large resistors (R_{t1}, R_{t2}) are added to represent the off-state leakage losses of the valves in the current source converters. Although these losses are negligible, without these resistors the AC side currents would be fully determined by the current source converter AC output currents and, thus, the relationship between the AC side currents and control action of the two current source converters could not be included in the system dynamic model.

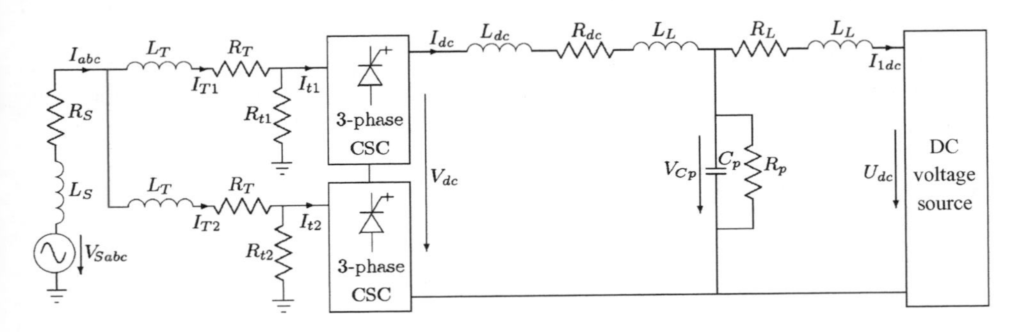

Figure 5.6 The equivalent circuit of the grouped two current source converters system.

In the diagram of Figure 5.7 the AC current phasors of the two current source converters are denoted by I_{T1} and I_{T2} respectively and the combined AC current phasor (I_{abc}) is the vectorial sum of these two currents, i.e. $I_{abc} = I_{T1} + I_{T2}$.

As the amplitudes of $|I_{T1}|$ and $|I_{T2}|$ are exactly the same, and determined by the converter DC side current, the following relations can be written for the combined AC current amplitude $|I_{abc}|$ and its phase angle:

$$|I_{abc}| = |I_{T1} + I_{T2}| = 2|I_{T1}|\cos\left(\frac{\theta_1 - \theta_2}{2}\right) \tag{5.80}$$

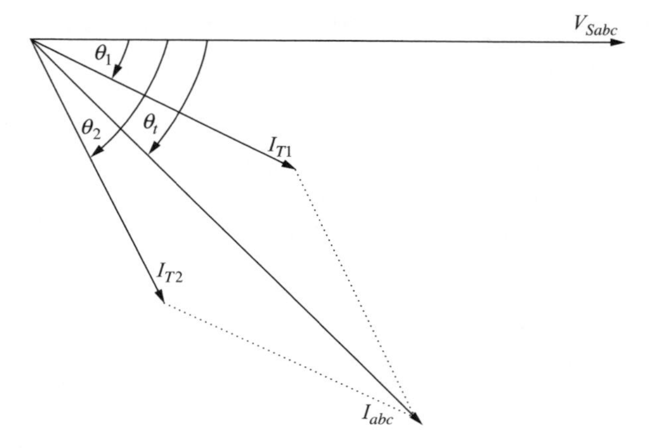

Figure 5.7 The phasor diagram of the two voltage source converters system.

$$\theta = \frac{\theta_1 + \theta_2}{2} \tag{5.81}$$

These equations show that the amplitude of I_{abc} can be controlled by $\cos\left(\frac{\theta_1 - \theta_2}{2}\right)$, and its phase angle by $\left(\frac{\theta_1 + \theta_2}{2}\right)$. Thus the combined effect of the two current source converter groups provides phase and amplitude current controllability independently, whereas a PWM converter uses the modulation index for the magnitude and phase shift for the phase angle control of the current.

In practice the two current source converters are identical, namely $R_{t1} = R_{t2} = R_t$ and the interfacing transformers have the same impedance (represented by a series connection of the resistance and inductance R_T and L_T). Thus the AC side voltage balance equations are:

$$\begin{aligned}
\mathbf{V_{Sabc}} &= L_S \frac{d\mathbf{I_{abc}}}{dt} + R_S \mathbf{I_{abc}} + L_T \frac{d\mathbf{I_{T1}}}{dt} + R_T \mathbf{I_{T1}} + R_t(\mathbf{I_{T1}} - \mathbf{I_{t1}}) \\
&= L_S \frac{d(\mathbf{I_{T1}} + \mathbf{I_{T2}})}{dt} + R_S(\mathbf{I_{T1}} + \mathbf{I_{T2}}) + L_T \frac{d\mathbf{I_{T1}}}{dt} + R_T \mathbf{I_{T1}} + R_t(\mathbf{I_{T1}} - \mathbf{I_{t1}})
\end{aligned}$$

$$\begin{aligned}
\mathbf{V_{Sabc}} &= L_S \frac{d\mathbf{I_{abc}}}{dt} + R_S \mathbf{I_{abc}} + L_T \frac{d\mathbf{I_{T2}}}{dt} + R_T \mathbf{I_{T2}} + R_t(\mathbf{I_{T2}} - \mathbf{I_{t2}}) \\
&= L_S \frac{d(\mathbf{I_{T1}} + \mathbf{I_{T2}})}{dt} + R_S(\mathbf{I_{T1}} + \mathbf{I_{T2}}) + L_T \frac{d\mathbf{I_{T2}}}{dt} + R_T \mathbf{I_{T2}} + R_t(\mathbf{I_{T2}} - \mathbf{I_{t2}})
\end{aligned}$$

where

$\mathbf{I_{abc}} = [I_a I_b I_c]^T$ is the three-phase current vector,

$\mathbf{V_{Sabc}} = [V_{Sa} V_{Sb} V_{Sc}]^T$ is the three-phase AC source voltage vector,

$\mathbf{I_{T1}} = [I_{T1a} I_{T1b} I_{T1c}]^T$ and $\mathbf{I_{T2}} = [I_{T2a} I_{T2b} I_{T2c}]^T$ are the three-phase current vectors of the two interface transformers,

$\mathbf{I_{t1}} = [I_{t1a} I_{t1b} I_{t1c}]^T$ and $\mathbf{I_{t2}} = [I_{t2a} I_{t2b} I_{t2c}]^T$ are the three-phase output current vectors of the two current source converters.

Under fundamental switching control, the AC output currents of the two current source converters are directly related to the common DC current, i.e.

$$\mathbf{I_{t1}} = \sqrt{2} k_i I_{dc} \left[\sin(\omega t + \phi_1) \sin\left(\omega t - \frac{2\pi}{3} + \phi_1\right) \sin\left(\omega t + \frac{2\pi}{3} + \phi_1\right) \right]^T$$

$$\mathbf{I_{t2}} = \sqrt{2} k_i I_{dc} \left[\sin(\omega t + \phi_2) \sin\left(\omega t - \frac{2\pi}{3} + \phi_2\right) \sin\left(\omega t + \frac{2\pi}{3} + \phi_2\right) \right]^T$$

where k_i is a constant determined by the current source converter topology.

Adding and subtracting these two equations respectively the following expressions are obtained:

$$\frac{d[\mathbf{I_{T1}} + \mathbf{I_{T2}}]}{dt} = -\frac{2R_S + R_T + R_t}{2L_S + L_T} [\mathbf{I_{T1}} + \mathbf{I_{T2}}] + \frac{R_t}{2L_S + L_T} [\mathbf{I_{t1}} + \mathbf{I_{t2}}] + \frac{2}{2L_S + L_T} \mathbf{V_{Sabc}}$$

$$\frac{d[\mathbf{I_{T1}} - \mathbf{I_{T2}}]}{dt} = -\frac{R_T + R_t}{L_T} [\mathbf{I_{T1}} - \mathbf{I_{T2}}] + \frac{R_t}{L_T} [\mathbf{I_{t1}} - \mathbf{I_{t2}}]$$

Substituting $\mathbf{I_S} = \mathbf{I_{T1}} + \mathbf{I_{T2}}$ and $\mathbf{I_N} = \mathbf{I_{T1}} - \mathbf{I_{T2}}$ into the equations above, the AC side expressions become:

$$\frac{d\mathbf{I_S}}{dt} = -a_1 \mathbf{I_S} + b_1 [\mathbf{I_{t1}} + \mathbf{I_{t2}}] + d\mathbf{V_{Sabc}} \tag{5.82}$$

$$\frac{d\mathbf{I_N}}{dt} = -a_2 \mathbf{I_N} + b_2 [\mathbf{I_{t1}} - \mathbf{I_{t2}}] \tag{5.83}$$

where $a_1 = \frac{2R_S + R_T + R_t}{2L_S + L_T}$, $a_2 = \frac{R_T + R_t}{L_T}$, $b_1 = \frac{R_t}{2L_S + L_T}$, $b_2 = \frac{R_t}{L_T}$, $d = \frac{2}{2L_S + L_T}$.

The DC side equations are

$$I_{dc} - I_{1dc} = \frac{V_{cp}}{R_p} + C_p \frac{dV_{cp}}{dt} \tag{5.84}$$

$$V_{dc} - V_{cp} = R_{dc} I_{dc} + (L_{dc} + L_L) \frac{dI_{dc}}{dt} \tag{5.85}$$

$$V_{cp} - U_{dc} = R_L I_{1dc} + L_L \frac{dI_{1dc}}{dt} \tag{5.86}$$

The following AC and DC side power balance results:

$$\begin{aligned} I_{dc} V_{dc} = {} & \mathbf{I_S^T V_{Sabc}} - \mathbf{L_S I_S^T} \frac{\mathbf{dI_S}}{\mathbf{dt}} - \mathbf{R_S I_S^T I_S} - \mathbf{L_T I_{T1}^T} \frac{\mathbf{dI_{T1}}}{\mathbf{dt}} - \mathbf{L_T I_{T2}^T} \frac{\mathbf{dI_{T2}}}{\mathbf{dt}} \\ & - \mathbf{R_T}(\mathbf{I_{T1}^T I_{T1}} + \mathbf{I_{T2}^T I_{T2}}) - P_{R_{t1}} - P_{R_{t2}} \end{aligned}$$

where $P_{R_{t1}}$ and $P_{R_{t2}}$ are the power losses consumed by the resistors R_{t1} and R_{t2} respectively.

Considering that $P_{R_{t1}}$ and $P_{R_{t2}}$ are negligible and that $\mathbf{I_{T1}}$ and $\mathbf{I_{T2}}$ are nearly the same as the two current source converters' AC currents ($\mathbf{I_{t1}}$ and $\mathbf{I_{t2}}$), the active power of the two current source converters on the DC side can be expressed as

$$\begin{aligned} P_{CSCs} &= I_{dc}V_{dc} \approx \mathbf{I_S^T V_{Sabc}} - L_S\mathbf{I_S^T}\frac{d\mathbf{I_S}}{dt} - R_S\mathbf{I_S^T I_S} - L_T\mathbf{I_{t1}^T}\frac{d\mathbf{I_{t1}}}{dt} - L_T\mathbf{I_{t2}^T}\frac{d\mathbf{I_{t2}}}{dt} - R_T(\mathbf{I_{t1}^T I_{t1}} + \mathbf{I_{t2}^T I_{t2}}) \\ &= \mathbf{I_S^T V_{Sabc}} - \mathbf{L_S I_S^T}\frac{\mathbf{dI_S}}{\mathbf{dt}} - \mathbf{R_S I_S^T I_S} - \frac{L_T}{2}\frac{d}{dt}(\mathbf{I_{t1}^T I_{t1}} + \mathbf{I_{t2}^T I_{t2}}) - R_T(\mathbf{I_{t1}^T I_{t1}} + \mathbf{I_{t2}^T I_{t2}}) \end{aligned}$$

Substituting the expression $V_{dc} = \frac{1}{I_{dc}}P_{CSCs}$ into equation 5.85 and rewriting equations 5.82 to 5.86 as

$$\frac{d\mathbf{I_S}}{dt} = -a_1\mathbf{I_S} + b_1[\mathbf{I_{t1}} + \mathbf{I_{t2}}] + d\mathbf{V_{Sabc}} \tag{5.82a}$$

$$\frac{d\mathbf{I_N}}{dt} = -a_2\mathbf{I_N} + b_2[\mathbf{I_{t1}} - \mathbf{I_{t2}}] \tag{5.83a}$$

$$\frac{dI_{dc}}{dt} = \frac{1}{L_{DC}}\left[\frac{1}{I_{dc}}P_{CSCs} - V_{cp} - R_{dc}I_{dc}\right] \tag{5.84a}$$

$$\frac{dV_{cp}}{dt} = \frac{1}{C_p}\left(I_{dc} - I_{1dc} - \frac{V_{cp}}{R_p}\right) \tag{5.85a}$$

$$\frac{dI_{1dc}}{dt} = \frac{1}{L_L}(V_{cp} - U_{dc} - R_L I_{2dc}) \tag{5.86a}$$

where $L_{DC} = L_{dc} + L_L$, the state equations in dq-frame can be obtained by performing the orthogonal transformation to equations and , i.e.:

$$\frac{dI_{Sd}}{dt} = -a_1 I_{Sd} + \omega I_{Sq} + b_1[I_{t1d} + I_{t2d}] + dV_{Sd} \tag{5.87}$$

$$\frac{dI_{Sq}}{dt} = -\omega I_{Sd} - a_1 I_{Sq} + b_1\left[I_{t1q} + I_{t2q}\right] + dV_{Sq} \tag{5.88}$$

$$\frac{dI_{dc}}{dt} = \frac{1}{L_{DC}}\left[\frac{1}{I_{dc}}P_{CSCs} - V_{cp} - R_{dc}I_{dc}\right] \tag{5.89}$$

$$\frac{dV_{cp}}{dt} = \frac{1}{C_p}\left(I_{dc} - I_{1dc} - \frac{V_{cp}}{R_p}\right) \tag{5.90}$$

$$\frac{dI_{1dc}}{dt} = \frac{1}{L_L}(V_{cp} - U_{dc} - R_L I_{1dc}) \tag{5.91}$$

$$\frac{dI_{Nd}}{dt} = -a_2 I_{Nd} + \omega I_{Nq} + b_2[I_{t1d} - I_{t2d}] \tag{5.92}$$

$$\frac{dI_{Nq}}{dt} = -\omega I_{Nd} - a_2 I_{Nq} + b_2\left[I_{t1q} - I_{t2q}\right] \tag{5.93}$$

where

$$P_{CSCs} = \mathbf{I}_{\mathbf{Sdq}}^{\mathbf{T}}\mathbf{V}_{\mathbf{Sdq}} - L_S\mathbf{I}_{\mathbf{Sdq}}^{\mathbf{T}}\frac{d}{dt}\mathbf{I}_{\mathbf{Sdq}} - R_S\mathbf{I}_{\mathbf{Sdq}}^{\mathbf{T}}\mathbf{I}_{\mathbf{Sdq}} - \frac{L_T}{2}\frac{d}{dt}(\mathbf{I}_{\mathbf{t1dq}}^{\mathbf{T}}\mathbf{I}_{\mathbf{t1dq}} + \mathbf{I}_{\mathbf{t2dq}}^{\mathbf{T}}\mathbf{I}_{\mathbf{t2}})$$
$$-R_T(\mathbf{I}_{\mathbf{t1dq}}^{\mathbf{T}}\mathbf{I}_{\mathbf{t1dq}} + \mathbf{I}_{\mathbf{t2dq}}^{\mathbf{T}}\mathbf{I}_{\mathbf{t2dq}})$$

The last two equations, 5.92 and 5.93, show that there is a stable self-governed subsystem inside the system, as the state variables I_{Nd} and I_{Nq} have no interaction with the other equations. They are only influenced by the two current source converters control variables $\mathbf{I}_{\mathbf{t1dq}}$ and $\mathbf{I}_{\mathbf{t2dq}}$. Therefore the whole system can be divided into two subsystems. The first subsystem consists of two converter bridges and two interfacing transformers and the second subsystem consists of the AC source (connected to the AC terminal of the first subsystem) and the DC side circuit (connected to the DC terminal of the first subsystem). Under the control action defined by the control variables $\mathbf{I}_{\mathbf{t1dq}}$ and $\mathbf{I}_{\mathbf{t2dq}}$ in the time domain, the state variables of the two subsystems will be driven to their specified states; however, the state variables of the first are not a function of the state variables of the second subsystem. To study the dynamic and steady operation of the system related to the AC and DC side variables, only the second subsystem needs to be considered and once the behaviour of this part is understood, that of the first subsystem can be determined. Therefore the state equations used for different applications do not need to include the two equations describing the behaviours of the first subsystem.

The generalized equations 5.87 to 5.91 can be directly used for the control of the active and reactive powers, DC voltage and current, and AC side voltage and current of an HVDC terminal. The number of system equations can be reduced for other applications. For STATCOM application, as $C_p = \infty$, $I_{1dc} = 0$, $R_L = 0$, $L_L = 0$ and $U_{dc} = V_{cp} = 0$, the state equations are:

$$\frac{dI_{Sd}}{dt} = -a_1 I_{Sd} + \omega I_{Sq} + b_1[I_{t1d} + I_{t2d}] + dV_{Sd} \tag{5.94}$$

$$\frac{dI_{Sq}}{dt} = -\omega I_{Sd} - a_1 I_{Sq} + b_1\left[I_{t1q} + I_{t2q}\right] + dV_{Sq} \tag{5.95}$$

$$\frac{dI_{dc}}{dt} = \frac{1}{L_{dc}I_{dc}}P_{CSCs} - \frac{R_{dc}I_{dc}}{L_{dc}} \tag{5.96}$$

For the back-to-back application the DC voltage source block in Figure 5.6 is an equivalent mirror of the AC system and current source converter on the left-hand side. As $C_p = 0$, $R_p = \infty$, $I_{1dc} = I_{dc}$, $V_{Cp} = U_{dc}$, $L_L = 0$ and $R_L = 0$, all terms in equations 5.90 and 5.91 are zero; subscript 'A' is added to the left-side AC system and current source converter in equations 5.87 to 5.88 and subscript 'B' is added to the right-side AC system and current source converter variables in equations 5.87 to 5.88. As $V_{cp} = U_{dc}$ and since $-U_{dc}I_{dc}$ is the active power transferred from the right-side system, substituting $-V_{cp}$ in equation 5.89 by $\frac{1}{I_{dc}}P_{BCSCs}$ the state equations for the back-to-back application become:

$$\frac{dI_{ASd}}{dt} = -a_{A1}I_{ASd} + \omega_A I_{ASq} + b_{A1}[I_{At1d} + I_{At2d}] + d_A V_{ASd} \tag{5.97}$$

$$\frac{dI_{ASq}}{dt} = -\omega_A I_{ASd} - a_{A1}I_{ASq} + b_{A1}\left[I_{At1q} + I_{At2q}\right] + d_A V_{ASq} \tag{5.98}$$

$$\frac{dI_{BSd}}{dt} = -a_{B1}I_{BSd} + \omega_B I_{BSq} + b_{B1}[I_{Bt1d} + I_{Bt2d}] + d_B V_{BSd} \tag{5.99}$$

$$\frac{dI_{BSq}}{dt} = -\omega_B I_{BSd} - a_{A1}I_{BSq} + b_{B1}\left[I_{Bt1q} + I_{Bt2q}\right] + d_B V_{BSq} \tag{5.100}$$

$$\frac{dI_{dc}}{dt} = \frac{1}{L_{dc}I_{dc}}[P_{ACSCs} + P_{BCSCs}] - \frac{R_{dc}I_{dc}}{L_{dc}} \tag{5.101}$$

These equations can also be used for a UPFC system by making $\omega_A = \omega_B = \omega$.

Using the following expressions for $\mathbf{I_{t1dq}}, \mathbf{I_{t2dq}}$ (the control variables of the grouped current source converters):

$$\mathbf{I_{t1dq}} = \mathbf{MI_{t1}} = [I_{t1d}I_{t1q}]^T = I_{dc}\sqrt{3}k_i[\cos(\phi_1)\sin(\phi_1)]^T$$

$$\mathbf{I_{t2dq}} = \mathbf{MI_{t2}} = [I_{t2d}I_{t2q}]^T = I_{dc}\sqrt{3}k_i[\cos(\phi_2)\sin(\phi_2)]^T$$

and

$$\begin{aligned}\mathbf{I_{t1dq}} + \mathbf{I_{t2dq}} &= [(I_{t1d} + I_{t2d})(I_{t1q} + I_{t2q})]^T \\ &= I_{dc}\sqrt{3}k_i[\cos(\phi_1) + \cos(\phi_2)\sin(\phi_1) + \sin(\phi_2)]^T = I_{dc}[v_1 v_2]^T\end{aligned}$$

the generalized state equations of the grouped current source converters can be expressed in the affine form:

$$\dot{\mathbf{X}} = \mathbf{F}(\mathbf{X}) + \mathbf{G}(\mathbf{X})\mathbf{U}$$

As P_{CSCs} in equations 5.89, 5.96 and 5.101 is related to $\frac{d}{dt}\mathbf{I_{Sdq}}$ and $\frac{d}{dt}\left(\mathbf{I^T_{t1dq}I_{t1dq}} + \mathbf{I^T_{t2dq}I_{t2dq}}\right)$, the following derivation is needed to relate P_{CSCs} to $\mathbf{I_{Sdq}}$ and I_{dc}:

$$\begin{aligned}P_{CSCs} &= \mathbf{I^T_{Sdq}V_{Sdq}} - L_S\mathbf{I^T_{Sdq}}\frac{d}{dt}\mathbf{I_{Sdq}} - R_S\mathbf{I^T_{Sdq}I_{Sdq}} - \frac{L_T}{2}\frac{d}{dt}\left(\mathbf{I^T_{t1dq}I_{t1dq}} + \mathbf{I^T_{t2dq}I_{t2dq}}\right) \\ &\quad - R_T(\mathbf{I^T_{t1dq}I_{t1dq}} + \mathbf{I^T_{t2dq}I_{t2dq}}) \\ &= -6k_i^2L_TI_{dc}\frac{dI_{dc}}{dt} + (1 - dL_S)(I_{Sd}V_{Sd} + I_{Sq}V_{Sq}) + (a_1L_S - R_S)\left(I^2_{Sd} + I^2_{Sq}\right) \\ &\quad - 6k_i^2R_TI^2_{dc} - b_1L_SI_{Sd}I_{dc}v_1 - b_1L_SI_{Sq}I_{dc}v_2\end{aligned}$$

Substituting this expression of P_{CSCs} into equation 5.89 produces the following:

$$\frac{dI_{dc}}{dt} = \frac{1}{L_{DC}+6k_i^2L_T}\left\{\frac{1}{I_{dc}}\left[(1-dL_S)(I_{Sd}V_{Sd}+I_{Sq}V_{Sq})+(a_1L_S-R_S)\left(I_{Sd}^2+I_{Sq}^2\right)\right]\right.$$

$$\left. -V_{cp}-\left(R_{dc}+6k_i^2R_T\right)I_{dc}-b_1L_SI_{Sd}v_1-b_1L_SI_{Sq}v_2\right\}$$

$$= c_1\frac{(I_{Sd}V_{Sd}+I_{Sq}V_{Sq})}{I_{dc}}+c_2\frac{\left(I_{Sd}^2+I_{Sq}^2\right)}{I_{dc}}-c_3V_{cp}-c_4I_{dc}-c_5I_{Sd}v_1-c_5I_{Sq}v_2 \quad (5.102)$$

where

$$c_1 = \frac{L_T}{(L_{DC}+6k_i^2L_T)(2L_S+L_T)}$$

$$c_2 = \frac{(R_T+R_t)L_S-R_SL_T}{(L_{DC}+6k_i^2L_T)(2L_S+L_T)}$$

$$c_3 = \frac{1}{L_{DC}+6k_i^2L_T}$$

$$c_4 = \frac{R_{DC}+6k_i^2R_T}{L_{DC}+6k_i^2L_T}$$

$$c_5 = \frac{R_TL_S}{(L_{DC}+6k_i^2L_T)(2L_S+L_T)}$$

Substituting P_{CSCs} into equation 5.96

$$\frac{dI_{dc}}{dt} = c_1\frac{(I_{Sd}V_{Sd}+I_{Sq}V_{Sq})}{I_{dc}}+c_2\frac{\left(I_{Sd}^2+I_{Sq}^2\right)}{I_{dc}}-c_4I_{dc}-c_5I_{Sd}v_1-c_5I_{Sq}v_2 \quad (5.103)$$

For the back-to-back application,

$$P_{ACSCs} = -6k_{Ai}^2L_{AT}I_{dc}\frac{dI_{dc}}{dt}+(1-d_AL_{AS})(I_{ASd}V_{ASd}+I_{ASq}V_{ASq})+(a_{A1}L_{AS}-R_{AS})\left(I_{ASd}^2+I_{ASq}^2\right)$$
$$-6k_{Ai}^2R_{AT}I_{dc}^2-b_{A1}L_{AS}I_{ASd}I_{dc}v_{A1}-b_{A1}L_{AS}I_{ASq}I_{dc}v_{A2}$$

$$P_{BCSCs} = -6k_{Bi}^2L_{BT}I_{dc}\frac{dI_{dc}}{dt}+(1-d_BL_{BS})(I_{BSd}V_{BSd}+I_{BSq}V_{BSq})+(a_{B1}L_{AS}-R_{BS})\left(I_{BSd}^2+I_{BSq}^2\right)$$
$$-6k_{Bi}^2R_{BT}I_{dc}^2-b_{B1}L_{BS}I_{BSd}I_{dc}v_{A1}-b_{B1}L_{BS}I_{BSq}I_{dc}v_{B2}$$

Substituting P_{ACSCs} and P_{BCSCs} into equation 5.101

$$\frac{dI_{dc}}{dt} = c_{A1}\frac{(I_{ASd}V_{ASd}+I_{ASq}V_{ASq})}{I_{dc}}+c_{A2}\frac{\left(I_{ASd}^2+I_{ASq}^2\right)}{I_{dc}}-c_{AB}I_{dc}-c_{A5}I_{ASd}v_{A1}-c_{A5}I_{ASq}v_{A2}$$
$$+c_{B1}\frac{(I_{BSd}V_{ASd}+I_{BSq}V_{BSq})}{I_{dc}}+c_{B2}\frac{\left(I_{BSd}^2+I_{BSq}^2\right)}{I_{dc}}-c_{B5}I_{ASd}v_{B1}-c_{B5}I_{BSq}v_{B2} \tag{5.104}$$

where

$$c_{A1}=\frac{L_{AT}}{\left(L_{dc}+6k_{Ai}^2L_{AT}+6k_{Bi}^2L_{BT}\right)\left(2L_S+L_{AT}\right)} \qquad c_{B1}=\frac{L_{BT}}{\left(L_{dc}+6k_{Ai}^2L_{AT}+6k_{Bi}^2L_{BT}\right)\left(2L_{BS}+L_{BT}\right)}$$

$$c_{A2}=\frac{(R_{AT}+R_{At})L_{AS}-R_{AS}L_{AT}}{\left(L_{dc}+6k_{Ai}^2L_{AT}+6k_{Bi}^2L_{BT}\right)\left(2L_{As}+L_{AT}\right)} \qquad c_{B2}=\frac{(R_{BT}+R_{Bt})L_{BS}-R_{BS}L_{BT}}{\left(L_{dc}+6k_{Ai}^2L_{AT}+6k_{Bi}^2L_{BT}\right)\left(2L_{BS}+L_{BT}\right)}$$

$$c_{A5}=\frac{R_{At}L_{AS}}{(L_{dc}+6k_{Ai}^2L_{AT}+6k_{Bi}^2L_{BT})(2L_{AS}+L_{AT})} \qquad c_{B5}=\frac{R_{Bt}L_{BS}}{\left(L_{dc}+6k_{Ai}^2L_{AT}+6k_{Bi}^2L_{BT}\right)(2L_{BS}+L_{BT})}$$

$$c_{AB}=\frac{R_{dc}+6k_{Ai}^2R_{AT}+6k_{Ai}^2R_{BT}}{L_{dc}+6k_{Ai}^2L_{AT}+6k_{Bi}^2L_{BT}}$$

Therefore the state equations of the grouped current source converters in the affine form for the control of an HVDC terminal are:

$$\begin{pmatrix} \frac{dI_{Sd}}{dt} \\ \frac{dI_{Sq}}{dt} \\ \frac{dI_{dc}}{dt} \\ \frac{dV_{cp}}{dt} \\ \frac{dI_{1dc}}{dt} \end{pmatrix} = \begin{pmatrix} -a_1I_{Sd}+\omega I_{Sq}+dV_{Sd} \\ -\omega I_{Sd}-a_1I_{Sq}+dV_{Sq} \\ c_1\frac{(I_{Sd}V_{Sd}+I_{Sq}V_{Sq})}{I_{dc}}+c_2\frac{\left(I_{Sd}^2+I_{Sq}^2\right)}{I_{dc}}-c_3V_{cp}-c_4I_{dc} \\ \frac{1}{C_p}\left(I_{dc}-I_{1dc}-\frac{V_{cp}}{R_p}\right) \\ \frac{1}{L_L}(V_{cp}-U_{dc}-R_LI_{1dc}) \end{pmatrix}$$
$$+\begin{pmatrix} b_1I_{dc} & 0 \\ 0 & b_1I_{dc} \\ -c_5I_{Sd} & -c_5I_{Sq} \\ 0 & 0 \\ 0 & 0 \end{pmatrix}\begin{pmatrix} v_1 \\ v_2 \end{pmatrix} \tag{5.105}$$

For the grouped current source converters STATCOM application the state equations in affine form are:

$$\begin{pmatrix} \frac{dI_{Sd}}{dt} \\ \frac{dI_{Sq}}{dt} \\ \frac{dI_{dc}}{dt} \end{pmatrix} = \begin{pmatrix} -a_1 I_{Sd} + \omega I_{Sq} + dV_{Sd} \\ -\omega I_{Sd} - a_1 I_{Sq} + dV_{Sq} \\ c_1 \frac{I_{Sd}V_{Sd} + I_{Sq}V_{Sq}}{I_{dc}} + c_2 \frac{I_{Sd}^2 + I_{Sq}^2}{I_{dc}} - c_4 I_{dc} \end{pmatrix} + \begin{pmatrix} b_1 I_{dc} & 0 \\ 0 & b_1 I_{dc} \\ -c_5 I_{Sd} & -c_5 I_{Sq} \end{pmatrix} \begin{pmatrix} v_1 \\ v_2 \end{pmatrix} \tag{5.106}$$

For the grouped current source converters back-to-back application the state equations in affine form are:

$$\begin{pmatrix} \frac{dI_{ASd}}{dt} \\ \frac{dI_{ASq}}{dt} \\ \frac{dI_{BSd}}{dt} \\ \frac{dI_{BSq}}{dt} \\ \frac{dI_{dc}}{dt} \end{pmatrix} = \begin{pmatrix} -a_A I_{ASd} + \omega_A I_{ASq} + b_{A1} V_{Asd} \\ -\omega_A I_{ASd} - a_A I_{ASq} + b_{A1} V_{ASq} \\ -a_B I_{BSd} + \omega_B I_{BSq} + b_{B1} V_{BSd} \\ -\omega_B I_{BSd} - a_B I_{BSq} + b_{B1} V_{BSq} \\ \frac{c_{A1}(I_{ASd}V_{ASd} + I_{ASq}V_{ASq})}{I_{dc}} + \frac{c_{A2}(I_{ASd}^2 + I_{ASq}^2)}{I_{dc}} + \frac{c_{B1}(I_{BSd}V_{BSd} + I_{BSq}V_{BSq})}{I_{dc}} + \frac{c_{B2}(I_{BSd}^2 + I_{BSq}^2)}{I_{dc}} - c_{AB} I_{dc} \end{pmatrix}$$

$$+ \begin{pmatrix} b_{A1} I_{dc} & 0 & 0 & 0 \\ 0 & b_{A1} I_{dc} & 0 & 0 \\ 0 & 0 & b_{B1} I_{dc} & 0 \\ 0 & 0 & 0 & b_{B1} I_{dc} \\ -c_{A5} I_{ASd} & -c_{A5} I_{ASq} & -c_{B5} I_{BSd} & -c_{B5} I_{BSq} \end{pmatrix} \begin{pmatrix} v_{A1} \\ v_{A2} \\ v_{B1} \\ v_{B2} \end{pmatrix} \tag{5.107}$$

These equations present the affine non-linearity as there are state variables in the input matrix *G(X)*. They could be linearized, and a controller designed, around a specific operation state, but they would only be valid within a small region. However, in a practical current source converter the DC side current I_{dc} may be controlled within a small margin or in a slow changing large region, but the corresponding AC side current (determined by the dq-components I_d and I_q) needs to be controlled between zero and the rated level gradually or sharply. The challenge is therefore to find a design method that is valid globally and independent of the operating states.

As explained earlier for the case of the affine voltage source converter, the non-linear controller of the current source converter can also be designed based on the non-linear transformation of the system state variables and the input variables to convert the affine non-linear system into a linear system. Therefore the transformed state variables can be controlled by the transformed input variables to vary in desired patterns by using well developed linear control techniques. The reversed transformation from the transformed input variables to the original input variables enables the control of the CSC system via the practical control variables and achieves the desired performance. Again in this case the main difficulty is in synthesizing the non-linear transformations, but for the grouped CSC systems this can be achieved based on the electrical technology. The following derivations present the transformations and procedures involved.

To convert the grouped current source converters STATCOM equation 5.106 into linear form the required transformed state variables are the energy stored in the AC and DC side inductors and their derivatives as well as the DC current, i.e.

$$z_3 = \frac{1}{2b_1}\left(I_{Sd}^2 + I_{Sq}^2\right) + \frac{1}{2c_5} I_{dc}^2$$

$$\begin{aligned} z_2 = \frac{dz_3}{dt} &= \frac{1}{b_1}\left(I_{Sd}\frac{dI_{Sd}}{dt} + I_{Sq}\frac{dI_{Sq}}{dt}\right) + \frac{1}{c_5} I_{dc}\frac{dI_{dc}}{dt} \\ &= \left(\frac{c_2}{c_5} - \frac{a_1}{b_1}\right)\left(I_{Sd}^2 + I_{Sq}^2\right) + \left(\frac{d}{b_1} + \frac{c_1}{c_5}\right)\left(V_{Sd}I_{Sd} + V_{Sq}I_{Sq}\right) - \frac{c_4}{c_5} I_{dc}^2 \end{aligned}$$

$$z_1 = I_{dc}$$

The transformed state equations for the grouped current source converters STATCOM application are

$$\begin{pmatrix} \dfrac{dz_1}{dt} \\ \dfrac{dz_2}{dt} \\ \dfrac{dz_3}{dt} \end{pmatrix} = \begin{pmatrix} w_1 \\ w_2 \\ z_2 \end{pmatrix} = \begin{pmatrix} c_1 \dfrac{I_{Sd}V_{Sd} + I_{Sq}V_{Sq}}{I_{dc}} + c_2 \dfrac{I_{Sd}^2 + I_{Sq}^2}{I_{dc}} - c_4 I_{dc} \\ h_{Csta}(X) \\ \dfrac{d}{dt}\left[\dfrac{1}{2b_1}\left(I_{Sd}^2 + I_{Sq}^2\right) + \dfrac{1}{2c_5} I_{dc}^2\right] \end{pmatrix}$$

$$+ \begin{pmatrix} -c_5 I_{Sd} & -c_5 I_{Sq} \\ g_{C1sta}(X) & g_{C2sta}(X) \\ 0 & 0 \end{pmatrix}\begin{pmatrix} v_1 \\ v_2 \end{pmatrix} \qquad (5.108)$$

where

$$h_{Cxta}(X) = \left(\frac{2a_1^2}{b_1} - \frac{2a_1c_2 + 2c_2c_4}{c_5}\right)(I_{Sd}^2 + I_{Sq}^2) + \frac{2C_4^2}{c_5}I_{dc}^2$$
$$+ \left(\frac{2c_2d - a_1c_1 - c_1c_4}{c_5} - \frac{3a_1d}{b_1}\right)(V_{Sd}I_{Sd} + V_{Sq}I_{Sq})$$
$$+ w\left(\frac{d}{b_1} + \frac{c_1}{c_5}\right)(V_{Sd}I_{Sq} - V_{Sq}I_{Sd}) + \left(\frac{d^2}{b_1} + \frac{c_1d}{c_5}\right)(V_{Sd}^2 + V_{Sq}^2)$$

$$g_{C1sta}(X) = \left(d + \frac{c_1b_1}{c_5}\right)I_{dc}V_{Sd} + \left(\frac{2b_1c_2}{c_5} - 2a_1 + 2c_4\right)I_{dc}I_{Sd}$$

$$g_{C2sta}(X) = \left(d + \frac{c_1b_1}{c_5}\right)I_{dc}V_{Sq} + \left(\frac{2b_1c_2}{c_5} - 2a_1 + 2c_4\right)I_{dc}I_{Sq}$$

They can be derived from

$$\frac{dz_2}{dt} = \frac{d}{dt}\left[\left(\frac{c_2}{c_5} - \frac{a_1}{b_1}\right)(I_{Sd}^2 + I_{Sq}^2) + \left(\frac{d}{b_1} + \frac{c_1}{c_5}\right)(V_{Sd}I_{Sd} + V_{Sq}I_{Sq}) - \frac{c_4}{c_5}I_{dc}^2\right].$$

The terms on the left in equation 5.108 are the linearized Z state variables and the top two rows of the right side of the equations define the relationship between the transformed input variables $[w_1 w_2]^T$ and the original input variables $[v_1 v_2]^T$.

If $\Delta_{Csta} = -c_5I_{Sd}g_{C2sta}(X) + c_5I_{Sq}g_{C1sta}(X) = c_5I_{dc}(d + \frac{c_1b_1}{c_5})(V_{Sd}I_{Sq} - V_{Sq}I_{Sd}) \neq 0$, i.e. $I_{dc} \neq 0$ and $\frac{V_{Sq}}{V_{Sd}} \neq \frac{I_{Sq}}{I_{Sd}}$, the original input vector $[v_1 v_2]^T$ can be derived from:

$$\begin{pmatrix} v_1 \\ v_2 \end{pmatrix} = \begin{pmatrix} -c_5I_{Sd} & -c_5I_{Sd} \\ g_{C1sta}(X) & g_{C2sta}(X) \end{pmatrix}^{-1} \begin{pmatrix} w_1 - c_1\dfrac{I_{Sd}V_{Sd} - I_{Sq}V_{Sq}}{I_{dc}} - c_2\dfrac{I_{Sd}^2 + I_{Sq}^2}{I_{dc}} + c_4I_{dc} \\ w_2 - h_{Csta}(X) \end{pmatrix} \tag{5.109}$$

To transform the grouped current source converters state equation 5.105 for use in HVDC terminal control, the choice of transformed state variables are the energy stored in the AC and DC side inductors and their derivatives, the DC current, the T-network capacitor voltage and the inductor current, i.e.:

$$z1 = I_{dc}$$

$$z_2 = \frac{dz_3}{dt} = \frac{1}{b_1}\left(I_{Sd}\frac{dI_{Sd}}{dt} + I_{Sq}\frac{dI_{Sq}}{dt}\right) + \frac{1}{c_5}I_{dc}\frac{dI_{dc}}{dt}$$

$$= \left(\frac{c_2}{c_5} - \frac{a_1}{b_1}\right)(I_{Sd}^2 + I_{Sq}^2) + \left(\frac{d}{b_1} + \frac{c_1}{c_5}\right)(V_{Sd}I_{Sd} + V_{Sq}I_{Sq}) - \frac{c_4}{c_5}I_{dc}^2 - \frac{c_3}{c_5}V_{cp}I_{dc}$$

$$z_3 = \frac{1}{2b_1}(I_{Sd}^2 + I_{Sq}^2) + \frac{1}{2c_5}I_{dc}^2$$

$$z_4 = V_{cp}$$

$$z_5 = I_{1dc}$$

and the transformed state equations for HVDC terminal control application are

$$\begin{pmatrix} \frac{dz_1}{dt} \\ \frac{dz_2}{dt} \\ \frac{dz_3}{dt} \\ \frac{dz_4}{dt} \\ \frac{dz_5}{dt} \end{pmatrix} = \begin{pmatrix} w_1 \\ w_2 \\ z_2 \\ \frac{1}{C_p}\left(z_1 - \frac{1}{R_p}z_4 - z_5\right) \\ \frac{1}{L_L}(z_4 - R_L z_5 - U_{dc}) \end{pmatrix} = \begin{pmatrix} c_1\frac{(I_{Sd}V_{Sd} + I_{Sq}V_{Sq})}{I_{dc}} + c_2\frac{(I_{Sd}^2 + I_{Sq}^2)}{I_{dc}} - c_3V_{cp} - c_4I_{dc} \\ h_{Cter}(X) \\ \frac{d}{dt}\left[\frac{1}{2b_1}(I_{Sd}^2 + I_{Sq}^2) + \frac{1}{2c_5}I_{dc}^2\right] \\ \frac{1}{C_p}\left(I_{dc} - \frac{V_{cp}}{R_p} - I_{1dc}\right) \\ \frac{1}{L_L}(V_{cp} - R_L I_{1dc} - U_{dc}) \end{pmatrix}$$

$$+ \begin{pmatrix} -c_5 I_{sd} & -c_5 I_{sq} \\ g_{C1ter}(X) & g_{C2ter}(X) \\ 0 & 0 \\ 0 & 0 \\ 0 & 0 \end{pmatrix}\begin{pmatrix} v_1 \\ v_2 \end{pmatrix} \qquad (5.110)$$

where

$$h_{Cter}(X) = \left(\frac{2a_1^2}{b_1} - \frac{2a_1c_2 + c_2c_4}{c_5} - \frac{c_2c_3V_{cp}}{c_5I_{dc}}\right)(I_{Sd}^2 + I_{Sq}^2)$$

$$+ \left(\frac{2c_2d - a_1c_1 - c_1c_4}{c_5} - \frac{3a_1d}{b_1} - \frac{c_1c_3V_{cp}}{c_5I_{dc}}\right)(V_{Sd}I_{Sd} + V_{Sq}I_{Sq})$$

$$+w\left(\frac{d}{b_1}+\frac{c_1}{c_3}\right)(V_{Sd}I_{Sq}-V_{Sq}I_{Sd})+\left(\frac{d^2}{b_1}+\frac{c_1 d}{c_5}\right)(V_{Sd}^2+V_{Sq}^2)+\left(\frac{2c_4^2}{c_5}-\frac{c_3}{c_5 C_p}\right)I_{dc}^2$$

$$+\left(\frac{2c_3 c_4}{c_5}+\frac{c_3}{c_5 R_p C_p}\right)V_{cp}I_{dc}+\frac{c_3}{c_5 C_p}I_{dc}I_{1dc}$$

$$g_{C1ter}(X)=\left(d+\frac{c_1 b_1}{c_5}\right)I_{dc}V_{Sd}+\left(\frac{2b_1 c_2}{c_5}-2a_1+2c_4\right)I_{dc}I_{Sd}+c_3 V_{cp}I_{Sd}$$

$$g_{C2ter}(X)=\left(d+\frac{c_1 b_1}{c_5}\right)I_{dc}V_{Sq}+\left(\frac{2b_1 c_2}{c_5}-2a_1+2c_4\right)I_{dc}I_{Sq}+c_3 V_{cp}I_{Sq}$$

They can be derived from:

$$\frac{dz_2}{dt}=\frac{d}{dt}\left[\left(\frac{c_2}{c_5}-\frac{a_1}{b_1}\right)(I_{Sd}^2+I_{Sq}^2)+\left(\frac{d}{b_1}+\frac{c_1}{c_5}\right)(V_{Sd}I_{Sd}+V_{Sq}I_{Sq})-\frac{c_4}{c_5}I_{dc}^2-\frac{c_3}{c_5}V_{cp}I_{dc}\right]$$

The transformed linear Z state equations can be written in the standard linear form i.e.:

$$\begin{pmatrix}\frac{dz_1}{dt}\\ \frac{dz_2}{dt}\\ \frac{dz_3}{dt}\\ \frac{dz_4}{dt}\\ \frac{dz_5}{dt}\end{pmatrix}=\begin{pmatrix}0&0&0&0&0\\0&0&0&0&0\\0&1&0&0&0\\ \frac{1}{C_p}&0&0&-\frac{1}{R_pC_p}&-\frac{1}{C_p}\\0&0&0&\frac{1}{L_L}&-\frac{R_L}{L_L}\end{pmatrix}\begin{pmatrix}z_1\\z_2\\z_3\\z_4\\z_5\end{pmatrix}+\begin{pmatrix}1&0&0\\0&1&0\\0&0&0\\0&0&0\\0&0&-\frac{1}{L_L}\end{pmatrix}\begin{pmatrix}w_1\\w_2\\U_{dc}\end{pmatrix}\quad(5.111)$$

and the relationship between the transformed input vector $[w_1 w_2]^T$ and the original input vector $[v_1 v_2]^T$ is given by the top two rows of equation 5.110, i.e.:

$$\begin{pmatrix}w_1\\w_2\end{pmatrix}=\begin{pmatrix}c_1\frac{(I_{Sd}V_{Sd}+I_{Sq}V_{Sq})}{I_{dc}}+c_2\frac{(I_{Sd}^2+I_{Sq}^2)}{I_{dc}}-c_3V_{cp}-c_4I_{dc}\\ h_{Cter}(X)\end{pmatrix}$$

$$+\begin{pmatrix}-c_5 I_{Sd}&-c_5 I_{Sq}\\ g_{C1ter}(X)&g_{C2ter}(X)\end{pmatrix}\begin{pmatrix}v_1\\v_2\end{pmatrix}$$

If $\Delta_{Cter} = -c_5 I_{Sd} g_{C2ter}(X) + c_5 I_{Sq} g_{C1ter}(X) = c_5 I_{dc}\left(d + \frac{c_1 b_1}{c_5}\right)(V_{Sd} I_{Sq} - V_{Sq} I_{Sd}) \neq 0$, i.e. $I_{dc} \neq 0$ and $V_{Sq}/V_{Sd} \neq I_{Sq}/I_{Sd}$,

$$\begin{pmatrix} v_1 \\ v_2 \end{pmatrix} = \begin{pmatrix} -c_5 I_{Sd} & -c_5 I_{Sq} \\ g_{C1ter}(X) & g_{C2ter}(X) \end{pmatrix}^{-1} \begin{pmatrix} w_1 - c_1 \dfrac{(I_{Sd}V_{Sd} + I_{Sq}V_{Sq})}{I_{dc}} - c_2 \dfrac{(I_{Sd}^2 + I_{Sq}^2)}{I_{dc}} + c_3 V_{cp} + c_4 I_{dc} \\ w_2 - h_{Cter}(X) \end{pmatrix} \quad (5.112)$$

To transform the grouped current source converters state equation 5.107 for back-to-back application, the transformed state variables needed are the energy stored in the DC and both AC side inductors as well as their derivatives, the right-side AC currents I_{ASd}, I_{ASq} and the left-side AC current I_{BSd}. The transformations are:

$$z_1 = I_{ASd}$$

$$z_2 = I_{ASq}$$

$$z_3 = I_{BSd}$$

$$\begin{aligned} z_4 &= \frac{dz_5}{dt} = \frac{c_{A5}}{b_{A1}}\left(I_{ASd}\frac{d}{dt}I_{ASd} + I_{ASq}\frac{d}{dt}I_{ASq}\right) + \frac{c_{B5}}{b_{B1}}\left(I_{BSd}\frac{d}{dt}I_{BSd} + I_{BSq}\frac{d}{dt}I_{BSq}\right) + I_{dc}\frac{d}{dt}I_{dc} \\ &= \left(c_{A2} - \frac{a_A c_{A5}}{b_{A1}}\right)(I_{ASd}^2 + I_{ASq}^2) + \left(c_{B2} - \frac{a_B c_{B5}}{b_{B1}}\right)(I_{BSd}^2 + I_{BSq}^2) - c_{AB} I_{dc}^2 \\ &\quad + (c_{A1} + c_{A5})(V_{ASd} I_{ASd} + V_{ASq} I_{ASq}) + (c_{B1} + c_{B5})(V_{BSd} I_{BSd} + V_{BSq} I_{BSq}) \end{aligned}$$

$$z_5 = \frac{c_{A5}}{2b_{A1}}(I_{ASd}^2 + I_{ASq}^2) + \frac{c_{B5}}{2b_{B1}}(I_{BSd}^2 + I_{BSq}^2) + 0.5 I_{dc}^2$$

and $\frac{dz_4}{dt}$ is derived as follows:

$$\begin{aligned} \frac{dz_4}{dt} &= \frac{d}{dt}\left[\left(c_{A2} - \frac{a_A c_{A5}}{b_{A1}}\right)(I_{ASd}^2 + I_{ASq}^2) + \left(c_{B2} - \frac{a_B c_{B5}}{b_{B1}}\right)(I_{BSd}^2 + I_{BSq}^2) - c_{AB} I_{dc}^2 \right. \\ &\quad \left. + (c_{A1} + c_{A5})(V_{ASd} I_{ASd} + V_{ASq} I_{ASq}) + (c_{B1} + c_{B5})(V_{BSd} I_{BSd} + V_{BSq} I_{BSq})\right] \\ &= \left[-2a_A\left(c_{A2} - \frac{a_A c_{A5}}{b_{A1}}\right) - 2c_{AB} c_{A2}\right](I_{ASd}^2 + I_{ASq}^2) + \left[-2a_B\left(c_{B2} - \frac{a_B c_{B5}}{b_{B1}}\right) - 2c_{AB} c_{B2}\right] \\ &\quad \times (I_{BSd}^2 + I_{BSq}^2) + \left[2b_{A1}\left(c_{A2} - \frac{a_A c_{A5}}{b_{A1}}\right) - 2c_{AB} c_{A1} - a_A(c_{A1} + c_{A5})\right](V_{ASd} I_{ASd} + V_{ASq} I_{ASq}) \\ &\quad + \left[2b_{B1}\left(c_{B2} - \frac{a_B c_{B5}}{b_{B1}}\right) - 2c_{AB} c_{B1} - a_B(c_{B1} + c_{B5})\right](V_{BSd} I_{BSd} + V_{BSq} I_{BSq}) + 2c_{AB}^2 I_{dc}^2 \end{aligned}$$

$$+\omega_A(c_{A1}+c_{A5})(V_{ASd}I_{ASq}-V_{ASq}I_{ASd})+\omega_B(c_{B1}+c_{B5})(V_{BSd}I_{BSq}-V_{BSq}I_{BSd})$$

$$+b_{A1}(c_{A1}+c_{A5})(V_{ASd}^2+V_{ASq}^2)+b_{B1}(c_{B1}+c_{B5})(V_{BSd}^2+V_{BSq}^2)$$

$$+\left[b_{A1}(c_{A1}+c_{A5})I_{dc}V_{ASd}+2b_{A1}\left(c_{A2}-\frac{a_Ac_{A5}}{b_{A1}}\right)I_{dc}I_{ASd}\right]v_{A1}$$

$$+\left[b_{A1}(c_{A1}+c_{A5})I_{dc}V_{ASq}+2b_{A1}\left(c_{A2}-\frac{a_Ac_{A5}}{b_{A1}}\right)I_{dc}I_{ASq}\right]v_{A2}$$

$$+\left[b_{B1}(c_{B1}+c_{B5})I_{dc}V_{BSd}+2b_{B1}\left(c_{B2}-\frac{a_Bc_{B5}}{b_{B1}}\right)I_{dc}I_{BSd}\right]v_{B1}$$

$$+\left[b_{B1}(c_{B1}+c_{B5})I_{dc}V_{BSq}+2b_{B1}\left(c_{B2}-\frac{a_Bc_{B5}}{b_{B1}}\right)I_{dc}I_{BSq}\right]v_{B2}$$

$$= h_{Cbtb}(X)+g_{C1btb}(X)v_{A1}+g_{C2btb}(X)v_{A2}+g_{C3btb}(X)v_{B1}+g_{C4btb}(X)v_{B2}$$

The transformed state equations for back-to-back application are

$$\begin{pmatrix}\dfrac{dz_1}{dt}\\[2mm]\dfrac{dz_2}{dt}\\[2mm]\dfrac{dz_3}{dt}\\[2mm]\dfrac{dz_4}{dt}\\[2mm]\dfrac{dz_5}{dt}\end{pmatrix}=\begin{pmatrix}w_{A1}\\w_{A2}\\w_{B1}\\w_{B2}\\z_4\end{pmatrix}=\begin{pmatrix}-a_AI_{ASd}+\omega_AI_{ASq}+b_{A1}V_{ASd}\\-\omega_AI_{ASd}-a_AI_{ASq}+b_{A1}V_{ASq}\\-a_BI_{BSd}+\omega_BI_{BSq}+b_{B1}V_{BSd}\\h_{Cbtb}(X)\\z_4\end{pmatrix}$$

$$+\begin{pmatrix}b_{A1}I_{dc}&0&0&0\\0&b_{A1}I_{dc}&0&0\\0&0&b_{B1}I_{dc}&0\\g_{C1btb}(X)&g_{C2btb}(X)&g_{C3btb}(X)&g_{C4btb}(X)\\0&0&0&0\end{pmatrix}\begin{pmatrix}v_{A1}\\v_{A2}\\v_{B1}\\v_{B2}\end{pmatrix}\tag{5.113}$$

The state equations of the Z variables can be written in the following standard linear form:

$$\begin{pmatrix} \frac{dz_1}{dt} \\ \frac{dz_2}{dt} \\ \frac{dz_3}{dt} \\ \frac{dz_4}{dt} \\ \frac{dz_5}{dt} \end{pmatrix} = \begin{pmatrix} 0 & 0 & 0 & 0 & 0 & 0 \\ 0 & 0 & 0 & 0 & 0 & 0 \\ 0 & 0 & 0 & 0 & 0 & 0 \\ 0 & 0 & 0 & 0 & 0 & 0 \\ 0 & 0 & 0 & 0 & 1 & 0 \end{pmatrix} \begin{pmatrix} z_1 \\ z_2 \\ z_3 \\ z_4 \\ z_5 \end{pmatrix} + \begin{pmatrix} 1 & 0 & 0 & 0 \\ 0 & 1 & 0 & 0 \\ 0 & 0 & 1 & 0 \\ 0 & 0 & 0 & 1 \\ 0 & 0 & 0 & 0 \end{pmatrix} \begin{pmatrix} w_{A1} \\ w_{A2} \\ w_{B1} \\ w_{B2} \end{pmatrix} \tag{5.114}$$

and the relationship between the transformed input vector $[w_1 w_2]^T$ and the original input vector $[u_1 u_2]^T$ is given by the top four rows of equation 5.46:

$$\begin{pmatrix} w_{A1} \\ w_{A2} \\ w_{B1} \\ w_{B2} \end{pmatrix} = \begin{pmatrix} -a_A I_{ASd} + \omega_A I_{ASq} + b_{A1} V_{ASd} \\ -\omega_A I_{ASd} - a_A I_{ASq} + b_{A1} V_{ASq} \\ -a_B I_{BSd} + \omega_B I_{BSq} + b_{B1} V_{BSd} \\ h_{Cbtb}(X) \end{pmatrix} + \begin{pmatrix} b_{A1} I_{dc} & 0 & 0 & 0 \\ 0 & b_{A1} I_{dc} & 0 & 0 \\ 0 & 0 & b_{B1} I_{dc} & 0 \\ g_{C1btb}(X) & g_{C2btb}(X) & g_{C3btb}(X) & g_{C4btb}(X) \end{pmatrix} \begin{pmatrix} v_{A1} \\ v_{A2} \\ v_{B1} \\ v_{B2} \end{pmatrix}$$

If $\Delta_{Cbtb} = b_{A1}^2 b_{B1} I_{dc}^3 g_{C4btb}(X) = b_{A1}^2 b_{B1} I_{dc}^4 \left[b_{B1}(c_{B1} + c_{B5}) V_{BSq} + 2 b_{B1}\left(c_{B2} - \frac{a_B c_{B5}}{b_{B1}}\right) I_{BSq}\right] \neq 0$, i.e. $I_{dc} \neq 0$ and $2(c_{B2} b_{B1} - a_B c_{B5}) I_{BSq} \neq b_{B1}(c_{B1} + c_{B5}) V_{BSq}$,

$$\begin{pmatrix} v_{A1} \\ v_{A2} \\ v_{B1} \\ v_{B2} \end{pmatrix} = \begin{pmatrix} b_{A1} I_{dc} & 0 & 0 & 0 \\ 0 & b_{A1} I_{dc} & 0 & 0 \\ 0 & 0 & b_{B1} I_{dc} & 0 \\ g_{C1btb}(X) & g_{C2btb}(X) & g_{C3btb}(X) & g_{C4btb}(X) \end{pmatrix}^{-1} \times \begin{pmatrix} w_{A1} + a_A I_{ASd} - \omega_A I_{ASq} - b_{A1} V_{ASd} \\ w_{A2} + \omega_A I_{ASd} + a_A I_{ASq} - b_{A1} V_{ASq} \\ w_{B1} + a_B I_{BSd} - \omega_B I_{BSq} - b_{B1} V_{BSd} \\ w_{B2} - h_{btb}(X) \end{pmatrix} \tag{5.115}$$

5.8 Non-linear control of VSC and CSC systems

The state equations derived in previous sections have shown that self-commutated VSC and CSC are affine non-linear systems and that they can be linearized using feedback linearization. Both VSC and CSC can provide four-quadrant controllability of the active and reactive powers or of the AC voltages and currents. Moreover, in flexible power transmission and distribution, the four-quadrant controllability (i.e. the magnitude and phase angle) need to be changed quickly for the full range of active and reactive power. The control strategies and controller designs for the self-commutated VSC and CSC systems are much more complicated than those of line-commutated conversion, as the linearization around an operation point cannot be used to achieve stable and high performance in the steady and dynamic states in the four-quadrant full-rated regions.

In diode converter rectification there is no power control freedom to force the DC voltage and current to vary in alternative ways. Although thyristor-based line-commutated conversion provides active power control by adjusting the firing delay angle (α), it lacks reactive power controllability; thyristor-based LCC keeps the firing delay angle within the $0^{\circ} < \alpha < 180^{\circ}$ region, and for commutation safety the firing delay angle is further limited to the region $\alpha_{\min} < \alpha < (180\circ - \alpha_{\min})$ (where $\alpha_{\min}$ is about $15\circ$ to 30° depending on the application). The DC output voltage of a thyristor LCC is nearly proportional to the function $\cos(\alpha)$, and within the restricted operation region, $\cos(\alpha)$ can be linearized around $\alpha = 90\circ$ without significant difference. The linearization strategy has been proved to provide satisfactory active power control in many thyristor LCC applications.

On the other hand the non-linear nature and wide range of operation requirements of self-commutated VSC and CSC impose considerable difficulty in control system design. Four-quadrant power control requires the power converter AC output voltage and/or current amplitude to vary within the full-rated range and phase angle in the region between $-180\circ < \phi < 180\circ$ and the control variables are all related to DC voltage or current multiplied by the functions $\sin(\phi)$ and $\cos(\phi)$. In the four quadrant region, both the functions $\sin(\phi)$ and $\cos(\phi)$ and their derivatives change significantly in magnitude and sign.

The non-linear control of the affine systems, also called feedback linearization, offers the possibility of designing a non-linear control system for the self-commutated VSC and CSC systems to achieve high global performance. The self-commutated VSC and CSC systems can be described by two sets of state equations, namely a set of the state equations (derived in sections 5.4–5.5 for VSC systems and in section 5.6 for CSC systems) and a set of output equations with multi-inputs and multi-outputs related to control objectives. They are:

$$\dot{\mathbf{X}} = \mathbf{F}(\mathbf{X}) + \mathbf{G}(\mathbf{X})\mathbf{U}$$

$$\mathbf{Y} = \mathbf{H}(\mathbf{X})$$

$$\dot{\mathbf{Z}} = \mathbf{A}\mathbf{Z} + \mathbf{B}\mathbf{W}\ \mathbf{Z} = \boldsymbol{\Phi}(\mathbf{X})\mathbf{W} = \mathbf{L}(\mathbf{X}, \mathbf{U})$$

where

X is the state variable vector of n-dimension in dq-frame,

U is the independent control input vector of m-dimension in dq-frame,

Z is the linearized state variable vector of n-dimension,

W is the linearized input variable vector of m-dimension,

Y is the independent controlled output vector of m-dimension defined in dq-frame.

The non-linear control system is expected to generate a control input vector in the time domain which keeps the system stable and forces the controlled output vector to operate at the referenced level, as well as provide the required steady and dynamic performances. If the dq-frame referenced state variable vector $\mathbf{X}^*$ (which is used to define the relation $Y^* = HX^*$), can be expressed by the referenced output, i.e. $\mathbf{X}^* = \mathbf{H}^{-1}\mathbf{Y}^*$, the required transformed state variable vector $\mathbf{Z}^*$ can be obtained by the transformation $\mathbf{Z}^* = \Phi(\mathbf{X}^*)$. Then a control system can be constructed based on linear state feedback, which is used to generate the required input vector **W** and the original input $\mathbf{U} = \mathbf{D}(\mathbf{W},\mathbf{X})$ to the converter system. Figure 5.8 shows the control system structure.

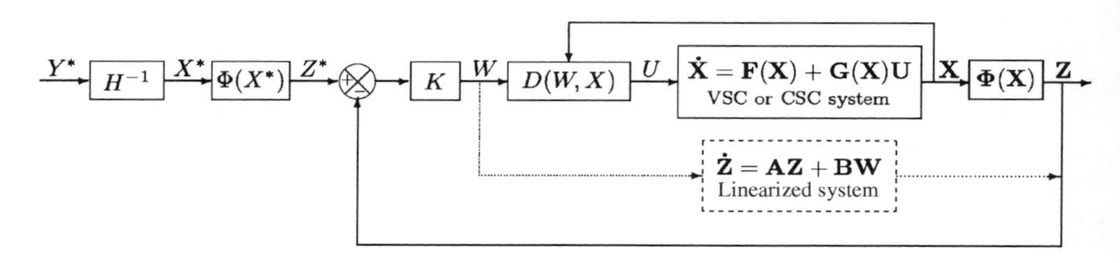

Figure 5.8 The control system structure of the VSC or CSC system.

In Figure 5.8 the referenced output vector Y^* is converted to the referenced state variable vector X^* in dq-frame by block H^{-1}, and X^* is converted by the transformation $Z^* = \Phi(X^*)$ to the required state reference of the linearized system. The closed-loop system from Z^* to Z can be treated as a full state feedback controlled linear system; K in the block represents the state error feedback diagonal matrix of n constants, which can be uniquely found based on the expected n poles of $(sI - A + BK)^{-1}$. The independent control input vector U to the converter system can be obtained by the input transformation $U = D(W,X)$ derived in the previous sections. Finally, the input vector U needs to be converted to phase angle orders and/or modulation indexes for the gate firing control logic.

The control system shown in Figure 5.8 indicates that the non-linear controller can be designed based on the well developed linear control system theory. The dashed-line part is used to design the state feedback matrix K, but is not involved in real time control. In real time control, the required control vector W for the linearized system is continuously converted to the converter control input vector U and sent to the converter gate firing logic to force the converter to operate on the desired state. The converter system state vector X in dq-frame is continuously transformed to the linearized state vector Z to perform the linearized state feedback. The linear transformation does not relate to a specific operating state and, therefore, the system performance is globally assured. Most of the required calculations and operations in the blocks of Figure 5.8 can be found in previous sections and only the conversion H^{-1} needs to be specifically derived for the different applications.

The control objectives are closely related to the specific application, and they decide the choice of output variables in the output vector Y. For power transmission application the outputs of interest are normally the active and reactive powers, the DC voltage (for VSC) and the DC current (for CSC). The active and reactive powers are non-linear functions of the converter system state variables and the required linearized state reference vector Z^*, corresponding to the active and reactive power orders, is derived below.

The converter system AC voltage and current are usually measured at the AC terminal connection of the converter to the power grid and the active and reactive powers are calculated from these measured AC voltage and current; the AC source voltage is used as the synchronizing signal for the converter gate firing logic. The measuring points in the equivalent circuits of Figures 5.1, 5.2, 5.4 and 5.5 are at the point between the source impedance and the interfacing transformer impedance. If the measured voltage vector is denoted V_O in the three-phase frame and V_{Odq} in the dq-frame, as well as the AC source current vector I_{abc} in the three-phase frame and I_{Sdq} in the dq-frame, the instantaneous terminal active (the active power absorbed by the convert system is defined as positive) and reactive (the inductive reactive power drawn by the converter system is defined as positive) powers are given by:

$$P_O = V_O^T I_{abc} = V_{Odq}^T I_{Sdq}$$

$$Q_O = (MV_O)^T N_3 (MI_{abc}) = V_{Odq}^T N I_{Sdq}$$

where M (the orthogonal transformation) is defined by equation 5.6,

$$N_3 = \begin{pmatrix} 0 & -1 & 0 \\ 1 & 0 & 0 \\ 0 & 0 & 0 \end{pmatrix} N = \begin{pmatrix} 0 & -1 \\ 1 & 0 \end{pmatrix} \tag{5.116}$$

the state variables $I_{Sdq} = [I_{Sd} I_{Sq}]^T$, if $V_{Odq}^T V_{Odq} \neq 0$, i.e. the converter AC terminal voltage amplitude is not zero and can be expressed in terms of V_{Odq}, P_O and Q_O, i.e.

$$I_{Sdq} = \begin{bmatrix} V_{Odq}^T \\ V_{Odq}^T N \end{bmatrix}^{-1} \begin{bmatrix} P_O \\ Q_O \end{bmatrix} = (V_{Odq}^T V_{Odq})^{-1} \left[V_{Odq} N^T V_{Odq} \right] \begin{bmatrix} P_O \\ Q_O \end{bmatrix} \tag{5.117}$$

The state equations have been derived in previous sections for voltage and current source converters of two independent control inputs systems. Except for the back-to-back application there are four independent control inputs systems, but they contain two converter and AC systems, each of them with two independent inputs. The two inputs can be used to control the two output variables of the AC system, while the four inputs can be used to control two independent outputs of the two interconnected AC systems.

For STATCOM application the main control objective is the reactive power measured at the converter AC terminal, while the DC voltage (in the VSC case) or DC current (in the CSC case) at the converter DC terminal is chosen as the output variable to stabilize the converter operation. The VSC-STATCOM system output vector is $Y = [Q_O V_{dc}]^T$ and the CSC-STATCOM system output vector is $Y = [Q_O I_{dc}]^T$. In the STATCOM application there is no active power exchange between the converter AC and DC side, and thus, the active power at the converter terminal (P_0) is zero if the losses are neglected.

For the VSC-STATCOM application, implemented either by PWM or grouped voltage source converter, the state equation in dq-frame is defined by equation 5.36, and if the reactive power and DC voltage references are Q_O^* and V_{dc}^*, the required linearized state reference vector $Z^* = \Phi(X^*)$ is:

$$\begin{pmatrix} z_1^* \\ z_2^* \\ z_3^* \end{pmatrix} = \begin{pmatrix} V_{dc}^* \\ V_{Sd}I_{Sd}^* + V_{Sq}I_{Sq}^* - R_{ST}\left[(I_{Sd}^*)^2 + (I_{Sq}^*)^2\right] - \dfrac{(V_{dc}^*)^2}{R_{dc}} \\ 0.5\left\{L_{ST}\left[(I_{Sd}^*)^2 + (I_{Sq}^*)^2\right] + C_{dc}(V_{dc}^*)^2\right\} \end{pmatrix} \tag{5.118}$$

$I_{Sdq}^* = \left[I_{Sd}^* I_{Sq}^*\right]^T$ and the top two components of the state vector $X^* = \left[I_{Sd}^* I_{Sq}^* V_{dc}^*\right]^T$ can be calculated by using equation 5.117:

$$\begin{aligned} I_{Sdq}^* &= (V_{Odq}^T V_{Odq})^{-1}[V_{Odq} N^T V_{Odq}] \begin{bmatrix} 0 \\ Q_O^* \end{bmatrix} = (V_{Odq}^T V_{Odq})^{-1} N^T V_{Odq} Q_O^* \\ &= \left[(V_{Odq}^T V_{Odq})^{-1} V_{Oq} Q_O^* - (V_{Odq}^T V_{Odq})^{-1} V_{Od} Q_O^*\right]^T \end{aligned} \tag{5.119}$$

Therefore, for the VSC-STATCOM application, X^* can be calculated from

$$X^* = H^{-1}Y^* = \begin{pmatrix} I_{Sd}^* \\ I_{Sd}^* \\ V_{dc}^* \end{pmatrix} = \begin{pmatrix} (V_{Od}^2 + V_{Oq}^2)^{-1} V_{Oq} & 0 \\ -(V_{Od}^2 + V_{Oq}^2)^{-1} V_{Od} & 0 \\ 0 & 1 \end{pmatrix} \begin{pmatrix} Q_O^* \\ V_{dc}^* \end{pmatrix} \tag{5.120}$$

Similarly for the CSC-STATCOM application implemented by the grouped CSC, the state equation in dq-frame is defined by equation 5.106 and the non-linear transformation for the state reference vector $Z^* = \Phi(X^*)$ is:

$$\begin{pmatrix} z_1^* \\ z_2^* \\ z_3^* \end{pmatrix} = \begin{pmatrix} I_{dc}^* \\ \left(\dfrac{c_2}{c_5} - \dfrac{a_1}{b_1}\right)\left[(I_{Sd}^*)^2 + (I_{Sq}^*)^2\right] + \left(\dfrac{d}{b_1} + \dfrac{c_1}{c_5}\right)(V_{Sd}I_{Sd}^* + V_{Sq}I_{Sq}^*) - \dfrac{c_4}{c_5}(I_{dc}^*)^2 \\ \dfrac{1}{2b_1}\left[(I_{Sd}^*)^2 + (I_{Sq}^*)^2\right] + \dfrac{1}{2c_5}(I_{dc}^*)^2 \end{pmatrix} \tag{5.121}$$

$I_{Sdq}^* = \left[I_{Sd}^* I_{Sq}^*\right]^T$ and the top two components of the state vector $X^* = \left[I_{Sd}^* I_{Sq}^* V_{dc}^*\right]^T$ can be calculated by equation 5.119. Therefore, for the grouped CSC-STATCOM, X^* can be calculated from:

$$X^* = H^{-1}Y^* = \begin{pmatrix} I_{Sd}^* \\ I_{Sd}^* \\ I_{dc}^* \end{pmatrix} = \begin{pmatrix} (V_{Od}^2 + V_{Oq}^2)^{-1} V_{Oq} & 0 \\ -(V_{Od}^2 + V_{Oq}^2)^{-1} V_{Od} & 0 \\ 0 & 1 \end{pmatrix} \begin{pmatrix} Q_O^* \\ I_{dc}^* \end{pmatrix} \tag{5.122}$$

For a back-to-back interconnection the active powers of the two terminals are nearly the same (as the interconnected converter losses are negligible); active and reactive power control is chosen for one of the two converter AC systems and the reactive power and DC voltage for the other, and thus the output vector is $Y = [P_1 Q_1 Q_2 V_{dc}]^T$.

For the back-to-back VSC-interconnection implemented either by PWM or grouped voltage source converter the state equation in dq-frame is defined by equation 5.44, and the linearized state reference vector $Z^* = \Phi(X^*)$ is:

$$\begin{pmatrix} z_1^* \\ z_2^* \\ z_3^* \\ z_4^* \\ z_5^* \end{pmatrix} = \begin{pmatrix} I_{d1}^* \\ I_{q1}^* \\ I_{d2}^* \\ V_{Sd1}I_{d1}^* + V_{Sq1}I_{q1}^* - R_{ST1}\left[(I_{d1}^*)^2 + (I_{q1}^*)^2\right] + V_{Sd2}I_{d2}^* + V_{Sq2}I_{q2}^* - R_{ST2}\left[(I_{d2}^*)^2 + (I_{q2}^*)^2\right] - \dfrac{(V_{dc}^*)^2}{R_{dc}} \\ 0.5\{L_{ST1}\left[(I_{d1}^*)^2 + (I_{q1}^*)^2\right] + L_{ST2}\left[(I_{d2}^*)^2 + (I_{q2}^*)^2\right] + C_{dc}(V_{dc}^*)^2\} \end{pmatrix} \tag{5.123}$$

The dq-frame state reference vector $X^* = [I_{d1}^* I_{q1}^* I_{d2}^* I_{q2}^* V_{dc}^*]^T$ can be calculated from the referenced output vector $Y^* = [P_1^* Q_1^* Q_2^* V_{dc}^*]$ and the two measured voltage source converter AC terminal voltages V_{Odq1} and V_{Odq2}:

$$X^* = \begin{pmatrix} I_{d1}^* \\ I_{q1}^* \\ I_{d2}^* \\ I_{q2}^* \\ V_{dc}^* \end{pmatrix} = \begin{pmatrix} (V_{m1})^{-1}V_{Od1} & (V_{m1})^{-1}V_{Oq1} & 0 & 0 \\ (V_{m1})^{-1}V_{Oq1} & -(V_{m1})^{-1}V_{Od1} & 0 & 0 \\ -(V_{m2})^{-1}V_{Od2}(1-k_o) & 0 & (V_{m2})^{-1}V_{Oq2} & 0 \\ -(V_{m2})^{-1}V_{Oq2}(1-k_o) & 0 & -(V_{m2})^{-1}V_{Od2} & 0 \\ 0 & 0 & 0 & 1 \end{pmatrix} \begin{pmatrix} P_1^* \\ Q_1^* \\ Q_2^* \\ V_{dc}^* \end{pmatrix} \tag{5.124}$$

where

$V_{m1} = V_{Od1}^2 + V_{Oq1}^2$,

$V_{m2} = V_{Od1}^2 + V_{Oq1}^2$, and

k_o represents the percentage losses of the converter active power.

Similarly, for the back-to-back grouped current source converter interconnection implemented by the affine state equation in dq-frame (equation 5.107, its linearized state reference vector $Z^* = \Phi(X^*)$ is:

$$\begin{pmatrix} z_1^* \\ z_2^* \\ z_3^* \\ z_4^* \\ z_5^* \end{pmatrix} = \begin{pmatrix} I_{ASd}^* \\ I_{ASq}^* \\ I_{BSd}^* \\ f_4(X^*) \\ \frac{c_{A5}}{2b_{A1}}\left[(I_{ASd}^*)^2 + (I_{ASq}^*)^2\right] + \frac{c_{B5}}{2b_{B1}}\left[(I_{BSd}^*)^2 + (I_{BSq}^*)^2\right] + 0.5(I_{dc}^*)^2 \end{pmatrix} \tag{5.125}$$

where

$$f_4(X^*) = (c_{A2} - \frac{a_A c_{A5}}{b_{A1}})\left[(I_{ASd}^*)^2 + (I_{ASq}^*)^2\right] + (c_{B2} - \frac{a_B c_{B5}}{b_{B1}})\left[(I_{BSd}^*)^2 + (I_{BSq}^*)^2\right] - c_{AB}(I_{dc}^*)^2$$
$$+ (c_{A1} + c_{A5})(V_{ASd}I_{ASd}^* + V_{ASq}I_{ASq}^*) + (c_{B1} + c_{B5})(V_{BSd}I_{BSd}^* + V_{BSq}I_{BSq}^*)$$

The dq-frame state reference vector $X^* = [I_{ASd}^* I_{ASq}^* I_{BSd}^* I_{BSq}^* I_{dc}^*]^T$ can be calculated from the referenced output vector $Y^* = [P_A^* Q_A^* Q_B^* V_{dc}^*]$ and the two measured voltage source converter AC terminal voltages V_{OAdq} and V_{OBdq}:

$$X^* = \begin{pmatrix} I_{ASd}^* \\ I_{ASq}^* \\ I_{BSd}^* \\ I_{BSq}^* \\ I_{dc}^* \end{pmatrix} = \begin{pmatrix} (V_{mA})^{-1}V_{OAd} & (V_{mA})^{-1}V_{OAq} & 0 & 0 \\ (V_{mA})^{-1}V_{OAq} & -(V_{mA})^{-1}V_{OAd} & 0 & 0 \\ -(V_{mB})^{-1}V_{OBd}(1-k_o) & 0 & (V_{mB})^{-1}V_{OBq} & 0 \\ -(V_{mB})^{-1}V_{OBq}(1-k_o) & 0 & -(V_{mB})^{-1}V_{OBd} & 0 \\ 0 & 0 & 0 & 1 \end{pmatrix} \begin{pmatrix} P_A^* \\ Q_A^* \\ Q_B^* \\ I_{dc}^* \end{pmatrix} \tag{5.126}$$

where $V_{mA} = V_{Od}^2 + V_{Oq1}^2$, $V_{mB} = V_{Od1}^2 + V_{Oq1}^2$, and k_o represents the percentage losses of the converter active power.

For HVDC transmission application the active and reactive powers are usually chosen as the control objectives at the sending end, while the reactive power and DC voltage are chosen as the control objectives at the receiving end; thus the output vector for the HVDC link terminal control is either $Y = [P_O Q_O]^T$ or $Y = [Q_O V_{dc}]^T$. Observing Figures 5.1, 5.2 and 5.5 and equations 5.42 and 5.111, the complete system for the HVDC terminal control consists of two parts, namely the converter connected to the AC system and the DC transmission network connected to a DC voltage source. The first part is non-linear and the second part linear. The two parts are cascaded connected and the state variable V_{dc} for voltage source converter terminals, or I_{dc} for current source converter terminals, is the only input of the second part. Therefore the reference states to be supplied are only the first three components in equations 5.40 and 5.105.

For the VSC-HVDC terminal at the sending end, the output vector is $Y = [P_{SO} Q_{SO}]^T$; the first two components of the reference state vector $X^* = [I_{Sd}^* I_{Sq}^* V_{Sdc}^*]^T$ can be calculated by using equation 5.117; the component V_{Sdc}^* is set as $V_{Sdc}^* = V_{Rdc}^* + \Delta V$, where V_{Rdc}^* is the required DC bus voltage at the receiving end and ΔV the resistive voltage drop across the DC transmission line. The referenced state vector is:

$$X^* = \begin{bmatrix} I_{Sd}^* \\ I_{Sd}^* \\ V_{Sdc}^* \end{bmatrix} = \begin{bmatrix} \begin{bmatrix} (V_{Od}^2+V_{Oq}^2)^{-1}V_{Od}(V_{Od}^2+V_{Oq}^2)^{-1}V_{Oq} \\ (V_{Od}^2+V_{Oq}^2)^{-1}V_{Oq}-(V_{Od}^2+V_{Oq}^2)^{-1}V_{Od} \end{bmatrix} \begin{bmatrix} P_{SO}^* \\ Q_{SO}^* \end{bmatrix} \\ V_{Rdc}^*+\Delta V \end{bmatrix} \tag{5.127}$$

For the receiving end VSC-HVDC terminal the output vector is $Y=[Q_{RO}V_{Rdc}]^T$, the corresponding reactive power reference is Q_{RO}^* and the DC bus voltage reference V_{Rdc}^*; the active power reference can be set as $P_{RO}^*=-P_{SO}^*+\Delta P$. Therefore the referenced state vector becomes:

$$X^* = \begin{bmatrix} I_{Rd}^* \\ I_{Rd}^* \\ V_{Rdc}^* \end{bmatrix} = \begin{bmatrix} \begin{bmatrix} (V_{Od}^2+V_{Oq}^2)^{-1}V_{Od}(V_{Od}^2+V_{Oq}^2)^{-1}V_{Oq} \\ (V_{Od}^2+V_{Oq}^2)^{-1}V_{Oq}-(V_{Od}^2+V_{Oq}^2)^{-1}V_{Od} \end{bmatrix} \begin{bmatrix} -P_{SO}^*+\Delta P \\ Q_{RO}^* \end{bmatrix} \\ V_{Rdc}^* \end{bmatrix} \tag{5.128}$$

For CSC-HVDC the sending end terminal output vector is $Y=[P_{SO}Q_{SO}]^T$; the first two components of the reference state vector $X^*=[I_{Sd}^*I_{Sq}^*I_{dc}^*]^T$ can be calculated by using equation 5.117, the component I_{dc}^* is set as $I_{dc}^*=P_{SO}^*/(V_{Rdc}^*+\Delta V)$, where V_{Rdc}^* is the required DC bus voltage at the receiving end and ΔV the resistive voltage drop across the DC transmission line. The referenced state vector is:

$$X^* = \begin{bmatrix} I_{Sd}^* \\ I_{Sd}^* \\ I_{dc}^* \end{bmatrix} = \begin{bmatrix} \begin{bmatrix} (V_{Od}^2+V_{Oq}^2)^{-1}V_{Od}(V_{Od}^2+V_{Oq}^2)^{-1}V_{Oq} \\ (V_{Od}^2+V_{Oq}^2)^{-1}V_{Oq}-(V_{Od}^2+V_{Oq}^2)^{-1}V_{Od} \end{bmatrix} \begin{bmatrix} P_{SO}^*+\Delta P \\ Q_{SO}^* \end{bmatrix} \\ P_{SO}^*/(V_{Rdc}^*+\Delta V) \end{bmatrix} \tag{5.129}$$

For the receiving end CSC-HVDC terminal, the output vector is $Y=[Q_{RO}V_{Rdc}]^T$, the corresponding reactive power reference is Q_{RO}^* and the DC bus voltage reference V_{Rdc}^*; the active power reference can be set as $P_{RO}^*=-P_{SO}^*+\Delta P$. Therefore the referenced state vector becomes:

$$X^* = \begin{bmatrix} I_{Rd}^* \\ I_{Rd}^* \\ I_{dc}^* \end{bmatrix} = \begin{bmatrix} \begin{bmatrix} (V_{Od}^2+V_{Oq}^2)^{-1}V_{Od}(V_{Od}^2+V_{Oq}^2)^{-1}V_{Oq} \\ (V_{Od}^2+V_{Oq}^2)^{-1}V_{Oq}-(V_{Od}^2+V_{Oq}^2)^{-1}V_{Od} \end{bmatrix} \begin{bmatrix} P_{SO}^* \\ Q_{SO}^* \end{bmatrix} \\ (-P_{SO}^*/\Delta P)/V_{Rdc}^* \end{bmatrix} \tag{5.130}$$

5.9 Summary

This chapter has described a generalized theory for the modeling and state control of non-linear three-phase self-commutating AC–DC and DC–AC voltage source and current source conversion systems at the converter control level. Both the VSC and CSC cases use a general equivalent suitable for the derivation of the average model, and the general model can be adapted for the specific applications by setting the parameters of the appropriate components to zero or infinity.

The derivations are based on the generalized state–space averaging method and the principle of AC and DC power balance. The converter system models are derived using feedback linearization and, as a result, the non-linear converter controller can be designed based on linear control theory.

References

1. Verghese, G.C., Ilic-Spong, M. and Lang, J.H. (1986) Modeling and control challenges in power electronics. *Proceedings of 25th Conference on Decision and Control*, Athens, Greece, December, 1986, pp. 39–45.
2. Petitclair, P., Bacha, S. and Rognon J.P. (1996) Averaged modeling and nonlinear control of an ASVC (advanced static VAR compensator). *27th Annual IEEE Power Electronics Specialists Conference*, 1996. *PESC*, Volume 1, pp. 753–758.
3. Yao, Z., Kesimpar, P., Donescu, V., Uchevin, N. and Rajagopalan, V. (1998) Nonlinear control for STATCOM based on differential algebra. *29th Annual IEEE Power Electronics Specialists Conference*, 1998. *PESC*, Volume 1, pp. 329–334.
4. Yang, Y., Kazerani, M. and Quintana, V.H. (2003) Modeling, control and implementation of three-phase PWM converters. *IEEE Transactions on Power Electronics* **18** (3), 857–864.
5. Yazdani, A. and Iravani, R. (2006) Dynamic model and control of the NPC-based back-to-back HVDC system. *IEEE Transactions on Power Delivery*, **21** (1), 414–424.
6. Yazdani, A. and Iravani, R. (2006) An accurate model for the DC-side voltage control of the neutral point diode clamped converter. *IEEE Transactions on Power Delivery*, **21** (1), 185–193.
7. Isidori, A. (1989) *Nonlinear Control Systems*, 2nd edition. Springer-Verlag, New York.
8. Lu, B. and Ooi, B.T. (2007) Nonlinear control of voltage-source converter systems. *IEEE Transactions on Power Electronics*, **22** (4), 1186–1195.

6

PWM–HVDC Transmission

6.1 Introduction

As explained in the first chapter, both the transistor and thyristor types of self-commutating power switches can now be combined in series to form reliable high-voltage valves. For other than relatively small voltages and distances, the cost and complexity of compensating for the large charging currents involved make AC power transmission by cable impractical. Therefore, AC power transmission has mainly developed around the overhead line technology. On the other hand, no charging current is required in DC power transmission. This advantage, together with greater transmission efficiency, encouraged the development of a high-voltage DC technology, but the high cost of the line-commutated converter terminals required long distances before the DC option could compete with AC transmission.

The availability of self-commutation, with simpler converter structures and more control flexibility has, in the past decade, justified the use of PWM-based voltage source conversion for high-voltage DC cable power transmission, with existing ratings of up to 330 MW [1]. A similar converter technology has already been applied to provide flexible static compensation (STATCOM) for the control of reactive power.

Due to the limited voltage capability of the cable technology, for very large power transfers the transmission links require the parallel connection of cables to provide high current ratings. For instance the line-commutated cross Channel interconnection between England and France uses eight cables operating at ±270 kV. This scheme, which has been operating since 1986, consists of two 1000 MW bipoles.

Proposals have been made for very large intercontinental links, such as the transfer of electricity generated from hydro and solar power sources from Africa to Europe. Such schemes, however, are likely to involve long-distance overhead transmission and, thus, the voltage limitation of the submarine part of the link will prevent the use of efficient voltage levels. It is likely that in these cases more cost-effective solutions would result from a combination of ultra-high voltage (UHV) AC (for the overhead line) and HVDC (for the submarine connection).

Self-Commutating Converters for High Power Applications J. Arrillaga, Y. H. Liu, N. R. Watson and N. J. Murray

6.2 State of the DC cable technology

As the cable is not exposed to lightning strikes, storms, falling trees, etc., the probability of line short-circuits in DC power transmission is greatly reduced. This is an important consideration in VSC transmission because such faults require isolation of the link by circuit breakers at both ends to permit clearance of the reactive energy and the return to normal power flow is very slow.

There is no limiting length in DC cable transmission and, therefore, there is no need for intermediate compensating stations. Generally, cables have far less impact on the environment than overhead transmission. The magnetic fields are almost completely eliminated by adopting the bipolar system and they do not produce ground currents. Moreover, the lifetime of the cable insulating materials is better for DC than AC.

VSC transmission allows only one DC polarity and thus the cable does not need to be designed for polarity reversals. This greatly simplifies the cable design, permitting the use of polymeric insulation, instead of the conventional oil-impregnated paper insulation. The polymeric insulation material can withstand high forces and repeated flexing and is thus better suited to deep water installation. This is because this type of cable can use galvanized steel wire armour, whereas AC cables need to use non-magnetic, less strong armours. Present standard voltages for the polymeric insulated cables used in recent self-commutating HVDC schemes are 84 and 150 kV and the installation of a 500 MW ±200 kV submarime link between Britain and Ireland has recently been announced.

In these cables the polymeric insulation is triple-extruded together with the conductor screen and the insulation screen. This offers a very robust design, suited to land installation (using ploughing techniques), as well as submarine use under severe conditions. Extruded HVDC single core cables are installed close to each other in bipolar pairs with anti-parallel currents that cancel the magnetic field.

The main problems of cable transmission for long distances overland are laying costs and reliability, due to the large number of joints required. These are caused by the limited length of cable that can be transported by road. This is not a problem, however, with submarine cable which is normally manufactured on the coast and transferred in one piece directly into cable laying ships.

6.3 Basic self-commutating DC link structure

HVDC light, a pioneering PWM–VSC technology developed by ABB [2], combines the high controllability of PWM conversion with the more reliable and environmentally friendly cable transmission. It is at present the most flexible HVDC technology, in that it provides four quadrant converter controllability, while maintaining the DC voltage constant. Additional benefits of the technology are:

- Compact and light-weight design (based on a modular concept).
- Use of DC cables with solid dielectric insulation.
- Bidirectional continuous power transfer controllability, without the need for equipment switching.
- Short installation and commissioning times, low operation and maintenance costs (as the stations can be operated remotely) and low footprint requirement.

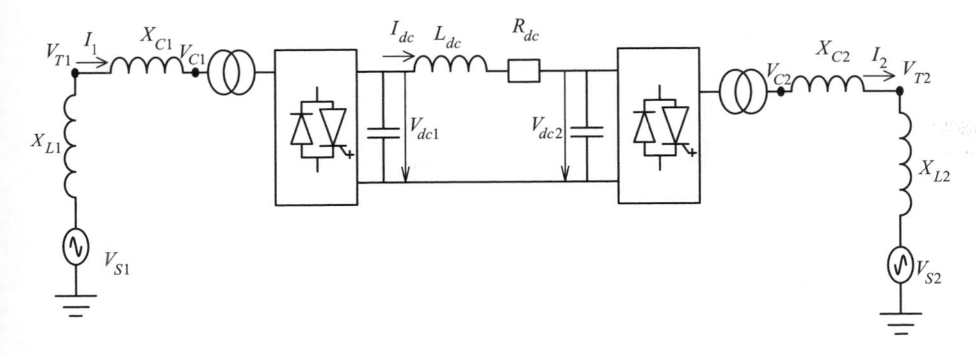

Figure 6.1 Basic VSC transmission link.

Given the present use of the early qualification of this technology as 'light' it may appear antagonistic to include it in a book discussing large power conversion. However, the indication is that the industry is continuing the development of this technology (now termed HVDC-Plus) for larger power ratings (in the region of 1000 MW and ±320 kV).

The basic configuration, shown in Figure 6.1, uses a two-level three-phase bridge converter at each end of the link. The converter valves consist of a number of series-connected IGBTs with anti-parallel diodes and each IGBT position is individually controlled and monitored via fibre optics. The valves, DC buses and DC capacitors are designed with low inductance to reduce the overvoltage across the valve at turn-off. Auxiliary power for the gate drive unit is generated from the voltage across the IGBT. Turn-on/-off of each IGBT is ordered via an optical link from the control equipment placed at ground potential.

The switching of the IGBT valves requires little energy due to the high impedance gate of the individual devices and this in turn permits the use of high-frequency switching, as required by PWM control. The main purpose of PWM is to achieve fast control of the amplitude and phase of the fundamental frequency voltage, which, in turn, permits control of the active and reactive power independently. However, as the rating of the converter is based on maximum currents and voltages, the reactive power capabilities of a converter must be treated against the active power capability [5]. The combined active/reactive power capabilities are illustrated in Figure 6.2. The capacitive limit in the figure is needed to impose a voltage limitation, i.e. if the voltage is reduced this limit increases.

Various modulation alternatives have been described in Chapter 2 to reduce the low harmonic orders. Moreover, the quality of the waveform needs to be further improved for power transmission application and this is normally achieved by the addition of a series reactor and some AC filters. The magnitude of the voltage harmonic components varies with the DC voltage, the switching frequency, the number of converter levels and the PWM strategy chosen.

The use of an optimal pulse width modulation (OPWM) technique provides harmonic elimination and reduces the converter losses. OPWM consists of two functions, their specific purposes being to calculate the time to the next sample instant and modulate the reference voltage vector. This technique concentrates the harmonics in a narrow bandwidth, and thus substantially reduces the size of the filters. Moreover, the converter losses are reduced by switching the valves less often when the current is high. Its effect is illustrated in Figure 6.3 for the case of a converter using a PWM technique designed for optimum harmonic cancellation. The figure shows the converter terminal phase-to-ground voltage (with the fundamental

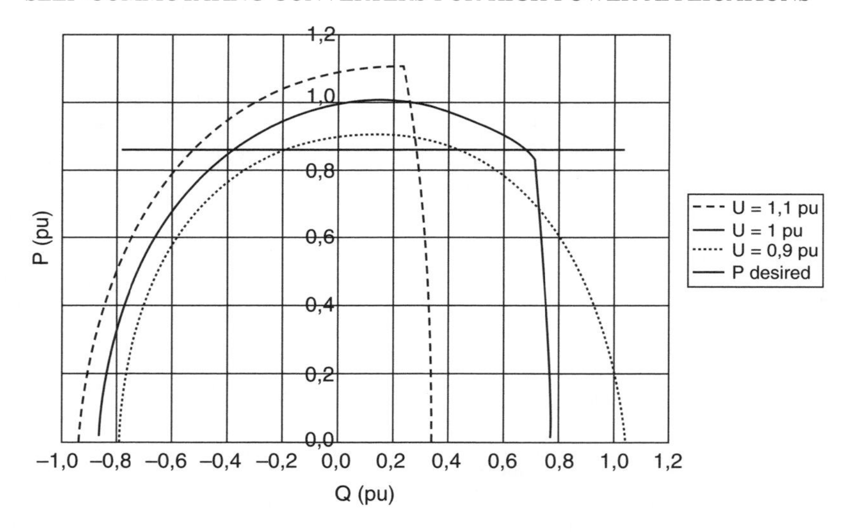

Figure 6.2 Active and reactive power converter capabilities.

component as a dotted line). The harmonic spectrum is given in terms of sequence components. The filter of a typical PWM–HVDC scheme contains two or three tuned branches and reduces the individual harmonics to 1% and the THD to 2.5 % with reference to the fundamental voltage measured at the point of common coupling.

On the DC side, the DC capacitor and the line smoothing reactor will normally be sufficient to limit the harmonic content. Some additional DC side filtering may be required, in the form of a common mode reactor (to eliminate the zero sequence harmonic content) and/or one or more single filters.

6.4 Three-level PWM structure

The first overland scheme using the three-level configuration, described in Chapter 3 (Section 3.3.1), is the Murray link commissioned in Australia in 2002. It is a 177 km long link of extruded polymer cable installed for the Victoria transmission system. The rated powers are 200 MW, + 140/−150 MVAr, the DC voltage ±150 kV and the AC voltages at the connecting points 132 kV and 220 kV. The scheme, shown in Figure 6.4, uses the three-level neutral point connection, as well as IGBT-based converters under PWM control with a switching frequency of 1350 Hz. The figure also shows the tuning frequencies for the AC and DC side filters. As the limits for the DC side harmonics in this scheme are very tight, a zero sequence reactor and a 9th/21st harmonic filter, were added as shown in the figure. The specific control modes are active power, reactive power and AC voltage. Although this link was initially unregulated, it has recently become regulated.

6.4.1 The cross sound submarine link

The Cross Sound Cable scheme is a bidirectional interconnection between the electricity market regions of New England and New York, and is the largest rated HVDC light submarine

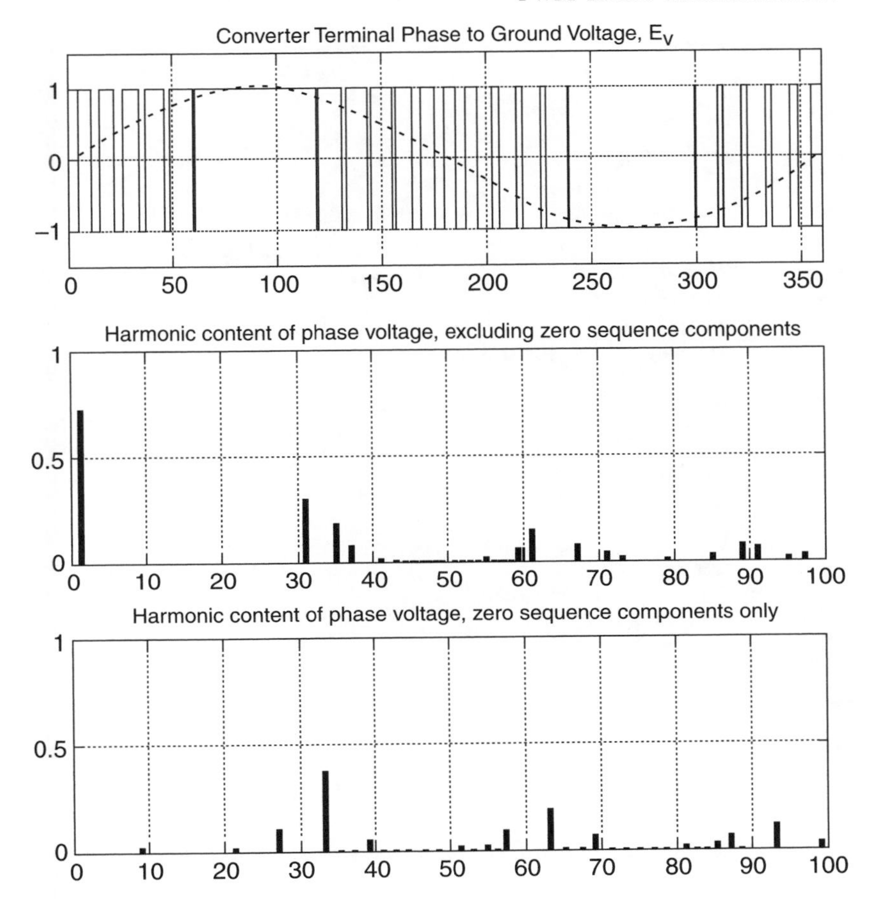

Figure 6.3 Voltage source converter PWM with optimum harmonic cancellation.

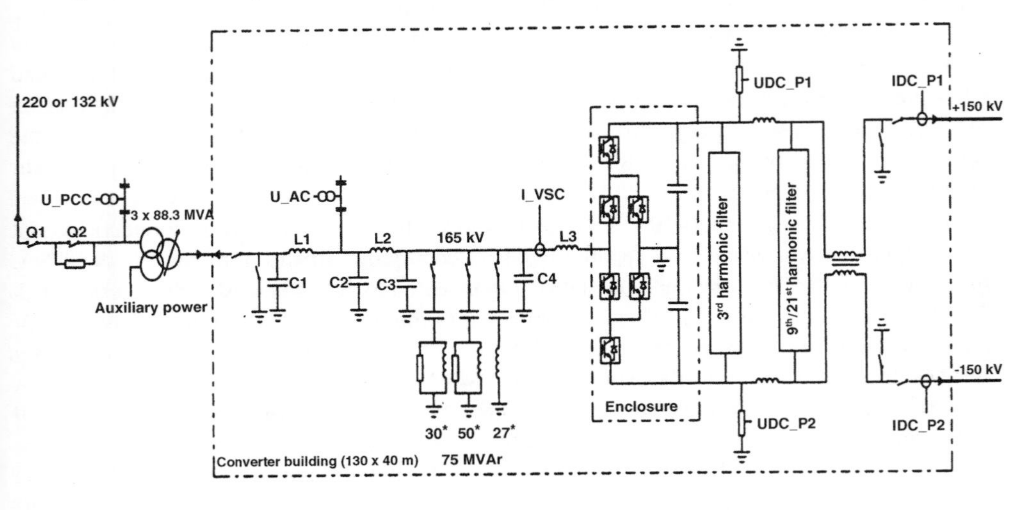

Figure 6.4 Single line diagram of one end of the Murray link.

link

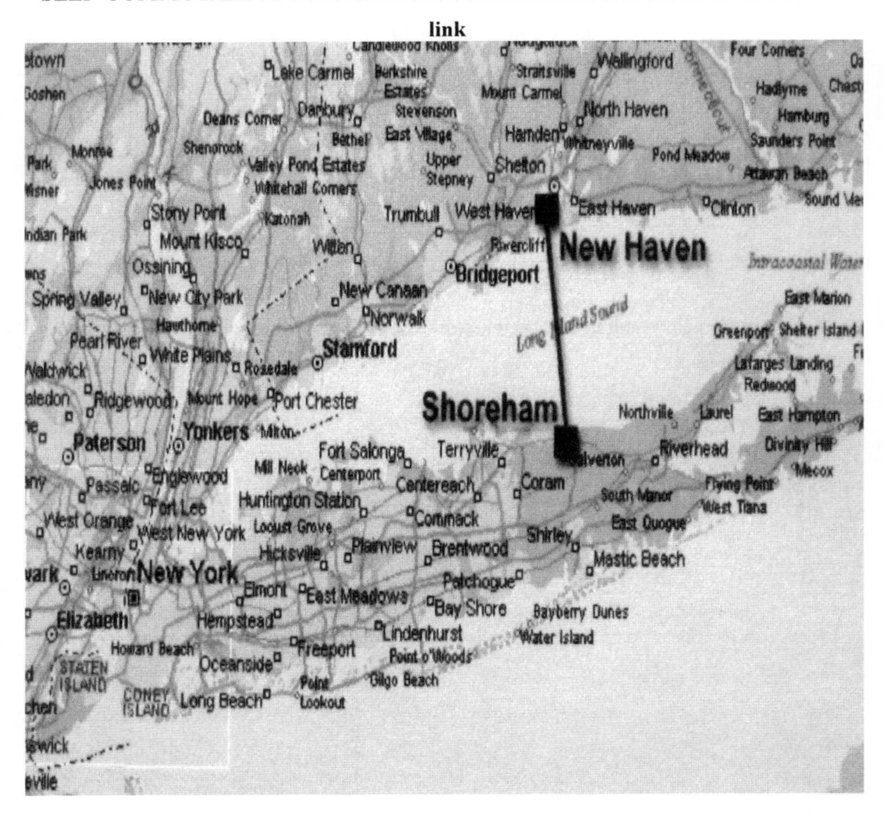

Figure 6.5 Geographical location of the Cross Sound Cable scheme.

scheme implemented so far [3]. It was commissioned in 2002 and had approximately 97.5% availability in its first year of operation. The geographical location is shown in Figure 6.5.

The two VSC–PWM stations are separated by 40 km of submarine cables operating at ±150 kV and 1200 A. The cables use solid dielectric, a 1300 mm^2 copper conductor, a lead alloy metal sheath, polyethylene protective sheath, galvanized steel armour wires and an outer protective layer of polypropylene yarn bonded with bitumen. The AC voltages at the connection points are 345 and 138 kV respectively, the DC voltage ±150 kV and the power rating 330 MW, ±75 MVAr.

A general description of the scheme is shown in Figure 6.6. The Connecticut and Long Island networks are two separately operated regions in the northeast of the USA. In Connecticut the DC link converter station is connected to the 345 kV AC transmission grid. The converter station at Long Island is connected to the Shoreham substation of the 138 kV transmission system. Power generation plants exist in close proximity of both converter stations. In the north terminal these include a 511 MVA generator at East Shore and a 461 MVA unit at Scovill Rock. On Long Island there are three 99 MVA generators near Shoreham as well as two units of 56 and 17.5 MVA respectively.

The use of three-level conversion halves the switching voltage amplitude and the PWM switching frequency with respect to those of two-level conversion used in earlier HVDC

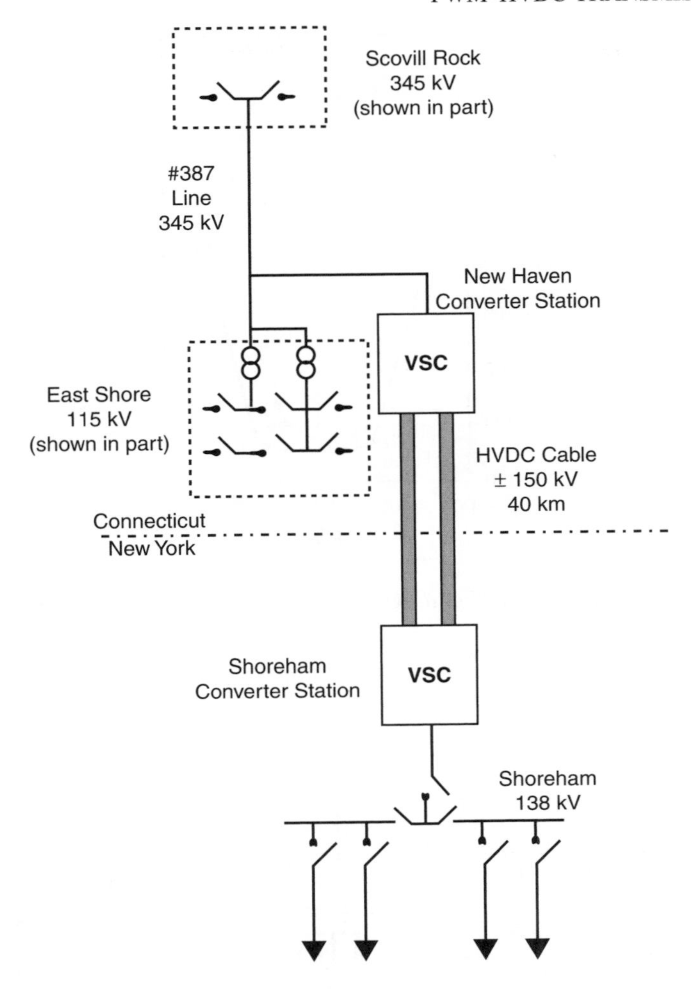

Figure 6.6 Diagram of the Cross Sound Cable interconnection.

projects. This configuration considerably reduces the switching losses without increasing the harmonic content. As the active and reactive powers are controlled independently, the converter stations can operate at any point within the PQ diagram shown in Figure 6.7.

A single line diagram of the converter stations is shown in Figure 6.8. The three-level PWM converter is connected via water series reactors ($L4$) to the 200 kV filter bus, which contains three shunt filters tuned to the 21st, 41st and 25th harmonics. The power line carrier filter equipment (consisting of $L1$, $L2$, $L3$, $C1$, $C2$, $C3$, $C4$) is installed on the secondary side of the converter transformer. The incoming breaker has associated pre-insertion resistors to minimize the transients at converter energization.

A third harmonic filter is connected on the DC side. This is due to the addition of a zero sequence third harmonic component to the sinusoidal reference in order to reduce the peak AC converter voltage As a result, for a given DC voltage, the fundamental frequency voltage produced by the voltage source converter is increased by about 15%.

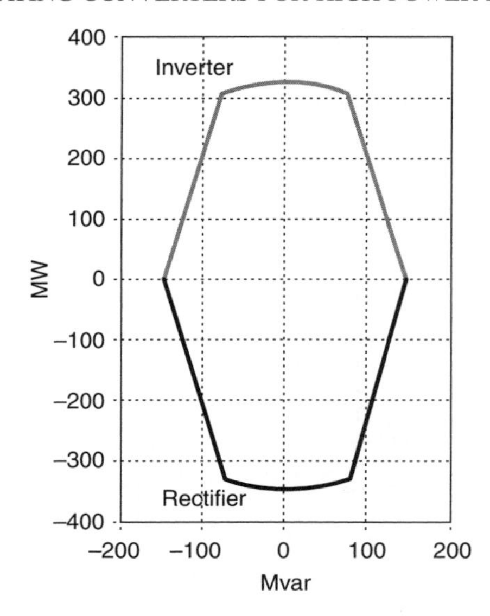

Figure 6.7 Steady state PQ diagram of the Cross Sound Cable terminals.

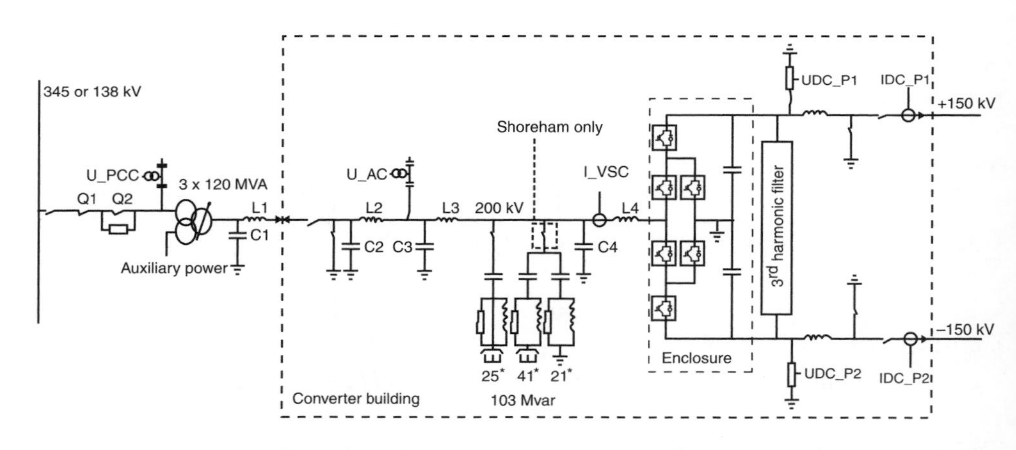

Figure 6.8 Diagram of the converter station.

A key feature of this scheme is to control the power transfers continuously from 0 to 330 MW, in accordance to scheduled transactions by those who have purchased rights to its capacity.

The scheme uses a single three-phase voltage source converter at each end. The converter is a three-level bridge using IGBT/diode packs, both for the main converter switches and for the neutral point clamping. The three-level concept reduces the rate of change of voltage, as the valves switch on and off only between + or −150 kV and 0 kV. The switching frequency is 1260 kV (corresponding to the 21st harmonic), which is half the frequency used in some two-level projects. Therefore the losses are correspondingly reduced, without increasing the harmonic content.

The active and reactive powers are controlled independently, although the latter is limited to one half of the former, as the main purpose of the link is to transfer active power. The ability to control reactive power may also be used to keep the AC voltage constant at the point of common coupling (PCC). By using PWM, the resulting converter terminal voltage is achieved by switching the valves according to a modulation scheme determined by several factors, such as calculated reference voltage, available DC voltage, harmonic generation and valve currents.

The controller includes a zero sequence third-harmonic component modulation (additional to the sinusoidal reference) to reduce the peak AC converter voltage and increase (by approximately 15%) the fundamental frequency component produced by the VSC process.

Non-linearities in the valve switchings create some low-order harmonics (mostly fifth and seventh) and thus a special controller is used to act on the PWM pattern to minimize the low-order harmonic currents at the PCC.

A sub-synchronous damping controller (SSDC) is also used, because both ends of the link contain power generating plant in the vicinity of the converters. The angular frequency deviation given by a PLL is band-pass filtered to extract the sensitive frequency range; this signal is then used in the SSDC and the limited output signal is added to the current orders.

Following the US blackout of August 2003, this link played an essential part in the restoration of power across Long Island; this event has showed the importance of enhancing the transmission infrastructure.

6.4.1.1 Control and protection

A diagram of the converter together with the pole controller is shown in Figure 6.9. The latter includes the converter AC current control, DC voltage control (DCVC), active power control (APC), reactive power control (RPC) and AC voltage control (ACVC). One converter controls the DC voltage and the other the active power. It is possible to operate either converter in reactive power control or in AC voltage control independently of which of them is controlling the active power.

The control is duplicated to increase availability with one control active and the other in stand by. The design allows interruption-free transfer between the two control systems.

Each control system consists of pole control and protection, as well as valve control units (VCUs). The latter receive the switching signals from the pole control and determine the switching orders sent to the valves.

The main object of the control system is to control the transferred active and reactive powers independently. The ability to control reactive power may be used either to keep the reactive power exchange or the AC voltage constant at the PCC independently from the DC voltage or active power control modes.

The pole control consists of a state feedback controller for the converter current control. The active and reactive currents are controlled independently. The controller is synchronized to the fundamental voltage using a phase-locked loop (PLL). The DC voltage, active power, reactive power and AC voltage controls calculate the reference currents to the converter current control. In steady state, one converter must operate in DC voltage control and the other in active power control. Although the converter protections are integrated in the same duplicated main computer system, they operate separately from the converter controls to ensure safe operation.

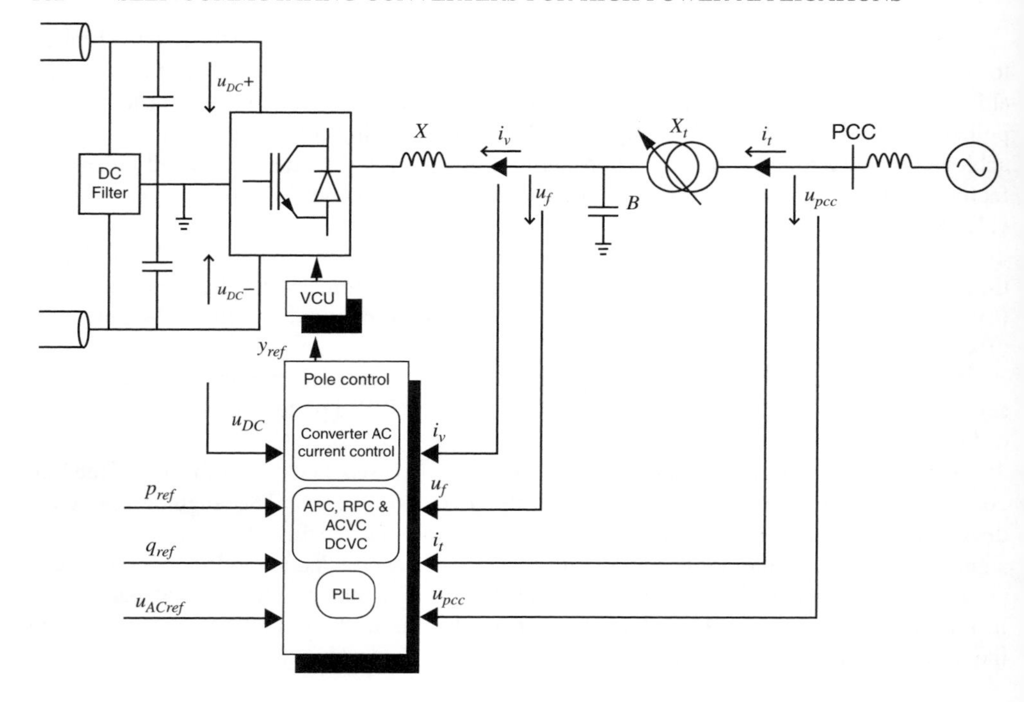

Figure 6.9 Diagram of the control functions used by the Cross Sound Cable project.

As the three-level converter is mid-point grounded, balancing of the DC pole voltage is needed to prevent ground current circulation. The bridge is balanced by using certain space vectors in the modulation process.

To prevent the amplification of some low-order harmonics caused by minor non-linearities in the valve switching, extra control is added to the PWM.

The AC voltage control includes a slope (or droop) similar to that used in generators automatic voltage regulators (AVRs). The reactive power capability is dependent on the active power transfer and, thus, a strong network may prevent getting the AC voltage set point due to the reactive power limit. This problem may be reduced by the use of a slope.

As indicated earlier, a sub-synchronous damping controller (SSDC) is used to prevent the amplification of possible sub-synchronous oscillations. With this aim, the angular frequency deviation given by a PLL is band-pass filtered to extract the sensitive frequency range; this signal is used in the SSDC and the limited output signal is added to the current orders.

6.4.1.2 Steady state behaviour

During commissioning the transmission losses were determined using the actual MW values from the 345 kV and 138 kV utility revenue meters. The actual measured losses are shown in Figure 6.10 to be lower than initially estimated, especially at high power transfers. The lower frequency switching of the three-level configuration is the main factor influencing the higher efficiency of the scheme compared to earlier built underground HVDC light systems.

Typical harmonic spectra for the voltage and current on the AC side are shown in Figure 6.11. The limits set for individual harmonics and THD (2% in New Haven and 3%

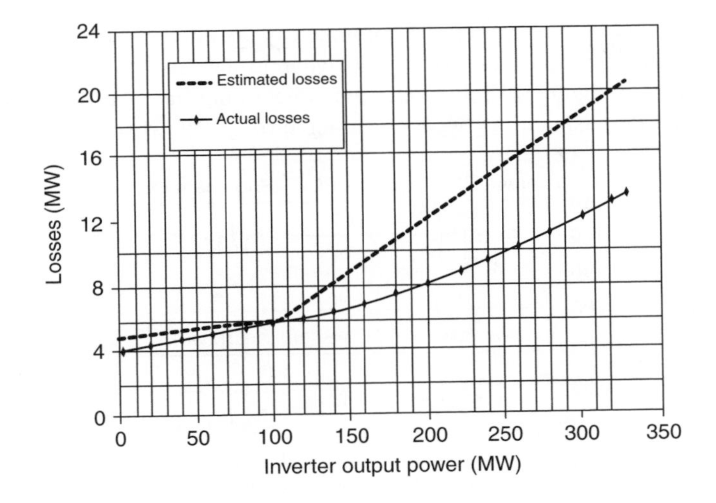

Figure 6.10 Cross Sound Cable losses, including station and cable.

in Shoreham) were easily met at both stations. In fact the harmonic levels of the voltage spectrum were also present when the converters were blocked. The requirement set for the IT product (12 A_{rms}) were initially more difficult to meet at Shoreham due to the levels of fifth and seventh harmonic voltage components. However, after optimizing the harmonic controller settings, the fifth and seventh harmonic currents are now kept below 2 A, as shown in the second graph of Figure 6.11.

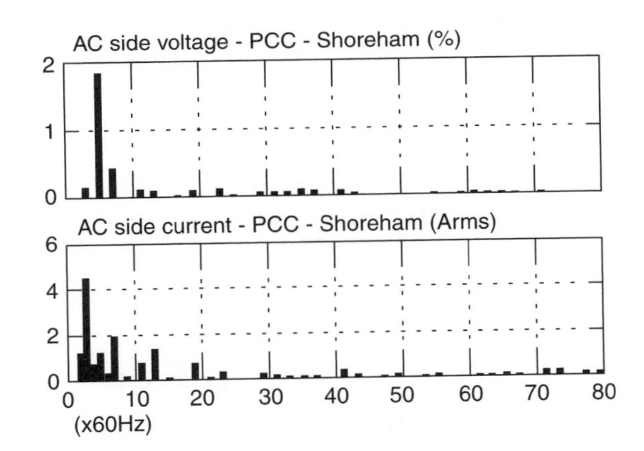

Figure 6.11 AC side harmonic content at Shoreham.

6.4.1.3 Response to faults

To minimize the transients during converter energization and deblocking, a pre-insertion resistor was added to the incoming circuit breaker and the AC filters circuit breakers were equipped with a synchronous closing function as well as a staggered sequence for the converter deblocking.

The Cross Sound Cable converters are continuously operated in AC voltage control at the connecting stations. This is also improving the dynamic stability of the system, as shown in Figures 6.12 and 6.13. Figure 6.12 illustrates the effect of multiple AC faults during a severe thunderstorm in Long Island. As well as riding through the disturbance, the converter

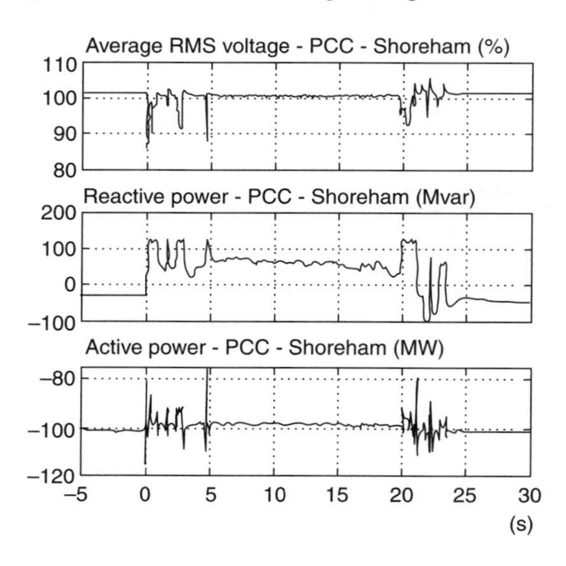

Figure 6.12 Dynamic response to AC faults in the Long Island system.

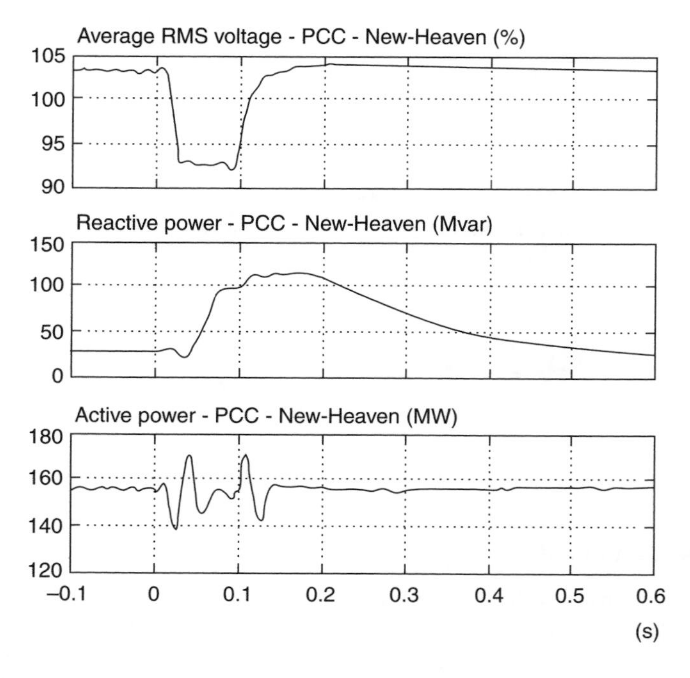

Figure 6.13 Dynamic response to an AC fault in the Connecticut system.

supported the voltage up to its reactive power limit of 125 MVAr, while keeping the active power practically constant. A similar behaviour is observed in Figure 6.13 for an AC fault in the Connecticut system.

6.5 PWM–VSC control strategies

A number of up to three degrees of freedom can be provided by the modulation process of the voltage source converter. They are:

- *Frequency control.* Controlling the frequency of the oscillator that determines the valve pulse firing sequence is essential when the voltage source converter is the only source of power, i.e. the voltage source converter link supplies power to an isolated load. When the voltage source converter is connected to an active power system the voltage source converter can participate in the system frequency control by regulating the power delivered to or taken from the AC system.
- *AC voltage control.* The AC voltage can be controlled by regulating the magnitude of the fundamental frequency component of the AC voltage produced by the voltage source converter on the converter side of the interface transformer, either by altering the DC capacitor voltage (in the case of multipulse and multilevel converters) or by varying the modulation index (in the case of PWM conversion). If the voltage source converter feeds into an isolated load, the AC voltage controller also provides automatic control of the power going into the load (this assumes that the sending end of the link is controlling the DC side voltage). The converter transformer may be provided with OLTC (on-load tap change control) with the purpose of keeping the bus voltage, and therefore the modulation index, within specified limits.
- *Active power control.* The control of active power transfer is achieved by regulating the phase angle of the fundamental frequency component of the converter generated AC voltage. Power is taken from or delivered to the AC system depending on the sign of this angle. The transfer of active power through the link requires simultaneous coordinated action at both ends of the link. Fast power transfer control can be used to damp electromechanical oscillations and for the improvement of transient stability following disturbances.
- *Reactive power control.* The reactive power generated or absorbed by the voltage source converter is controlled by the magnitude of the converter AC voltage source, which in PWM conversion is determined by the modulation index. The use of this function is important when the other converters in the transmission system are operating to maintain their respective AC voltages.
- *DC voltage control.* As the various converters in a DC link share a common DC voltage, at least one of them is required to control the DC capacitor voltage, a task achieved by regulating the small extra power required to charge or discharge the capacitor to maintain the specified DC voltage level. A proportional or proportional–integral controller will be used for this task. In the case of multipulse and multilevel conversion, DC voltage regulation is the only way for the voltage source converter to achieve AC voltage control.

- *AC current control.* Current control is often a desirable feature to ensure that the converter valves are not overloaded. This control can be achieved directly or through vector control, where the control of the current is an intermediate step in the control of other parameters such as the active and reactive powers.

6.6 DC link support during AC system disturbances

Within its rated capability, the PWM–VSC link can change its working point almost instantaneously, a property that can be used to enhance the AC network stability by providing the best mixture of active and reactive power following disturbances [4].

6.6.1 Strategy for voltage stability

The potential support of the voltage source converter is described here with reference to the simplified system of Figure 6.14(b), which represents the receiving end of a VSC transmission link. Figure 6.14(a) shows the crossing between the capability curve of the voltage source converter terminal and receiving end circle indicates the stable solution for a particular voltage level. If the link sending end voltage drops it is possible to establish immediately a new stable solution in the P-Q circle.

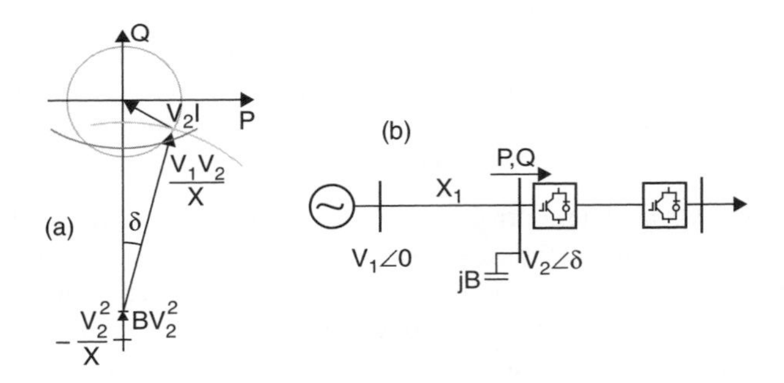

Figure 6.14 Power circle diagram (a) and series equivalent circuit (b).

6.6.2 Damping of rotor angle oscillation

With reference to a VSC–HVDC interconnection, provided that the two separate systems do not have the same frequency of oscillating modes, the healthy side can store or discharge energy to damp rotor oscillations at the disturbed end. Figure 6.15 shows such a case when the AC system is exposed to a fault in the vicinity of the generator. The fault is cleared after 100 ms and the corresponding rotor angle is plotted in Figure 6.15(b) for active power modulation and compared to what a pure reactive power modulation can achieve.

The difference in damping capability is clearly shown in the power-angle shown in Figure 6.15. This 'text-book' example indicates that active power modulation is four times more effective than reactive power for the damping of oscillations.

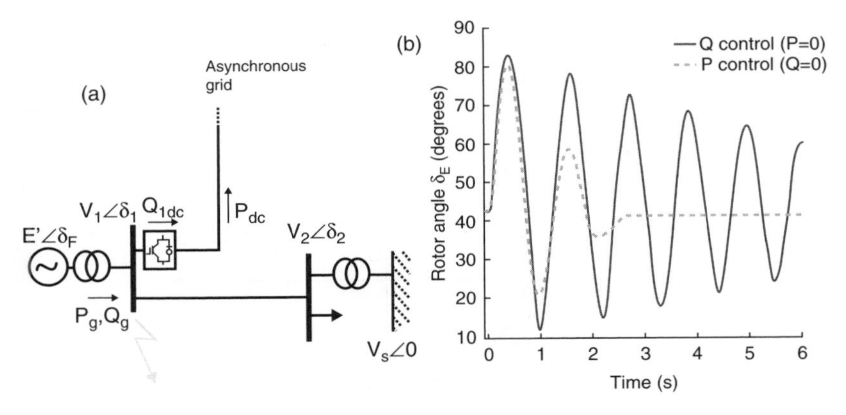

Figure 6.15 Rotor angle oscillation damping following a fault.

6.6.3 Converter assistance during grid restoration

Following a blackout on the AC system, the voltage source converter link will be able to assist recovery almost immediately by its inherent fast voltage controllability, regardless of the short-circuit capacity of the AC system. The extent of the benefit will, of course, depend on whether one or both ends of the link are exposed to the blackout.

The DC link can then energize some of the AC transmission lines at a lower voltage in order to avoid the overvoltages caused by the Ferranti effect, thus permitting the remote connection of transformers operate at a safer voltage level. Following the transformer connection, the DC link can ramp the AC voltage up to its nominal value. The voltage source converter link can also feed auxiliary power to the local plants at the remote end and thus provide them with a stable voltage and frequency to start on.

6.6.4 Contribution of the voltage source converter to the AC system fault level

As the converter AC current can be controlled, the converter contribution to the short-circuit power is expected to be small. This contribution varies in inverse proportion to the short-circuit ratio and is maximum when the converters operate at zero active power. The contribution also depends on the control strategy used. Under reactive power control, the short-circuit current contribution will be small, because the current order limit reduces with the voltage. In the AC voltage control mode, the contribution increases with decreasing active power if the current limit is not changed. Of course the increased short-circuit contribution is associated with an improved performance in voltage stability, i.e. the voltage dips during distant faults are reduced. During AC faults the current control will rapidly lower the fundamental frequency voltage of the bridge to reduce the DC current down to the pre-fault level.

The gate units are provided with a primary valve/bridge protection system that acts in nanoseconds and a back-up protection system that acts in a few microseconds, the latter based on the current flowing in the DC capacitors and phase reactors.

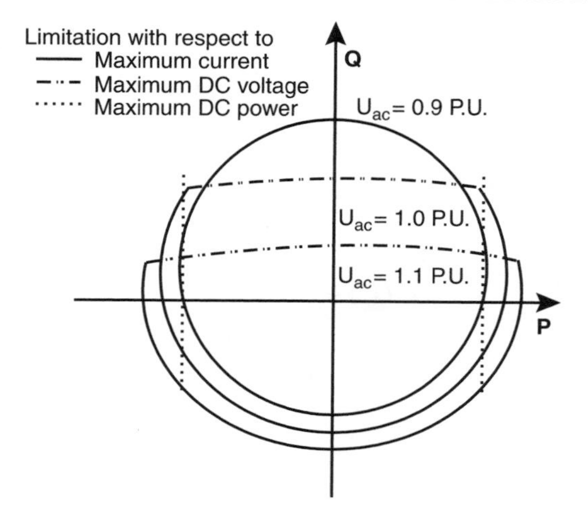

Figure 6.16 Active and reactive power converter capabilities.

6.6.5 Control capability limits of a PWM–VSC terminal

The capability of a PWM–VSC converter station to help the stability of the power system to which it is connected is limited, as shown in Figure 6.16, by the following three factors:

1. The maximum current through the IGBTs. This gives rise to a maximum MVA circle in the power plane where the maximum current and the actual AC voltage are multiplied. When the AC voltage drops, the MVA capability reduces accordingly.
2. The maximum DC voltage level. The reactive power is mainly dependent on the voltage difference between the AC voltage that the voltage source converter can generate and the grid AC voltage. If the latter is high, the difference between the maximum DC voltage and the AC voltage will be low. The reactive power capability is then moderate, but increases with decreasing AC voltage.
3. The maximum current capability of the DC cable.

When the voltage level reduces, the capability is determined by the maximum current level. The small bias in the Q-axis direction is due to the presence of a line reactor and filter capacitance within the VSC transmission system. When available, the converter transformers tap changers also play an important part in the control of the reactive power transfer.

The VSC link can alter the operating point anywhere within the capability chart practically instantaneously. Fast power reversal is also possible, which requires reversing the DC direction and not the voltage polarity. The cable technology used can handle such reversals without problems.

The practically instantaneous variation of the active and reactive power within the capability curve can be used to support the grid with the best mixture of active and reactive power during stressed conditions. In general, active power modulation will provide the best damping, but a mix of active and reactive power control will in some cases be the most practical solution. VSC transmission is able to support the AC grid with a suitable power factor to improve system stability.

6.7 Summary

PWM–VSC is the most flexible of all present converter control concepts and has already been successfully used for high voltage applications in hundreds of megawatts. The experience gathered from several PWM–HVDC schemes indicates that this technology is especially suited to cable power transmission. ABB has already announced the extension of the power and voltage ratings of the HVDC light configuration to 1100 MW and ±320 kV respectively, which should substantially reduce the number of parallel cables to be laid in future submarine interconnections. While perfectly suited for use in large submarine interconnections, this technology is unlikely to expand into UHV long-distance transmission.

From the operational viewpoint the use of PWM adds considerable flexibility to voltage source conversion. The modulation process develops independent fundamental and harmonic frequency voltage control from a constant DC source; this in turn provides four-quadrant power controllability and avoids the use of low-order harmonic filters.

In HVDC–CSC schemes, a change in the power direction requires DC voltage reversal, a condition that makes the control of multiterminal interconnections impractical. PWM-based VSC can achieve that purpose without altering the DC voltage magnitude and polarity. The ability to control the transfer of active power as well as the terminal voltage has made VSC–PWM conversion (based on IGBT switching) attractive for HVDC transmission.

Among the advantages of increased switching frequency are: simpler filtering arrangements, smaller DC ripple and smaller size of VSC substations. An additional benefit of PWM is that the size of the DC capacitor can be reduced. The size of the capacitor is decided to ensure that the voltage ripple is between 2 and 10%. Every time the valves are switched the current direction in the DC capacitor changes direction and this effect reduces the voltage ripple.

However, the frequent and fast switching of the AC bus voltage between two levels results in repetitive high-frequency transient stresses, which affect the design of the voltage source converter components and, unless appropriately filtered, will cause electromagnetic interference. Wound components are particularly vulnerable to repetitive, high-frequency transient stresses which cause premature ageing of the insulation materials. As well as increased voltage (dv/dt) stresses, PWM–VSC results in higher converter losses and reduced power capability for each device.

References

1. Arrillaga, J., Liu, Y.H. and Watson, N.R. (2007) *Flexible Power Transmission–The HVDC Options*, John Wiley & Sons, London.
2. Asplund, G., Eriksson, K. and Svensson, K. (1997) DC Transmission Based on Voltage Source Converters. *CIGRE SC-14 Colloquium,* Johannesburgh, S.A.
3. Ronstrom, L., Railing, B.D. Miller, J.J. et al. (2004) Cross Sound Cable project second generation VSC technology for HVDC. *CIGRE B4-102*, Paris.
4. Johansson, S.G., Asplund, G., Jansson, E. and Ruderwall, R. (2008) Power system stability benefits with DC transmission systems. *CIGRE B4-204*, Paris.

7

Ultra High-Voltage VSC Transmission

7.1 Introduction

VSC in the form of PWM has so far been the only option adopted by the industry for self-commutating HVDC transmission [1–5]. This option offers the following technical benefits:

1. The active and reactive powers can be controlled independently.
2. An excellent dynamic response can be achieved, which is an important condition to comply with the grid code requirements in the event of AC system faults.
3. A collapse network can be restored by means of what is referred to as the black-start capability.
4. Weak power systems can be interconnected without the need to install compensating equipment.

The switching device presently employed in self-commutating VSC schemes is the IGBT, the characteristics of which permit high-frequency switching, and it is, thus, perfectly suited to PWM conversion. The switching losses, however, are high and this is likely to restrict the voltage applicability of IGBT–PWM (at present the limit being 300 kV). Moreover, existing PWM–IGBT converter configurations are limited to three levels, which result in very large dV/dt's at the switching instances.

UHV DC is only justified for the transmission of very large powers at very long distances. These two factors are likely to require the use of overhead line transmission and this restricts the type of switching devices and converter configurations to be used. It is unlikely that the IGBT-based PWM technology will be considered for the proposed ±800 kV UHV DC technology.

From the multilevel configurations discussed in Chapters 3 and 4, the only option so far used by the industry is modular multilevel conversion (MMC). This multilevel configuration,

Self-Commutating Converters for High Power Applications J. Arrillaga, Y. H. Liu, N. R. Watson and N. J. Murray

that has recently been chosen for a submarine scheme in North America, provides four-quadrant power controllability without the high losses and dV/dt of PWM [6]. A structurally simpler multilevel alternative suitable for UHV DC application is the H-bridge multilevel voltage reinjection (MLVR) concept described in Chapter 4 (Section 4.5.2). However, to achieve independent reactive power controllability at the ends of a DC transmission system, MLVR-HB DC requires the special double group converter control described in this chapter.

7.2 Modular multilevel conversion

The MMC concept is described in Chapter 3 (Section 3.6). The structure of an MMC-based HVDC converter is shown in Figure 7.1. This configuration consists of a three-phase bridge, where each of the six legs is formed by the series connection of submodules (SM) and a reactor. The basic submodule components, shown on the right of the figure, are an IGBT half bridge and a DC storage capacitor.

The submodule can operate in three different states:

1. *The two IGBTs are off.* In this state, when the ACc power switch is closed, if the current flows from the positive DC pole in the direction of the AC terminal the capacitor is being charged; when the current flows in the opposite direction, the current bypasses the capacitor. This state is also adopted for all the submodules in the event of a serious failure.

2. *IGBT1 is on and IGBT 2 off.* Irrespective of the direction of current flow, the voltage of the storage capacitor is applied to the terminals of the submodule. Depending on the

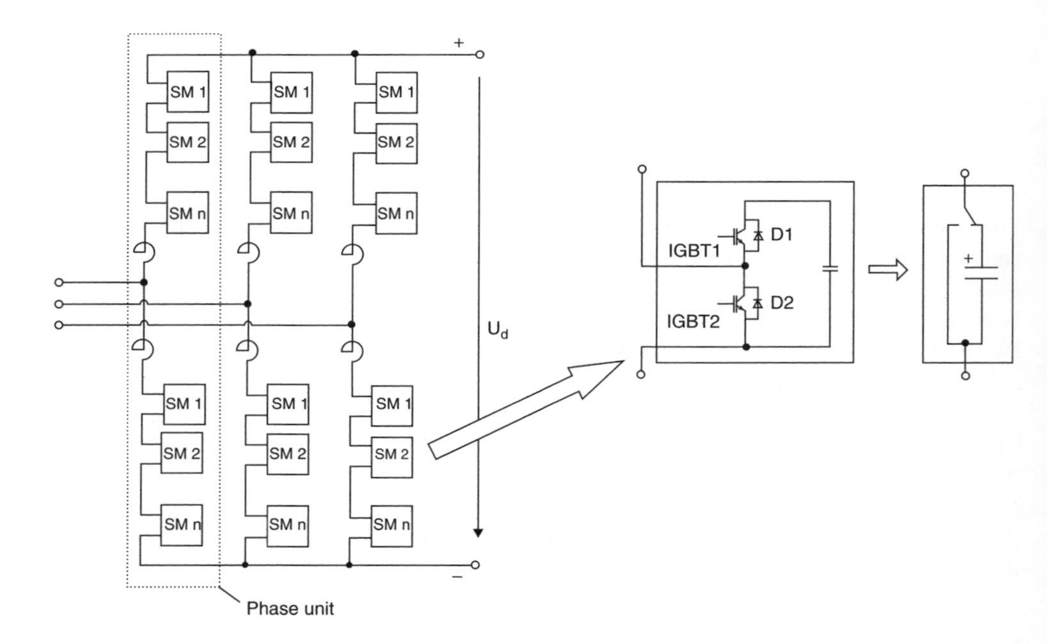

Figure 7.1 The MMC configuration. (Reproduced by permission of CIGRE.)

direction of flow, the current circulates through D1, charging the capacitor, or through IGBT1, discharging the capacitor.

3. *IGBT1 is off and IGBT2 on.* In this state the current either flows through IGBT2 or D2 depending on its direction. Thus no voltage is applied to the submodule terminals and the voltage across the capacitor remains unchanged.

Following the above procedure, each submodule of the converter leg is individually selected and controlled and, therefore, a converter leg represents a controllable voltage source. The total voltage of the two converter legs of the phase units is always equal to the DC voltage, a condition achieved, regardless of the submodule states, by the presence of diodes D1. Therefore, the DC voltage is maintained constant, while the required AC converter voltage is achieved by adjusting the ratio of the converter leg voltages via the state of the submodules. Altering the AC converter voltage without changing the DC voltage is an important advantage of this multilevel configuration, compared to the conventional ones, because it permits independent control of the reactive power at both ends of a DC link.

The series reactors of the phase units in Figure 7.1 are needed to connect the modules in parallel on the DC side because the three generated DC voltages of the phase modules cannot be exactly equal; thus without the reactors there would be relatively large balancing currents between the phase modules. These reactors also play their part in reducing the effects of faults inside or outside the converter. As a result, the current rates of rise can be limited to a few tens of amperes per microsecond even for critical faults, like DC terminal short circuits. Thus the IGBTs can then be turned off at perfectly acceptable current levels.

Under normal operating conditions only one sublevel per converter leg switches at any given time; therefore, the AC voltages can be adjusted in very small increments and both the AC side harmonics and the DC ripple are achieved.

A possible limitation of the MMC concept, as compared with the multilevel alternatives, is that the number of steps of the converter voltage (i.e. the number of active submodules) changes with the converter AC voltage. This may not be a problem for a small range of voltage variation, but should affect the higher harmonic orders for a large range of voltage control.

As well as the excellent steady state waveforms shown in Section 3.6, the MMC topology offers a strong dynamic behaviour, a condition essential for compliance with the grid code requirements, including fault ride-through capability following faults in the AC network. This is illustrated in Figure 7.2, which shows the response of a 400 MW system to an AC line-to-ground fault. It is clear that the fault can be handled without the need to block the converter.

Following a DC pole-to-pole short circuit the IGBTs are blocked within microseconds and a command to open the AC circuit breaker is sent. Because of the converter reactors, the IGBTs are turned off within their nominal current range. Unlike other VSC technologies, DC circuit breaker action is not required to avoid a discharge of the capacitors. Following fault recovery, reclosure of the AC circuit breaker will permit the energy transfer to continue.

MMC technology is particularly suited for HVDC cable transmission and Siemens has constructed two converter stations for a submarine HVDC link in the Bay of San Francisco to transmit 400 MW at $\pm$200 kV. The scheme is expected to increase the level of network security and reliability by providing voltage support as well as reduce system losses. Moreover, the provision of adequate protection against DC line fault makes MMC also suitable for overhead transmission.

While existing MMC technology uses IGBT switches, there seems to be no reason why the control concept should not be used with self-commutating thyristor-type switches, such as the

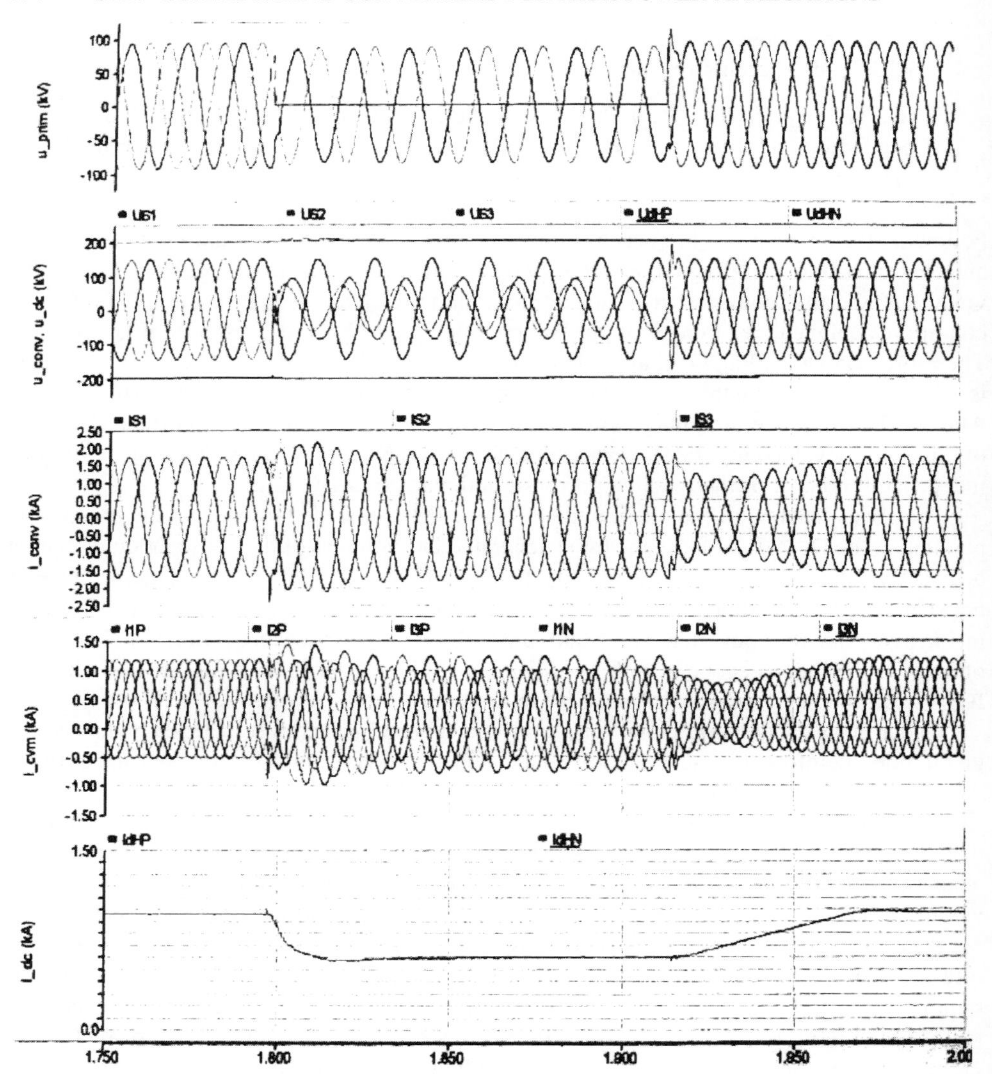

Figure 7.2 MMC dynamic response to an AC line-to-ground fault of a 400 MW system. From top to bottom the graphs show: AC line-to-ground voltages, AC phase-to-ground voltages, AC converter currents, converter arm currents, DC current. (Reproduced by permission of CIGRE.)

IGCT. This should reduce the switching losses and make the MMC technology better suited for UHV DC transmission

7.3 Multilevel H-bridge voltage reinjection

The multilevel cascaded H-bridge configuration is already well established in the area of reactive power compensation [7]. However in its present form is not suitable for HVDC transmission because the H-bridge high-voltage conversion process requires the use of many isolated DC sources in series. A multilevel configuration that uses the H-bridge concept to derive the voltage reinjection signal has been described in Section 4.5.2 [8, 9]. This configuration is suitable for UHV DC application, because it keeps the DC voltage nominally constant and produces the multilevel waveform with a simple reinjection structure, where the auxiliary switches only increase in direct proportion to the level number (as compared with diode clamping, where the number of extra switches required increases with the square of the level number).

7.3.1 Steady state operation of the MLVR-HB converter group

For power transmission application the multilevel converter stations only have one independent control variable, namely the phase displacement between the AC source and the converter output voltages, and this variable is used to control the active power flow between the AC source and the converter. Altering the reactive power injection requires a DC voltage change, an action that simultaneously affects the two sides of the link; thus, with present controls the reactive power adjustments at the receiving and sending ends can not be made independently from each other. The effect of this restriction on the voltage level and/or reactive power exchanges achievable at the terminals of the DC link is considered next to show that, other than in very weak systems, the MLVR-HB multilevel option is perfectly adequate for use in HVDC transmission.

The basic equivalent circuit of Figure 7.3 is used to describe the power transfer characteristics of a converter group. To simplify the description, the interface transformers are represented as ideal transformers in series with their leakage reactance (X_C). V_C is the converter output voltage (on the primary side of the ideal transformer) and V_T the converter terminal voltage (behind the interface reactance). The AC system is represented by a series reactance (X_L) and an ideal source (V_S). On the DC side V_{dc} represents the converter DC voltage and the line is modelled as a series inductive (L_{dc}) resistive (R_{dc}) circuit. The transfer of active power should ideally take place with minimum loss (i.e. with minimum AC current at each end). At one end (normally the receiving end) this is achieved by adjusting the converter-supplied reactive power component.

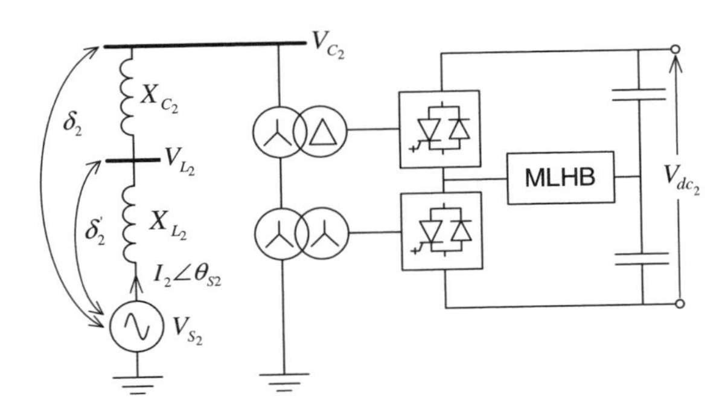

Figure 7.3 Simplified model of one terminal of a single group MLVR-HB HVDC link.

7.3.1.1 Current relationships

With reference to Figures 7.4 and 7.5, the magnitudes and phase angles of the AC side currents are:

$$I_1 = \frac{\sqrt{V_{S1}^2 + V_{C1}^2 - 2V_{C1}V_{S1}\cos\phi_1}}{\sqrt{3}(X_{L1} + X_{C1})} = \frac{\sqrt{V_{S1}^2 + V_{T1}^2 - 2V_{T1}V_{S1}\cos\phi_1'}}{\sqrt{3}X_{L1}} \tag{7.1}$$

$$\theta_{C1} = \cos^{-1}\left[\frac{V_{S1}}{\sqrt{V_{S1}^2 + V_{C1}^2 - 2V_{C1}V_{S1}\cos\phi_1}}\sin\phi_1\right] \text{ for } V_{C1} \rangle V_{S1}\cos\phi_1 \tag{7.2}$$

or

$$\theta_{C1} = -\cos^{-1}\left[\frac{V_{S1}}{\sqrt{V_{S1}^2 + V_{C1}^2 - 2V_{C1}V_{S1}\cos\phi_1}}\sin\phi_1\right] \text{ for } V_{C1} \le V_{S1}\cos\phi_1$$

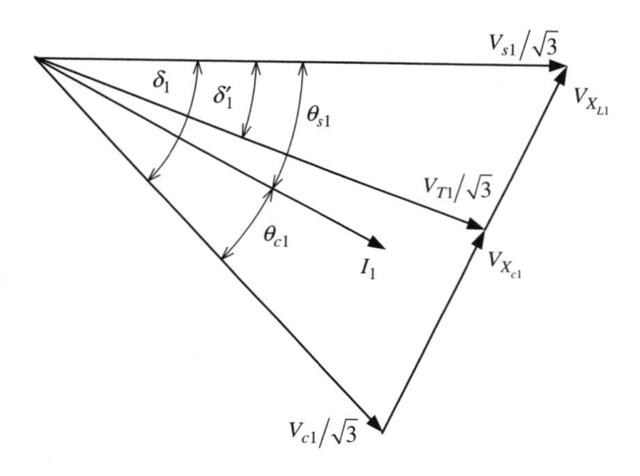

Figure 7.4 Phasor diagram for rectifier operation.

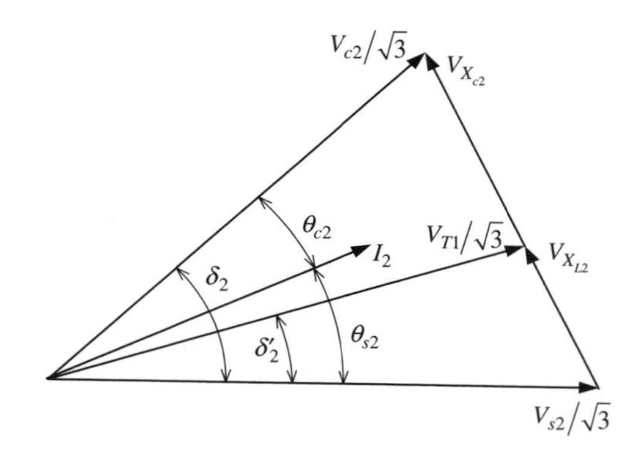

Figure 7.5 Phasor diagram for inverter operation.

and

$$\theta_{S1} = \phi_1 - \theta_{C1}$$

$$I_2 = \frac{\sqrt{V_{S2}^2 + V_{C2}^2 - 2V_{C2}V_{S2}\cos\phi_2}}{\sqrt{3}(X_{L2} + X_{C2})} = \frac{\sqrt{V_{S2}^2 + V_{T2}^2 - 2V_{T2}V_{S2}\cos\phi'_2}}{\sqrt{3}X_{L2}} \tag{7.3}$$

$$\theta_{S2} = \cos^{-1}\left[\frac{V_{C2}}{\sqrt{V_{S2}^2 + V_{C2}^2 - 2V_{C2}V_{S2}\cos\phi_2}}\sin\phi_2\right] \text{ for } V_{S2} \rangle V_{C2}\cos\phi_2$$

or

$$\theta_{S2} = -\cos^{-1}\left[\frac{V_{C2}}{\sqrt{V_{S2}^2 + V_{C2}^2 - 2V_{C2}V_{S2}\cos\phi_2}}\sin\phi_2\right] \text{ for } V_{S2} \leq V_{C2}\cos\phi_2 \tag{7.4}$$

and

$$\theta_{C2} = \phi_2 - \theta_{S2}$$

The AC currents' real and imaginary components (using their respective source voltages as a reference) are:

$$I_{\text{Re}1} = I_1\cos\theta_{S1} \quad I_{\text{Im}1} = I_1\sin\theta_{S1} \tag{7.5}$$

$$I_{\text{Re}2} = I_2\cos\theta_{S2} \quad I_{\text{Im}2} = I_2\sin\theta_{S2} \tag{7.6}$$

7.3.1.2 AC current minimization at the receiving end

It can be shown that the receiving end AC current is at its minimum value (for a specified value of P_2) when

$$\frac{V_{C2}}{V_{S2}} = \sqrt{1 + \frac{P_2^2(X_{L2} + X_{C2})^2}{V_{S2}^4}} \tag{7.7}$$

The component $\frac{P_2^2(X_{L2}+X_{C2})^2}{V_{S2}^4}$ in equation (7.7) written in terms of the short-circuit ratio is

$$\frac{P_2^2(X_{L2} + X_{C2})^2}{V_{S2}^4} = \left[\frac{P_2}{P_{2rated}}\right]^2 \left[\frac{V_{2rated}}{V_{S2}}\right]^4 \left[\frac{1}{SCR_2}\right]^2 \tag{7.8}$$

where V_{2rated} and P_{2rated} are the rated values of the voltage and power.

To achieve minimum AC current, equation (7.7) indicates that the ratio $\frac{V_{C2}}{V_{S2}}$ needs to increase when the active power demand increases. Moreover, their maximum increments are strongly dependent on the AC system impedance, i.e. a strong AC system requires a low increment and a weak system a high increment. Also a slight variation in the AC source voltage (V_{S2}) can cause a significant change of the ratio due to the 4 exponent.

For example, if the short-circuit ratio of the receiving end is 2.5, to transfer the rated active power with a source voltage adjustment of $V_{S2} = 0.96V_{2rated}$ the required ratio is $\frac{V_{C2}}{V_{S2}} = 1.091$, i.e. the converter AC output voltage would be adjusted to $V_{C2} = 1.047V_{2rated}$. Thus if a 5% AC source and DC voltage variation is considered acceptable, an appropriate control of the $\frac{V_{C2}}{V_{S2}}$ or $\frac{V_{T2}}{V_{S2}}$ ratio can provide transmission at unity power factor at the receiving end of the link.

As the two converters in a point-to-point DC link are directly connected, only one end of the link can be used to control the DC voltage. Therefore adjustment of the ratio converter to source voltage at the other end cannot be achieved by altering the DC voltage.

If the AC source voltage variation of the receiving end is limited to, say, $(1-\varepsilon)V_{2rated}$, then V_{C2} will vary between $(1-\varepsilon)V_{2rated}$ and $(1+\varepsilon)V_{2rated}$ (depending on P_2) and still achieve minimum AC current at this end. The required value of ε for this to happen can be derived from

$$\frac{1+\varepsilon}{1-\varepsilon} = \sqrt{1 + \frac{1}{(1-\varepsilon)^4(SCR)^2}} \tag{7.9}$$

which provides the following results:

Short-circuit ratio$_2$	1	2	2.5	3	4	5
ε	NC	0.0718	0.0436	0.0294	0.0161	0.0102

7.3.1.3 AC current minimization at the sending end

Similarly, the minimum AC current at the sending end (the side not controlling the DC voltage) occurs when the following relation is met:

$$\frac{V_{C1}}{V_{S1}} = \sqrt{1 + \left[\frac{P_1}{P_{1rated}}\right]^2 \left[\frac{V_{1rated}}{V_{S1}}\right]^4 \left[\frac{1}{SCR_1}\right]^2} \tag{7.10}$$

Comparing equations (7.10) and (7.7) for a specified active power and ignoring the DC transmission loss, the $\frac{V_{C1}}{V_{S1}}$ ratio requires a smaller change than $\frac{V_{C2}}{V_{S2}}$ if the sending end is stronger than the receiving end. Thus the DC voltage (or the converter terminal voltage) control should be carried out on the weaker side of the link. The converter on the strong side of the link will control the active power and a relatively small variation of the source voltage (according to equation (7.10) should be sufficient to transfer that power at unity power factor. Thus, with appropriate coordination between the AC sources and the DC voltage adjustment, the link should be able to operate efficiently.

For a specified value of the weaker system AC voltage, the stronger system AC source voltage can be set to ensure that the rated active power is delivered with the source operating at unity power factor. Ignoring the small difference between the DC voltages on both sides of the link, the unity power factor condition for the stronger system is given by equation (7.10), where $V_{1rated} = \frac{k_{C2}}{k_{C1}} V_{2rated}$.

When operating at the rated power, the weaker side maintains the relationship

$$\frac{V_{C2}}{V_{2rated}} = 1+\varepsilon \tag{7.11}$$

and by fixing the AC source at the sending end to $V_{S1} = (1-\varepsilon')V_{1rated}$, it will operate at unity power factor when ε' is determined from the expression

$$\frac{1+\varepsilon}{1-\varepsilon'} = \sqrt{1+\left[\frac{V_{1rated}}{V_{S1}}\right]^4\left[\frac{1}{SCR_1}\right]^2} \tag{7.12}$$

Some numerical results for the relationship between ε' and SCR_1 are listed below:

If $SCR_2 = 2$ and $\varepsilon = 0.0718$ for SCR_1 levels of 2, 2.5, 3, 4 and 5, the resulting values of ε' are 0.0718, 0.0177, −0.0105, −0.0377 and −0.0501 respectively.

If $SCR_2 = 2.5$ and $\varepsilon = 0.0718$ for SCR_1 levels of 2.5, 3, 4, and 5 the corresponding values of ε' are 0.0436, 0.0161, −0.0104 and −0.0225.

If $SCR_2 = 3$ and $\varepsilon = 0.0294$ for SCR_1 levels of 3, 4 and 5 the corresponding values of ε' are 0.0294, 0.0033 and −0.0086.

If the two AC voltage sources are fixed to $V_{S1} = (1-\varepsilon')V_{1rated}$ and $V_{S2} = (1-\varepsilon)V_{2rated}$, with the DC voltage controlled at the weaker end, only the weaker AC system will operate at unity power factor. The actual power factor at the source of the stronger system is given by

$$\cos[\theta_{S1}(p_1)] = \frac{\sin\phi_1}{\sqrt{1+\left[\frac{V_{C1}}{V_{S1}}\right]^2 - 2\frac{V_{C1}}{V_{S1}}\sqrt{1-\sin^2\phi_1}}} = \frac{\sin\delta_1}{\sqrt{1+\left[\frac{V_{C2}}{V_{2rated}}\right]^2\frac{1}{(1-\varepsilon')^2} - 2\frac{V_{C2}}{V_{2rated}}\frac{1}{1-\varepsilon'}\sqrt{1-\sin^2\phi_1}}} \tag{7.13}$$

where

$$\sin\phi_1 = \frac{P_1(X_{L1}+X_{C1})}{V_{C1}V_{S1}} = \frac{P_1/P_{1rated}}{\frac{V_{C1}}{V_{1rated}}\frac{V_{S1}}{V_{1rated}}(SCR_1)} = \frac{P_1/P_{1rated}}{\frac{V_{C2}}{V_{2rated}}(1-\varepsilon')(SCR_1)} \tag{7.14}$$

and

$$\frac{V_{C2}}{V_{2rated}} = (1-\varepsilon)\sqrt{1+\left[\frac{P_2}{P_{2rated}}\right]^2\left[\frac{1}{1-\varepsilon}\right]^4\left[\frac{1}{SCR_2}\right]^2} \tag{7.15}$$

By way of example, a link with short-circuit ratios of 2.5 and 3 at the weaker and stronger ends respectively will be able to maintain unity power factor operation at the weaker system end for the full range of power transfer, whereas the stronger system power factor will range from 0.998 to 1 for an active power range of between 0.5 and 1 pu. However, for very low power transfers (below 0.1) the strong system will operate with a low power factor; when the power transfer is zero the maximum reactive power capability is 8.1 %.

7.3.2 Addition of four-quadrant power controllability [10]

The AC voltage output of the multilevel configurations can only be adjusted by varying the DC voltage, and since this cannot be done independently on both sides of the link, the reactive power requirements at both terminals are interdependent. This is an important disadvantage of the multilevel concept, compared to PWM, when the DC link is expected to provide some reactive power as an ancillary service.

For large power transmission ratings, the DC link terminals normally use two or more series-connected converter groups. With reference to a double group configuration, this section describes a control strategy that achieves independent four-quadrant power controllability at the sending and receiving ends of the VSC-HVDC interconnection.

7.3.2.1 DC link configuration and operating principles

Figure 7.6 shows the basic structure of a MLVR-HB DC link, with the reinjection circuit represented by a black box connected between the DC capacitor centre point and the common point of the two component bridges of each of the 12-pulse groups. The link is based on the

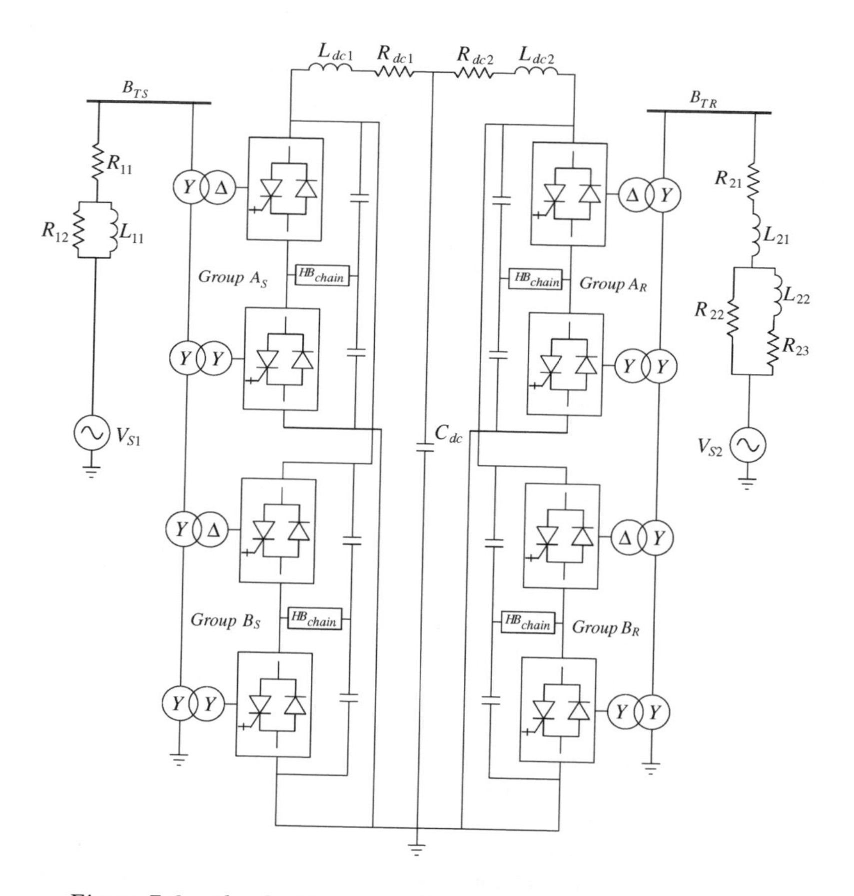

Figure 7.6 The double group MLVR-HB HVDC configuration.

CIGRE benchmark model, with some modifications made to suit the multilevel voltage source converter features; in particular, the addition of the reinjection circuit makes the conventional link filters unnecessary.

If the AC voltages of the two component groups at each end are denoted V_A and V_B and the total AC voltage $V_C(= V_A + V_B)$, the phasor diagram of Figure 7.7 illustrates their relationship with the AC source voltage (V_S) by ignoring the resistance between the converter terminals and their respective AC sources.

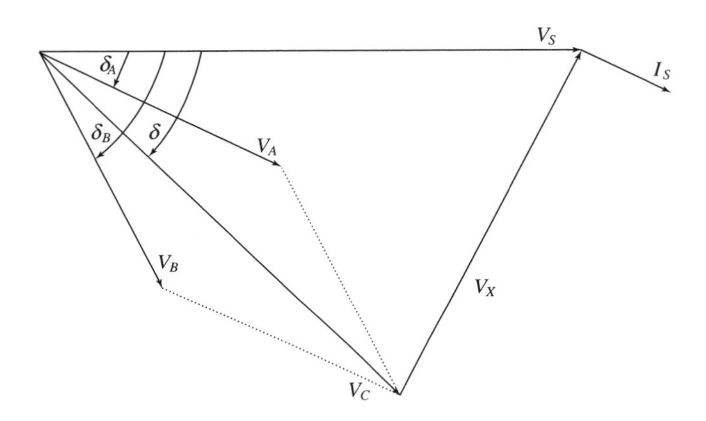

Figure 7.7 Phasor diagram with double-group firing-shift control.

If the DC voltages of the two converter groups are perfectly balanced, the amplitudes of V_A and V_B are exactly the same and the following relationships can be written:

$$|V_C| = 2|V_A|\cos\left(\frac{\delta_A - \delta_B}{2}\right) \tag{7.16}$$

$$\delta = \frac{\delta_A + \delta_B}{2} \tag{7.17}$$

So the amplitude of V_C is controlled by $(\delta_A - \delta_B)$, and its phase angle by $(\delta_A + \delta_B)$, and they are both independent of the DC voltage, as in the case of PWM conversion.

The active power flowing from V_S to V_C is

$$P = \frac{V_C V_S \sin(\delta)}{X} = \frac{V_A V_S(\sin\delta_A + \sin\delta_B)}{X} \tag{7.18}$$

and the reactive power injected by the converter into the AC system is

$$Q_C = \frac{V_C(V_C - V_S\cos\delta)}{X} = \frac{2V_A^2[1 + \cos(\delta_A - \delta_B)] - V_A V_S(\cos\delta_A + \cos\delta_B)}{X} \tag{7.19}$$

7.3.3 DC link control structure

In line with VSC procedure, the MLVR uses the phase angle difference between the voltages across the interface transformer to control the active power flow, while the voltage magnitude difference controls the (generated or absorbed) reactive power. As the converter side AC voltage of the interface transformer is directly determined by the converter DC voltage and the DC voltages are closely related, only one side has complete freedom to control the terminal voltage.

The terminal voltage of the converter interface transformer must be kept within a small margin to satisfy the local load. For use in the control system this voltage is more convenient, as the AC system source voltage would require long distance communication. If the transformer leakage reactance is small relative to the AC system inductance, controlling the link DC voltage and controlling the AC terminal voltage are very similar processes. In a well designed

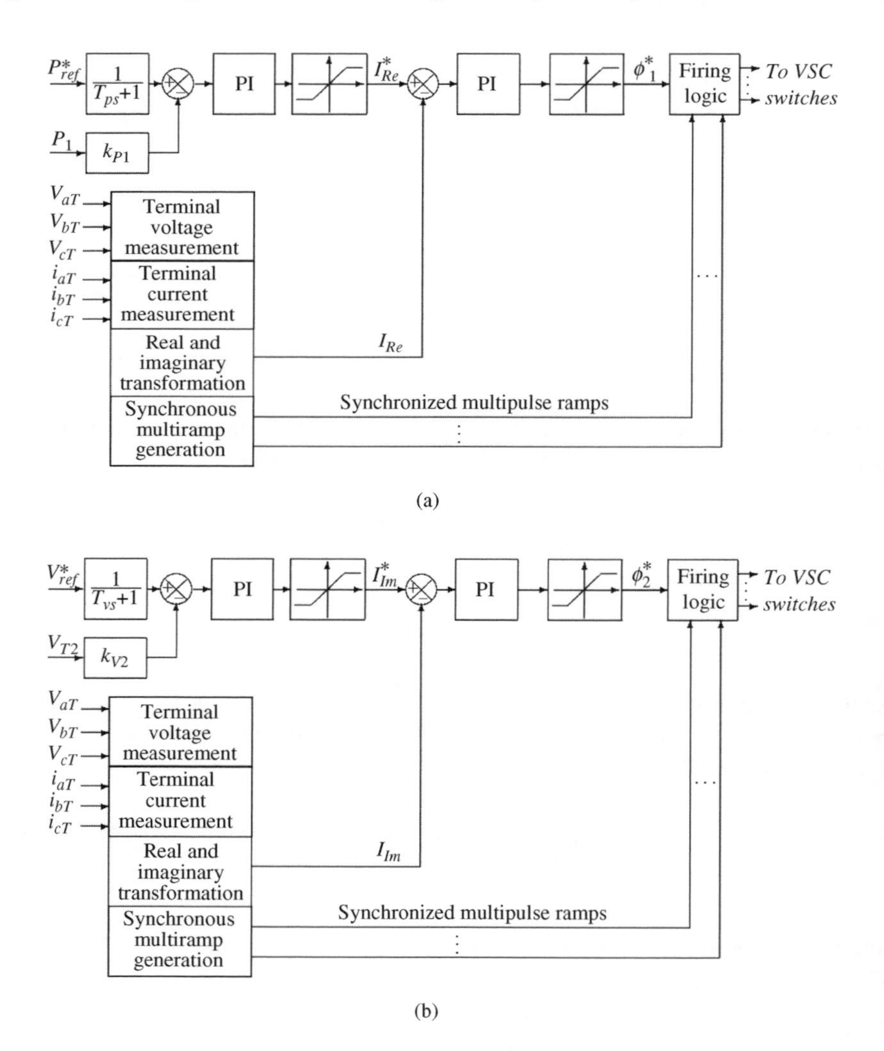

Figure 7.8 Control structures of the sending (a) and receiving (b) ends.

AC system, adjusting the DC link voltage by a narrow margin should normally be sufficient to minimize the AC side (and thus the DC side) currents to reduce the DC transmission losses.

In the simulation the transformer terminal voltage V_{T2} and DC voltage V_{dc2} on the weaker side of the link are used as the control variables to minimize the AC currents. This action is complemented by appropriate tuning of the two AC system source voltages. Under normal operating conditions, the real and imaginary components of the converter output currents directly determine the active and reactive powers and can thus be used as the control variables. Moreover, under abnormal conditions the direct control of the current provides safer operation than the use of power variables.

With reference to the HVDC interconnection, Figures 7.8(a) and (b) show the simplified control structures, without terminal voltage adjustment, for the sending and receiving end converters. The multipulse ramp signals of the converter firing logic are synchronized with the respective converter terminal voltages. The sending end real power P^*_{ref} and receiving end terminal voltage V^*_{ref} references are then compared with their respective measured values. The errors are sent to proprotional–integral (PI) controllers with saturation limits to derive the real and imaginary current components for the sending and receiving end controllers respectively. Finally, the two PI controllers generate the phase angle displacement commands using the real and imaginary current errors of the two converters. The saturation limits of the real and imaginary PI controllers ensure that the phase displacements are within safe operation ranges. The phase displacements are then sent to their respective converter logic to generate the gate firing signals.

The control structure required at the weaker end to control the DC voltage is very much the same as that needed to control the terminal AC voltage, the only difference being that the monitored voltage signal required will be V_{dc2} instead of V_{T2}. Equation (7.15) shows that in V_{C2} only the per unit active power $\frac{P_2}{P_{2rated}}$ varies, while ε and SCR_2 remain constant for a given system. Thus the reference voltage (V^*_{ref}) is generated from the monitored active power P_2 in real time.

7.3.4 Verification of reactive power control independence

The configuration of Figure 7.6 and the control structure described in Section 7.3.3 have been modelled in the PSCAD-EMTDC package. The specific features and parameters of the MLVR-VSC based configuration are:

- Total *absence* of filters.
- The interface transformers are connected in series and their individual leakage reactances are 7%.
- The AC system short-circuit ratios are chosen as 2.5 and 3 for the receiving and sending ends respectively; the required AC system parameters to achieve these short-circuit ratios are listed in Table 7.1.

Table 7.1 Sending and receiving AC system parameters

R_{11}	R_{12}	L_{11}	R_{21}	R_{22}	R_{23}	L_{21}	L_{22}
2.13Ω	3140Ω	0.100H	0.432Ω	43.2Ω	0.432Ω	0.0275H	0.0275H

- The rated voltages are 345 kV (at the sending end) and 230 kV (at the receiving end).
- As the short-circuit ratio$_2$ is set to 2.5, $\varepsilon = 0.0436$ and $V_{S2} = 230(1-\varepsilon) = 220\,kV$; then to compensate for the DC voltage difference caused by the DC line resistance, ε' is set to zero and therefore $V_{S1} = 345\,kV$.
- For convenience all variables displayed in the graphs are in per unit; the base parameters for the sending end are $1.673\sqrt{2}\,kA$ and 1000 MVA and those of the receiving end $2.673\sqrt{2}$ kA and 1000 MVA, while the DC side voltage and current bases are 500 kV and 2 kA respectively.
- The DC line parameters are listed in Table 7.2.

Table 7.2 DC Transmission line parameters

R_{dc1}	R_{dc2}	L_{dc1}	L_{dc2}	C_{dc}
2.5Ω	2.5Ω	$0.5968H$	$0.5968H$	$26.0\,\mu F$

The test system is controlled to follow sending end real power and receiving end terminal voltage orders. When the rated power is transferred from the 345 to the 230 kV AC system the two AC sources operate at optimal power factors.

The variables plotted from the EMTDC results are normalized in a per unit system. The base references for the sending end current, and real and reactive powers are $1.673\sqrt{2}$ kA, 1000 MW and 1000 MVA respectively. For the receiving end the corresponding bases are 2.673 kA, 1000 MW and 1000 MVA.

The active power is set as positive when it flows from the sending to the receiving end; the reactive power is positive when the converter current leads the voltage. The following symbols are used in the figures illustrating the results of the simulation:

P_{refS}– active power order at the sending end.

P_S– measured active power at the sending end converter terminal bus B_{TS}.

Q_S– measured reactive power at the sending end converter terminal bus B_{TS}.

P_R– measured active power at the receiving end converter terminal bus B_{TR}.

Q_R– measured reactive power at the receiving end converter terminal bus B_{TR}.

V_{TS}– measured *rms* voltage at the sending end converter terminal bus B_{TS}.

V_{TR}– measured *rms* voltage at the receiving end converter terminal bus B_{TR}.

V_{dcS}– measured DC voltage at the sending end converter DC terminal.

V_{dcR}– measured DC voltage at the receiving end converter DC terminal.

γ_{AS}– phase displacement between *Group AS* MLVR-VSC and its transformer terminal voltages.

γ_{BS}– phase displacement between *Group BS* MLVR-VSC and its transformer terminal voltages.

γ_{AR}– phase displacement between *Group AR* MLVR-VSC and its transformer terminal voltages.

γ_{BR}– phase displacement between *Group BR* MLVR-VSC and its transformer terminal voltages.

I_{aS}– measured phase A current of the sending end AC system.

V_{aTS}– measured phase A voltage at the sending converter terminal bus B_{TS}.

I_{aR}– measured phase A current of the receiving end AC system.

V_{aTR}– measured phase A voltage at the receiving converter terminal bus B_{TR}.

I_{dc}– measured DC current in the DC transmission line.

θ_S– sending end AC source power factor angle.

7.3.4.1 Simulation results

Figure 7.9 shows the response of the test system to a change of active power from zero to 1 pu in the absence of reactive power supply at either end, when the receiving end DC voltage order is set to 1.05 pu.

Figure 7.10 contains the waveforms of the AC currents and converter terminal voltages in the dynamic region (0.15–0.55 s) of the simulation shown in Figure 7.9 for the sending and receiving ends. Also shown are the converter terminal voltages, the DC voltages and the phase displacement commands to fire the converter switches.

Figure 7.11 demonstrates the control flexibility of the reactive powers generated by the sending and receiving end converter terminals. In a period of 6 s, the sending end reactive power order changes from 0 to 0.5 pu (at 1 s), from 0.5 to 0.3 (at 5 s), while the receiving end reactive order changes from 0 to −0.2 pu (at 0.1 s), from −0.2 to 0.2 pu (at 2 s) and from 0.2 to 0.5 pu (at 4 s). The *rms* voltages at the converter terminals show that they are strongly related to the converter reactive power generated.

7.3.5 Control strategies

Having shown that the reactive powers at the sending and receiving ends can be controlled independently, a control strategy is next tested to optimize the DC link operation. At the receiving end (the weaker system) the AC voltage is maintained constant by making the supply of reactive power vary in proportion to the amount of active power transfer. The stronger sending end AC system is controlled to operate at unity power factor in order to minimize the AC current and DC voltage and, thus, reduce the power losses.

7.3.5.1 Unity power factor control at the sending end

At the sending end the converter generates the appropriate reactive power to keep the displacement angle between the AC system voltage and current at 0° or 180° depending on the direction of the power transfer. This operating condition is represented in the phasor diagram of Figure 7.12, which shows that the voltage across the AC system reactance (V_X) is in quadrature with the AC system voltage (V_S) and current (I_S).

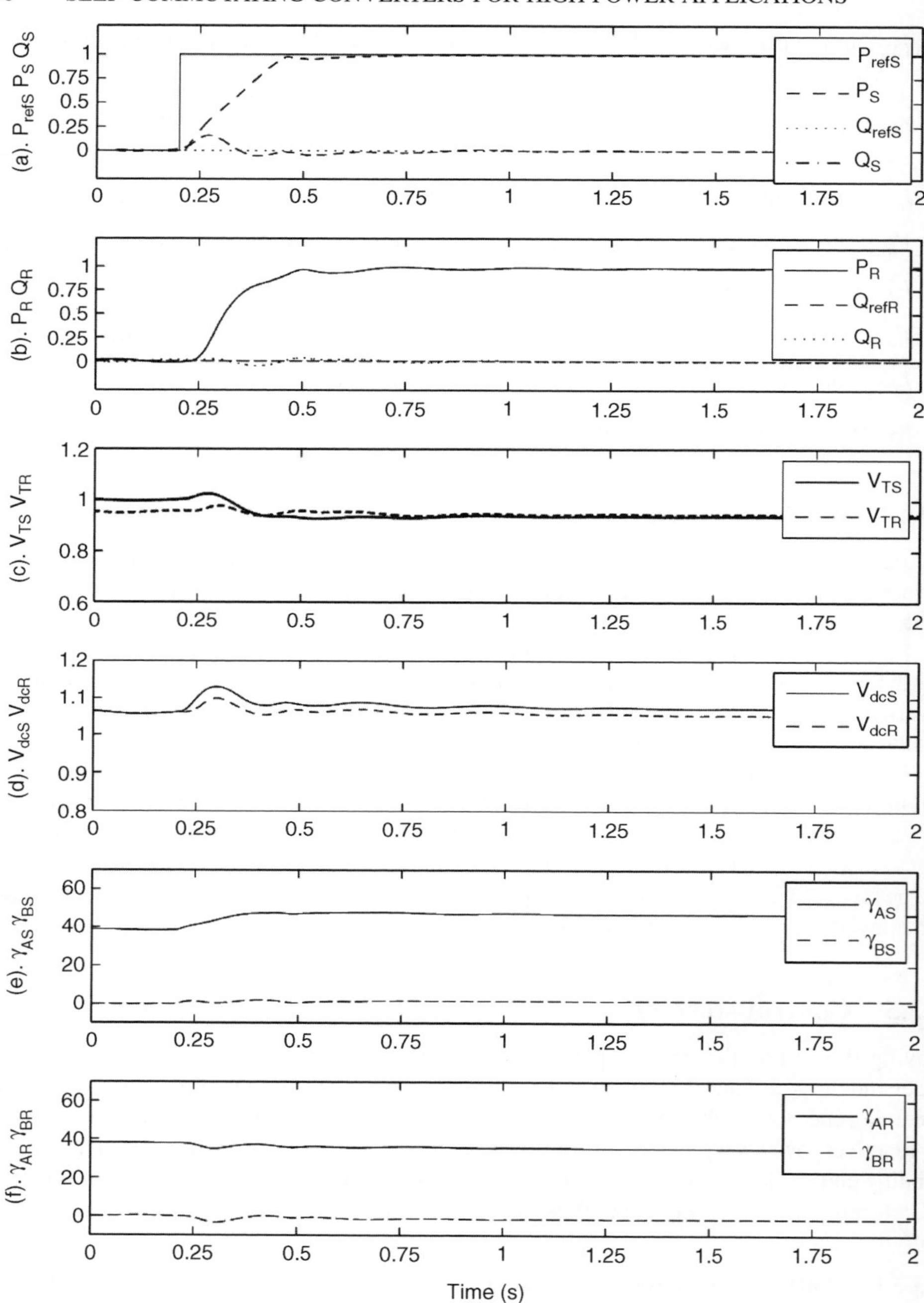

Figure 7.9 Active power response to a 1 pu active power step order.

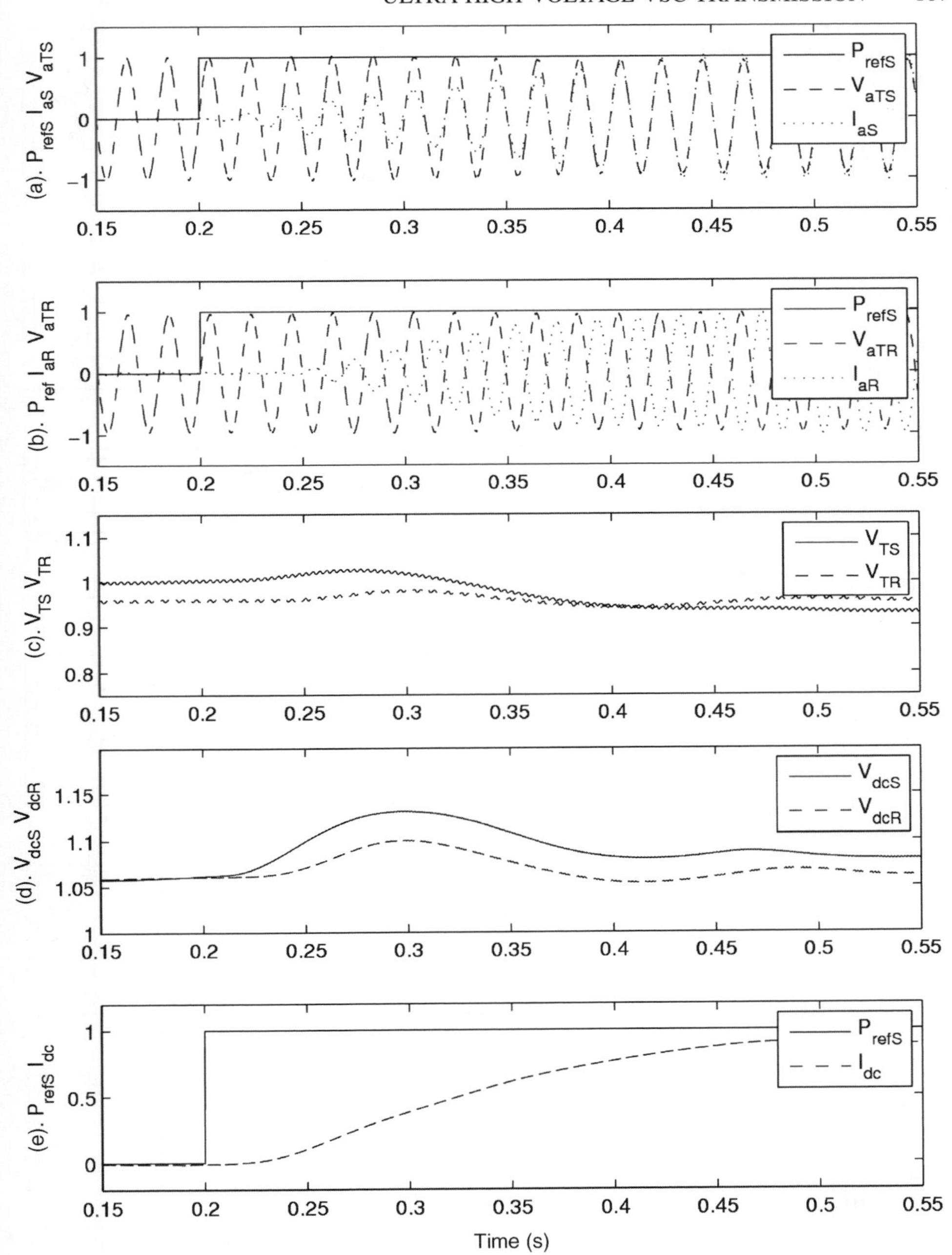

Figure 7.10 The voltage and current response to the 1 pu active power step order.

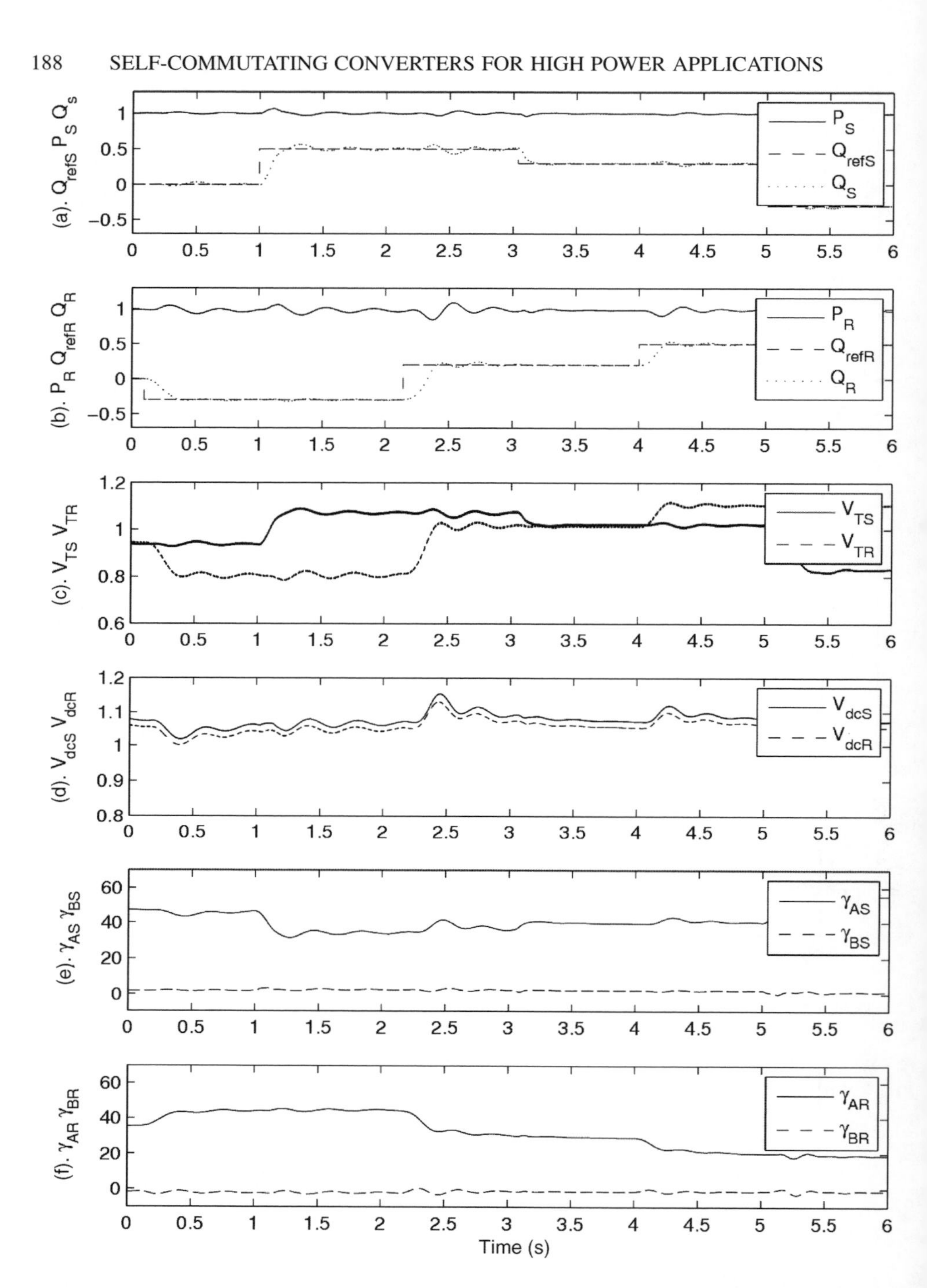

Figure 7.11 The reactive power responses during reactive power operation.

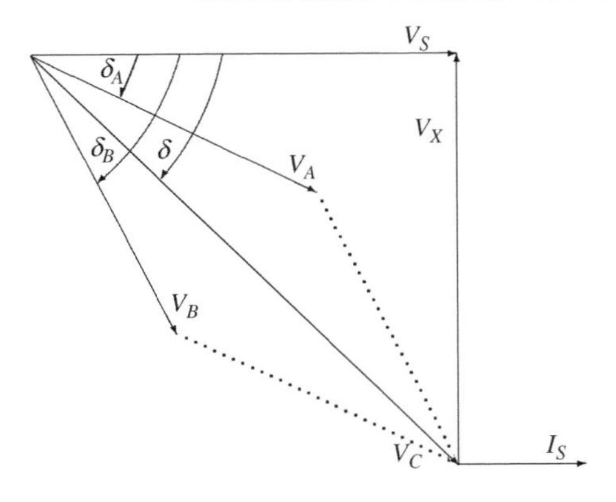

Figure 7.12 The phasor diagram of unity power factor operation.

Therefore, for a nominal power transfer of 1 or −1 pu, the converter terminal voltage is

$$V_{Trated} = \sqrt{V_S^2 + (XI_{Rated})^2} = \sqrt{V_S^2 + \left(\frac{V_S}{I_{SC}} I_{Rated}\right)^2} = V_S\sqrt{1 + \left(\frac{1}{SCR}\right)^2} \tag{7.20}$$

and the converter output voltage behind the leakage reactance of the interface transformer (represented by nominal reactance k_u) is

$$V_{Crated} = V_S\sqrt{1 + \left(\frac{1}{SCR} + k_u\right)^2} \tag{7.21}$$

Equation (7.21) indicates that the converter terminal voltage can vary from V_S by a small amount depending on the AC system short-circuit ratio. For instance for short-circuit ratio >3 the deviation is less than 5.4%. With a strong AC system and a small leakage reactance maintaining a unity power factor operation should have very little effect on the converter power rating. The use of a small leakage reactance is possible with the reinjection scheme because of the high quality of the voltage and current waveforms. This strategy requires the continuous detection of the displacement angle between voltage and current; however, this measurement is difficult to achieve accurately when the current level is very low and, under such an operating condition, the control system does not operate satisfactorily.

The control structure needs to be reliable and effective for all operating conditions, including a start from zero current; therefore, a zero imaginary control strategy needs to be used during low power transmission states. As the DC interconnection has to be able to control the power flow in both directions, the power angle can, therefore, be either 0° or 180°. The control action around the 0° or 180° is:

$$d\theta = \left[\frac{1}{\sqrt{P_S^2 + Q_S^2}} - \frac{Q_S^2}{\sqrt{(P_S^2 + Q_S^2)^3}} \right] dQ \quad \theta = 0^0$$

$$d\theta = -\left[\frac{1}{\sqrt{P_S^2 + Q_S^2}} - \frac{Q_S^2}{\sqrt{(P_S^2 + Q_S^2)^3}} \right] dQ \quad \theta = 180^0 \tag{7.22}$$

Therefore the reactive power control logic for unity power factor must take into account the power flow direction and level. The logic outputs shown in Figure 7.13 are used for power transfers from sending to receiving end (P_{S1}) and from receiving to sending end (P_{S2}) respectively; they operate based on a preset value (which is 0.05 pu in the simulation).

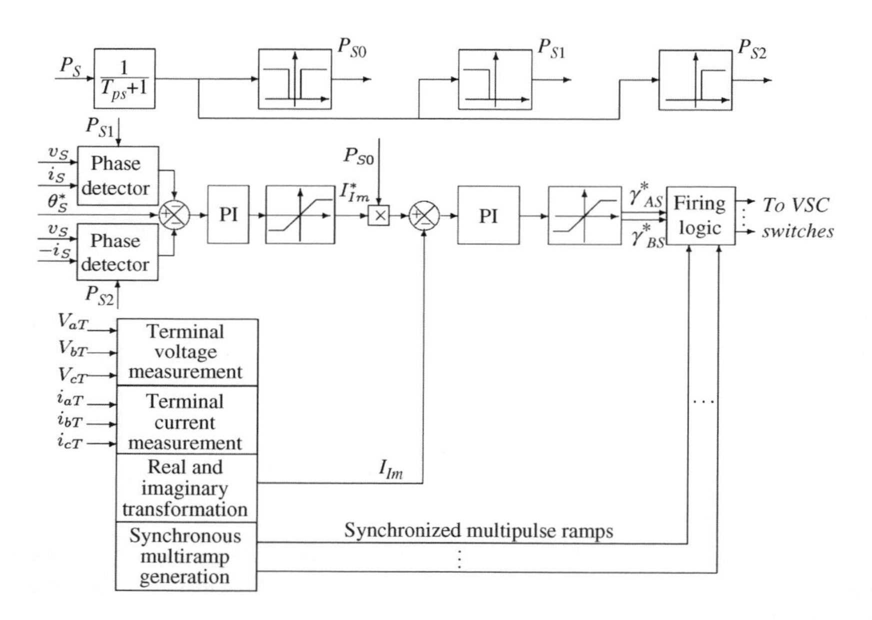

Figure 7.13 The reactive power control structure for unity power factor operation.

The logic unit P_{S0} decides whether to use zero imaginary current control or unity power factor control, whereas P_{S1} and P_{S2} decide which of the detected power angles is to be used as feedback signal to achieve unity power factor operation. For this purpose there are two power angle detectors, one measuring the angle between the source voltage (v_S) and the source current and the other between the source voltage and the reverse source current ($-i_S$). In both cases the measured power angle (θ_S) can be within $-180° \leq 180°$ for both power flow directions.

The difference between the detected power angle of the AC source and its zero setting is sent to a PI controller to determine the imaginary current order; if the active power absolute value is less than 0.05 pu, the order is forced to zero by the logic signal P_{S0} to provide a stable control signal and reduce the AC source current in order to minimize losses. The remaining part of the controller is the same as described in the previous section.

The single equivalent of Figure 7.6 is only available in the case of a generating station directly connected to the DC link. In the general case, information on V_S is not available at the converter location. The alternative, if the control is to be performed on the basis of terminal measurements only, is to force the line current to be in phase with the converter terminal voltage, a solution that provides minimum converter rating. In practice a compromise must be made between the extra converter rating needed to provide reactive power and the reduction of transmission losses. This requires carrying out extensive power flow simulation at the planning stage.

7.3.5.2 Constant terminal voltage control at the receiving end

When the converter is connected to a weak AC system, the use of a constant terminal voltage control strategy will benefit the power system considerably. If the converter terminal voltage is set to the AC source voltage, the reactive power consumed by the reactance between the AC system and converter will be shared equally by the AC system and converter; this will also reduce the reactive power consumption at low levels of active power transfers.

The control loop for constant terminal voltage shown in Figure 7.14 is very similar to the reactive power control loop described in the previous section. The main difference is the derivation of the imaginary current order, which in this case is generated from the difference between the converter terminal voltage and its set value.

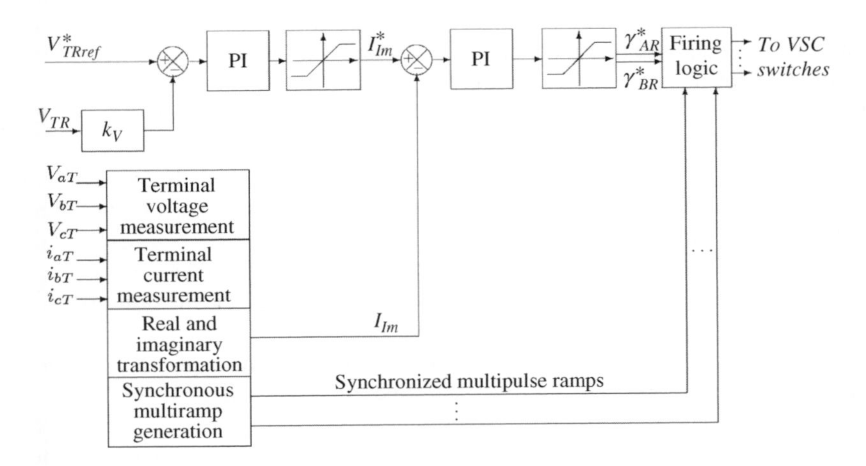

Figure 7.14 The reactive power control structure for constant terminal voltage operation.

7.3.5.3 Unity power factor and constant terminal voltage operation

The EMTDC simulation model contains four control loops. At the sending end, the active power control loop forces the HVDC link to transfer the specified active power and the reactive power control loop uses the detected power angle to keep the AC source operating at unity power factor. At the receiving end, the DC and AC voltage control loops maintain the converter terminal DC and AC voltages at present constant values for all the active power levels. This control strategy minimizes the AC current of the sending end and keeps the AC voltage of the receiving end converter terminal constant.

The high quality of the voltage and current waveforms of the MLVR scheme eliminates the need for harmonic filters and the flexible reactive power control capability enables the link to control the reactive power dynamically, without costly capacitors on the AC side.

To demonstrate the response of this strategy in the steady state and dynamic regions, tests are carried out with different power order settings as described below:

Case (i)

- From 0 to 0.05 s $P_{refS} = 0\,\text{pu}$.
- From 0.05 s to 1 s$P_{refS} = 0.3\,\text{pu}$.
- From 1 s to 2 s $P_{refS} = 0.6\,\text{pu}$.
- From 2 s to 3 s $P_{refS} = 1\,\text{pu}$.
- From 3 s to 4 s $P_{refS} = 0.6\,\text{pu}$.
- From 4 s to 5 s $P_{refS} = 0.3\,\text{pu}$.
- From 5 s to 6 s $P_{refS} = 0\,\text{pu}$.

Thee simulation results are shown in Figure 7.15. Figure 7.15(a) illustrates the active power order and power response measured at the sending end. As the latter is directly controlled, the response is smooth and no overshoot is observed. Figure 7.15(b) shows the active power response at the receiving end. As the latter is not directly controlled, but derived via terminal DC voltage constant control, the response to the various changes in active power has a small overshoot.

Figure 7.15(c) illustrates the sending end DC voltage and the converter terminal voltage. As the DC voltage is controlled at the receiving end (set at 1.05 pu), the sending end DC voltage varies with the active power level. In the dynamic regions the DC voltage varies to accelerate the changes in DC current along the transmission line; when this current reaches the set level, it returns to the steady state level, the maximum drift from the set value being 3%.

A similar situation occurs at the receiving end, where both the DC voltage and the converter terminal voltage are directly controlled (this is shown in Figure 7.15(d)). Except for a small variation in the dynamic region these two voltages are kept constant.
Figure 7.15(e) shows the sending end power factor and reactive power measured at the converter terminal. The graph indicates that the converter provides the required reactive power to maintain the sending end AC source operating at unity power factor. The power factor at the converter terminal is very close to unity, reducing slightly when the active power reaches the maximum (nominal) power. For the period between 5 and 6 s, where the active power has been set to zero, the value of the power factor angle of the AC source is not used as a control parameter, instead, the converter is being controlled by the zero imaginary current order.

Finally Figure 7.15(f) shows the receiving end power factor and reactive power measured at the converter terminal. It indicates that the converter provides the

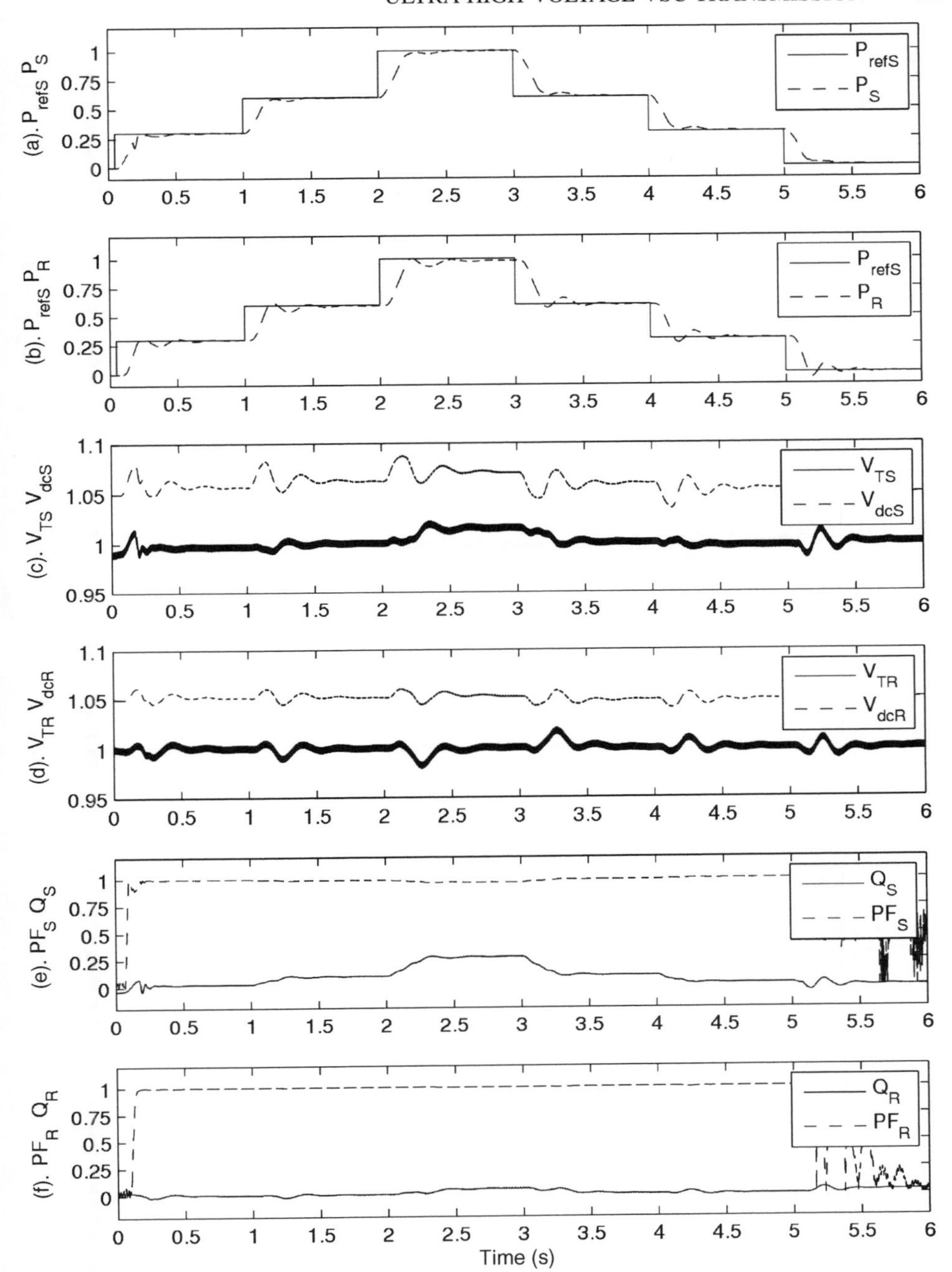

Figure 7.15 The dynamic response under the positive step active power order.

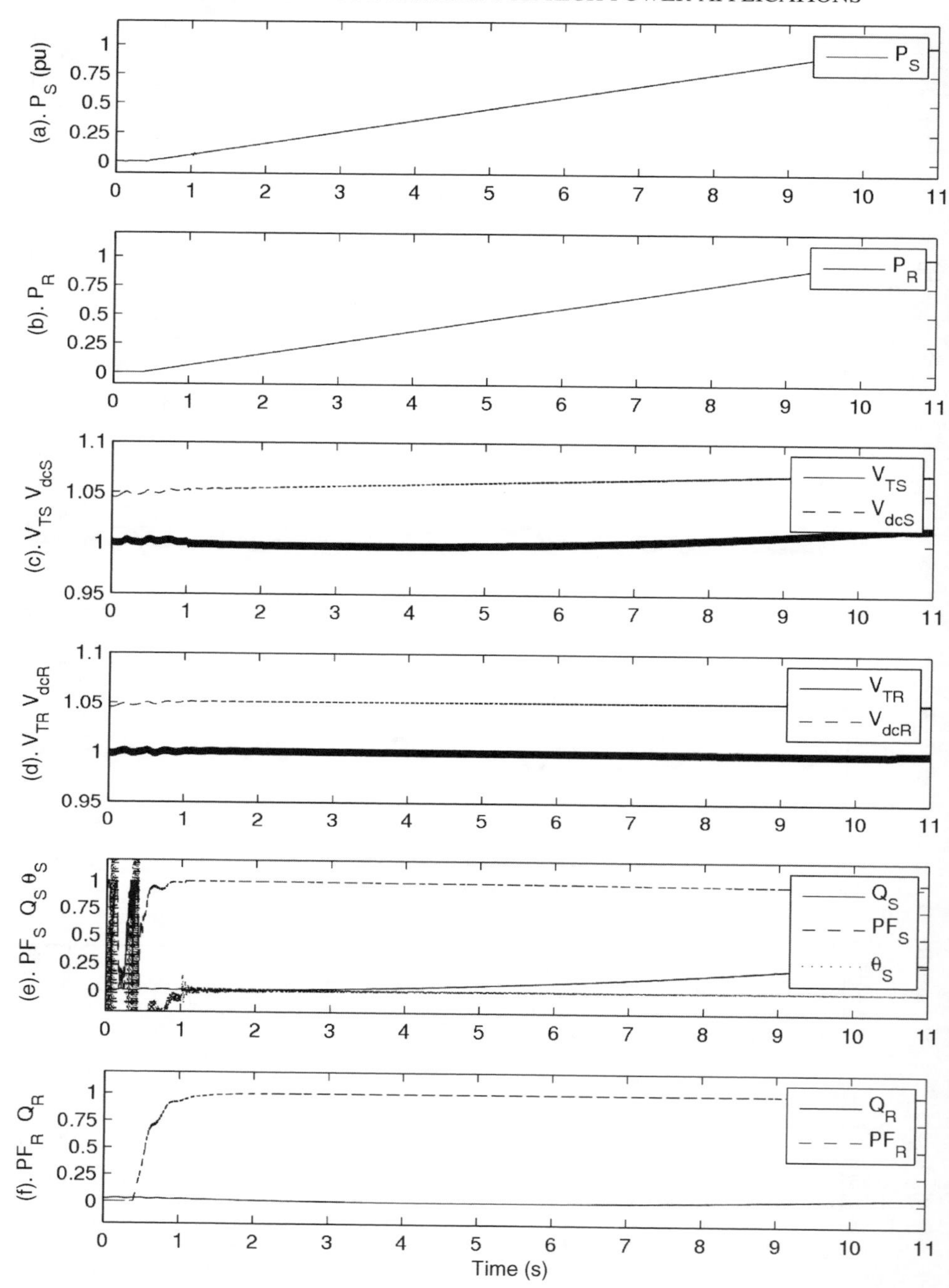

Figure 7.16 The dynamic response under the positive step active power order.

appropriate reactive power to maintain the terminal voltage constant. The power factor is very close to unity, reducing only slightly when the active power reaches 1 pu, as the converter has to generate more reactive power to maintain the terminal voltage constant.

Case (ii)

In this case the active power order is increased gradually from zero to 1 pu over a 10 s period and the simulation results, illustrated in Figure 7.16(a) and (b), indicate that the active powers at both ends follow the order very well.

Figure 7.16(c) and (d) show the converter terminal DC and AC voltages at the sending and receiving ends. Along with the gradual active power order changes (which start increasing at 0.5 s, reaching 1 pu at 10.5 s), the sending end converter terminal DC and AC voltages gradually increase to transfer the increasing power and generate appropriate reactive power to maintain the AC source operating at unity power factor. The DC and AC voltages at the receiving end are kept constant by their respective control loops.

Figure 7.16(e) shows the sending end converter power factor, the generated reactive power and the AC source power angle. In the region of low active power, the measured power angle of the sending end AC source is invalid (the sending end converter terminal is controlled to generate zero reactive power) when the active power level reaches 5% (under the unity power factor control strategy θ_S is kept at 0°). The sending end converter generates the required reactive power, which increases with the active power and when the latter reaches 1 pu the converter-generated reactive power is 0.28 pu. The converter terminal power factor keeps very close to unity, dropping only to 0.96 when the active power level is close to 1 pu.

Finally, Figure 7.16(f) shows the receiving end power factor and the reactive power generated required to maintain the terminal voltage constant. The converter contribution increases with the active power and when the latter reaches 1 pu the converter generates 0.07 pu of reactive power. The converter power factor is kept very close to unity except for the very low active power period.

7.4 Summary

A multilevel configuration based on the concept of cascaded H-bridge voltage reinjection has been shown to provide very effective conversion for HVDC transmission. In common with other reinjection configurations, the cascaded H-bridge solution removes the main problems of the existing PWM transmission technology, namely the large dv/dt across the converter valves and the high switching losses. The main converter valves also benefit from the soft switching condition naturally provided by the reinjection concept.

As compared with previous HVDC multilevel reinjection proposals, the cascaded H-bridge reinjection alternative permits the use of high levels because of the simplicity of the switching structure (which increases only in direct proportion to the level number) and its active DC capacitor balancing capability.

EMTDC simulation has shown that this type of MLVR provides a very good response to step changes in the active and reactive powers under a variety of control strategies.

References

1. Asplund, G., Eriksson, K. and Tollerz, O. (2000) Land and sea cable interconnections with hvdc light. *CEPSI Conference*, Manila, Philippines.
2. Asplund, G., Eriksson, K. and Svensson, K. (1997) DC transmission based on voltage source converters. *CIGRE SC-14 Colloquium*, Johannesburgh, S.A.
3. Iov, F., Sorensen, P., Hansen, A.D. and Blaabjerg, F. (2005) Modelling and control of VSC based DC connection for active stall wind farms to grid., *Proceedings of International Power Electronics Conference* Japan pp 1078–1084.
4. Mattsson, I., Ericsson, A., Railing, B.D., Miller, J.J., Williams, B., Moreau, G. and Clarke, C.D. (2004) Murraylink, the longest underground HVDC cable in the world. *CIGRE paper B4-103*, Paris.
5. Railing, B.D., Moreau, G. and Ronstrom, L. (2004) Cross Sound cable project second generation VSC technology for HVDC. *CIGRE paper B4-102*, Paris.
6. Dorn, J., Huang, H. and Retzmann, D. (2008) A new multi-level voltage-sourced converter topology for HVDC applications. *CIGRE 2008 Session*, paper B4-304.
7. Peng, F.Z., Lai, J.S., McKeever, J.W. and Van Coevering, J. (1996) A multilevel voltage source inverter with separate dc sources for static VAR generation. *IEEE Transactions on Industry Applications*, **32** (5), 1130–7.
8. Liu, Y.H., Arrillaga, J., and Watson, N.R. (2008) Cascaded H-bridge voltage reinjection – Part I: A new concept in multi-level voltage source conversion. *IEEE Transactions on Power Delivery*, **23** (2), 1175–82.
9. Liu, Y.H., Arrillaga, J. and Watson, N.R. (2008) Cascaded H-bridge voltage reinjection – Part II: Application to HVDC Transmission. *IEEE Transactions on Power Delivery*, **23** (2), 1200–1206.
10. Liu, Y.H., Arrillaga, J. and Watson, N.R. (2007) Addition of four-quadrant power controllability to multi-level VSC HVDC transmission. *IEE Proceedings on Generation Transmission & Distribution*, **1** (6), 872–8.

8

Ultra High-Voltage Self-Commutating CSC Transmission

8.1 Introduction

For almost 50 years HVDC has been used to transmit bulk power over long distances, the largest scheme being Itaipu, with nominal voltage and power ratings of ±600 kV and 6000 MW respectively. In the past decade India and China have built several schemes with ratings of up to ±500 kV and 3000 MW. The very large powers (of up to 6000 MW) and distances (typically 2000 km) of many DC links under consideration in China, India and Brazil, justify the need to increase the DC voltage level to ±800 kV. Although there is no operating experience at such voltage levels, the general opinion is that they do not represent an unreasonable risk and the manufacturers are ready for the task.

Two recent CIGRE contributions [1, 2] describe the research and development carried out so far and the issues requiring further 3attention, such as the HVDC system configuration, the AC system requirements, the interaction with the HVDC system and the testing levels to be used for the equipment involved.

There is no intention at present to deviate from the conventional robust thyristor technology, that, notwithstanding its dependence on the AC system for the valve commutations and the reactive power required, offers very reliable operating features, such as:

- fast recovery from DC line faults;
- ability to operate at reduced voltage when insulation is weakened by pollution or physical damage;
- monopolar operation during pole outages;

Self-Commutating Converters for High Power Applications J. Arrillaga, Y. H. Liu, N. R. Watson and N. J. Murray

- power modulation to stabilize the AC system;
- not responding to cascade outage crisis (typical in AC systems).

Potential problems with UHV are the effect of pollution on DC insulation, the fact that in the past transformer reliability has been below expectations, possible commutation failures and uncertainty in the expected harmonic levels.

It is only a question of time before line-commutated conversion is replaced by self-commutated conversion for UHV DC application. To try and help in that direction, this chapter analyses the applicability of the MLCR concept [3, 4] described in Chapter 4 to self-commutating CSC transmission.

8.2 MLCR-HVDC transmission

Multilevel current reinjection (MLCR) conversion has been shown in Chapter 4 to simplify the switching structure of earlier proposed multilevel configurations without losing any of their properties, and is likely to be the best self-commutating alternative for UHV DC transmission. This section describes the control and simulated performance of the basic MLCR-HVDC configuration.

8.2.1 Dynamic model

Figure 8.1 is a simplified diagram of an HVDC transmission system interconnecting two separate power systems represented by ideal sources (V_{S1}, V_{S2}), each in series with an impedance. The converter terminal voltages on the system side of the converter transformers are V_{T1} and V_{T2}. The AC output currents are specified by their real and imaginary components $(I_{\text{Re1}}, I_{\text{Im1}})$ and $(I_{\text{Re2}}, I_{\text{Im2}})$. The DC voltages at the converter terminals are V_{dc1} and V_{dc2} respectively and the common DC current I_{dc}.

The following formulation applies to the circuit of Figure 8.1:

$$P_1 = V_{dc1} I_{dc} = \frac{k_{v1} V_{T1} \cos\theta_1 A(\theta_1, \theta_2)}{L_m s + R} \tag{8.1}$$

$$\begin{aligned} Q_1 &= \sqrt{3} V_{T1} I_{T1} \sin\theta_1 = \sqrt{3} V_{T1} k_{i1} I_{dc} \sin\theta_1 \\ &= \frac{\sqrt{3} V_{T1} \sin\theta_1 k_{i1} A(\theta_1, \theta_2)}{L_m s + R} \end{aligned} \tag{8.2}$$

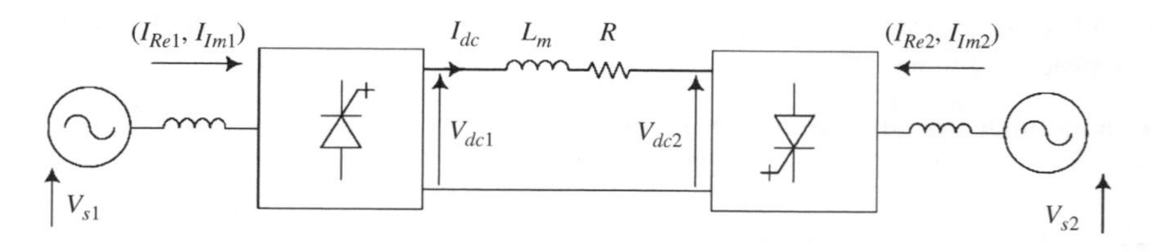

Figure 8.1 Simplified CSC HVDC transmission link.

$$P_2 = V_{dc2}I_{dc} = \frac{k_{v2}V_{T2}\cos\theta_2 A(\theta_1,\theta_2)}{L_m s + R} \tag{8.3}$$

$$\begin{aligned} Q_2 &= \sqrt{3}V_{T2}I_{T2}\sin\theta_2 = \sqrt{3}V_{T2}k_{i2}I_{dc}\sin\theta_2 \\ &= \frac{\sqrt{3}V_{T2}\sin\theta_2 k_{i2}A(\theta_1,\theta_2)}{L_m s + R} \end{aligned} \tag{8.4}$$

$$I_{\mathrm{Re1}} = \frac{k_{i1}\cos(\theta_1)A(\theta_1,\theta_2)}{L_m s + R} \tag{8.5}$$

$$I_{\mathrm{Im1}} = \frac{k_{i1}\sin(\theta_1)A(\theta_1,\theta_2)}{L_m s + R} \tag{8.6}$$

$$I_{\mathrm{Re2}} = \frac{k_{i2}\cos(\theta_2)A(\theta_1,\theta_2)}{L_m s + R} \tag{8.7}$$

$$I_{\mathrm{Im2}} = \frac{k_{i2}\sin(\theta_2)A(\theta_1,\theta_2)}{L_m s + R} \tag{8.8}$$

where
V_T is the fundamental *rms* value (phase to phase) of the converter terminal voltage.
θ is the phase angle difference between the AC current and converter terminal voltage (which, in the absence of commutation overlap is also the firing angle and the power factor angle).

$$A(\theta_1,\theta_2) = \left|k_{v1}V_{T1}\cos\theta_1 - k_{v2}V_{T2}\cos\theta_2\right| \tag{8.9}$$

P is the real power transfer at the converter terminals and Q the reactive power supplied to, or extracted from, the converter.

8.2.2 Control structure

Based on equations (8.1) to (8.8), Figure 8.2 shows a block diagram of the control system in terms of active and reactive powers. This diagram is modified in Figure 8.3 to make it more directly applicable to the control of the active and reactive current components.

As shown in these diagrams, the real and reactive powers vary with the DC current, I_{dc}, which in turn depends on the DC side voltages V_{dc1}, V_{dc2}. The DC voltage is a cosine function of θ, which can vary within the range $-180^0 \leq \theta \leq 180^0$. This makes the MLCR-CSC a very non-linear system. In practice, however, the DC voltages are kept within very narrow limits, such as shown in Figure 8.4. The sending end operates in a narrow band of $\theta_2 \leq \theta \leq \theta_1$ and the receiving end in the corresponding band of $\theta_4 \leq \theta \leq \theta_3$. Controlling the converter near unity power factor is difficult, as there is hardly any change in the cosine function in this area, which produces practically no change in the DC current.

Since the real and imaginary output current components at the sending and receiving ends of the link are interrelated, the real component of the sending end and the imaginary component of the receiving end can be set as the control parameters. Then the imaginary

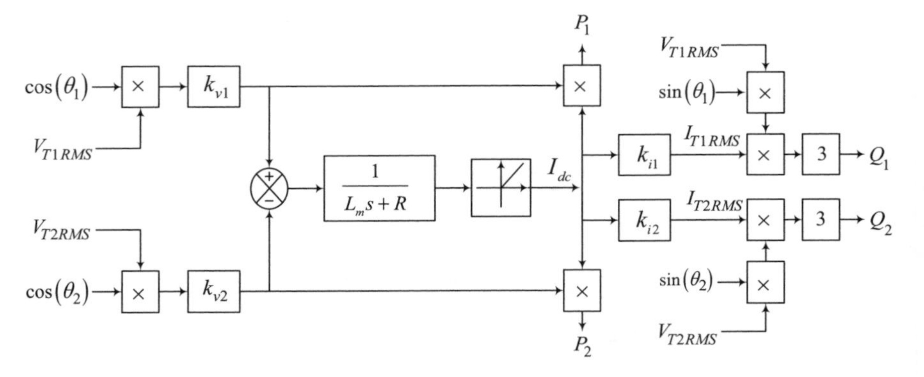

Figure 8.2 Block diagram for active and reactive power control.

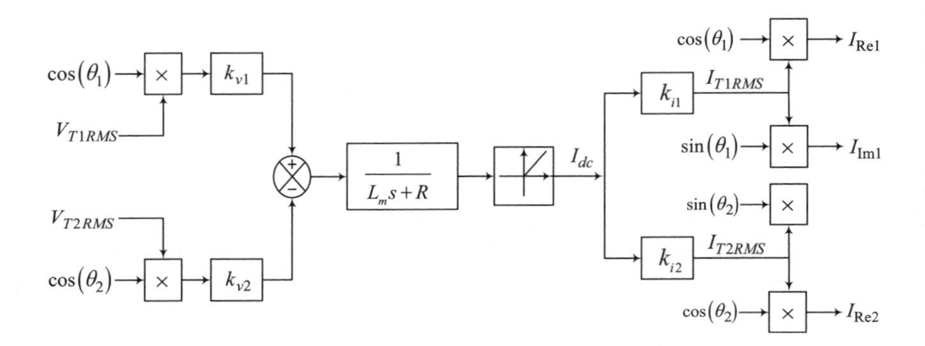

Figure 8.3 Block diagram for real and imaginary current control.

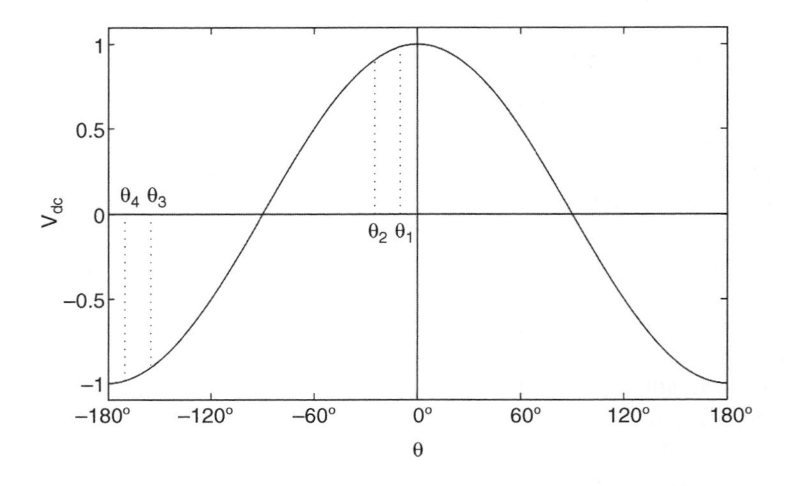

Figure 8.4 Operating regions of the converters at the sending and receiving ends.

component of the former and the real component of the latter are dependent on the operating state. The firing angles are placed on the negative side, enabling the two converters to supply some reactive power to their respective AC systems.

The control structure of Figure 8.5 shows that the measured output currents are transformed into their real and imaginary components using the measured source voltages as a reference. The latter are also used as a reference to synchronize the multipulse ramp signals sent to the converter valves firing logic.

The real P^* and reactive Q^* power references are divided by the source voltages to obtain the real and imaginary current orders. Finally, using the real and imaginary current errors, the PI controllers derive the $\Delta\theta_1, \Delta\theta_2$ signals to be added to the $-15°$ and $-165°$ settings to generate the firing instants to be sent to the CSC firing logic.

The direct current control characteristic of the CSC schemes permits the link to operate safely during normal and abnormal conditions. The AC side current can be controlled symmetrically under normal and AC fault conditions (even after losing one phase) and, therefore, MLCR-HVDC provides high reliability and fast recovery times.

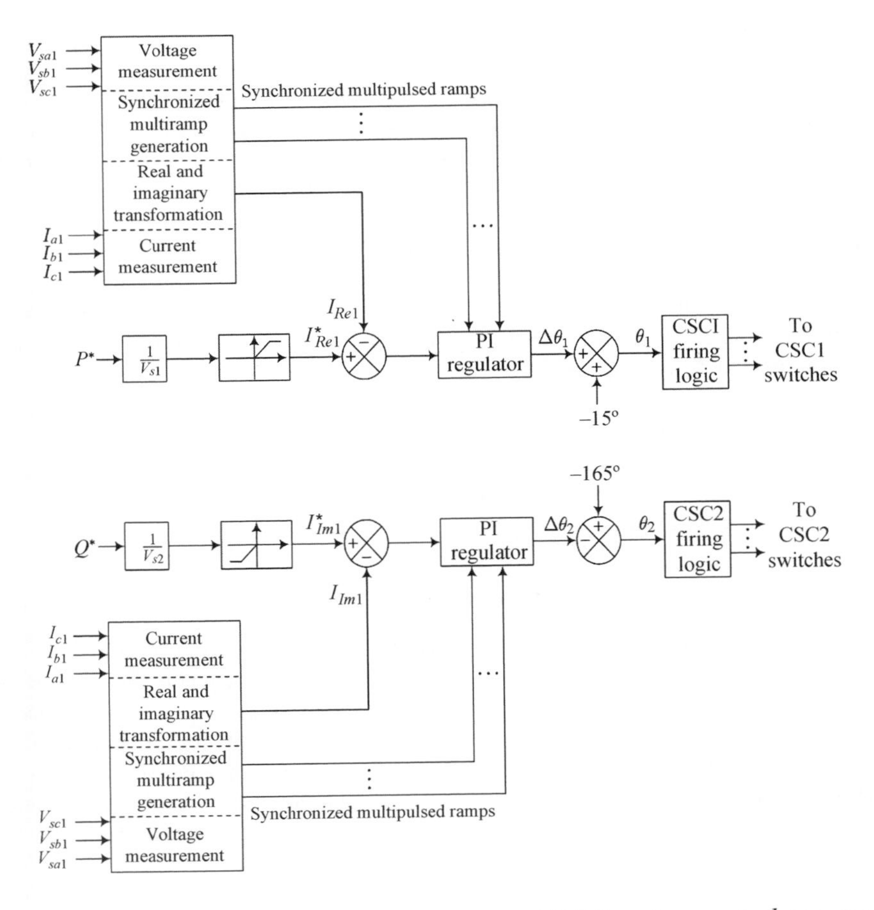

Figure 8.5 The MLCR current source converter HVDC system control structure.

8.3 Simulated performance under normal operation

The test system used for the EMTDC simulation consists of a two-terminal HVDC link, based on the simplified diagram of Figure 8.1, with each of the AC systems represented by a 500 kV (phase-to-phase) voltage source in series with a reactance calculated to give a short-circuit ratio of 2.5 for a DC power rating of 1000 MW. The DC line is modelled by a smoothing reactance of 2 H in series with a 5 Ω resistance.

8.3.1 Response to active power changes

The simulated dynamic response to step variations in the active power is demonstrated in Figure 8.6, which includes the following information:

- P_{ref} – real power order.
- Q_{ref}– reactive power order.
- P_1, P_2 – active power at the sending and receiving ends.
- Q_1, Q_2 – reactive power at the sending and receiving ends.

Initially the link is operating under a real power order of 1000 MW at the sending station and a reactive power order of −800 MVAr at the receiving station (i.e. generating 800 MVAr). After 100 ms the real power order is changed to 1300 MW, and at 250 ms the power order is returned to the original setting. In each case the results (Figure 8.6(a) and (b)) show that the system reaches a new steady state condition after approximately 50 ms, with P_1 experiencing an overshoot of about 30% of the step change, while P_2 shows no overshoot.

The effect of these changes on the reactive powers (Figure 8.6(c) and (d)) show a larger disturbance at the sending end (the power controlling station). Upon reaching the new steady state condition (after 50 ms) the reactive power at the sending station settles at a lower level (−500 instead of the original −685 MVAr according to Figure 8.6(c)). This drop in the supply of reactive power causes a corresponding reduction in the sending end system voltage (from 551.5 to 528.5 kV, or 4%). The voltage variation will, of course, depend on the magnitude of the active power change and the converter short-circuit ratio. In general, therefore, the assistance of on-load tap change (OLTC) may still be needed to keep the voltage within specified limits.

8.3.2 Response to reactive power changes

Figure 8.7 illustrates the response of the link to step changes in reactive power at the receiving station, while maintaining the active power setting constant. Initially Q is set at −800 MVAr, after 100 ms its value is reduced to −600 MVAr and at 250 ms it is returned to its original level. Again, Figure 8.7(a) and (b) show that it takes approximately 50 ms for the system to reach a new steady sate, and that Q_2 experiences a 25% overshoot whereas no overshoot is observed in Q_1. In this case the sending end station suffers a large reduction in the reactive power injection (from −685 down to −450 MVAr according to Figure 8.7(b)), thereby causing the ac system voltage to drop by about 3.7% (from 551.5 to 531 kV).

Figure 8.8 illustrates the result of incorporating OLTC control following a changes in the power setting. To this effect, again the receiving system is subjected to a reduction in the reactive

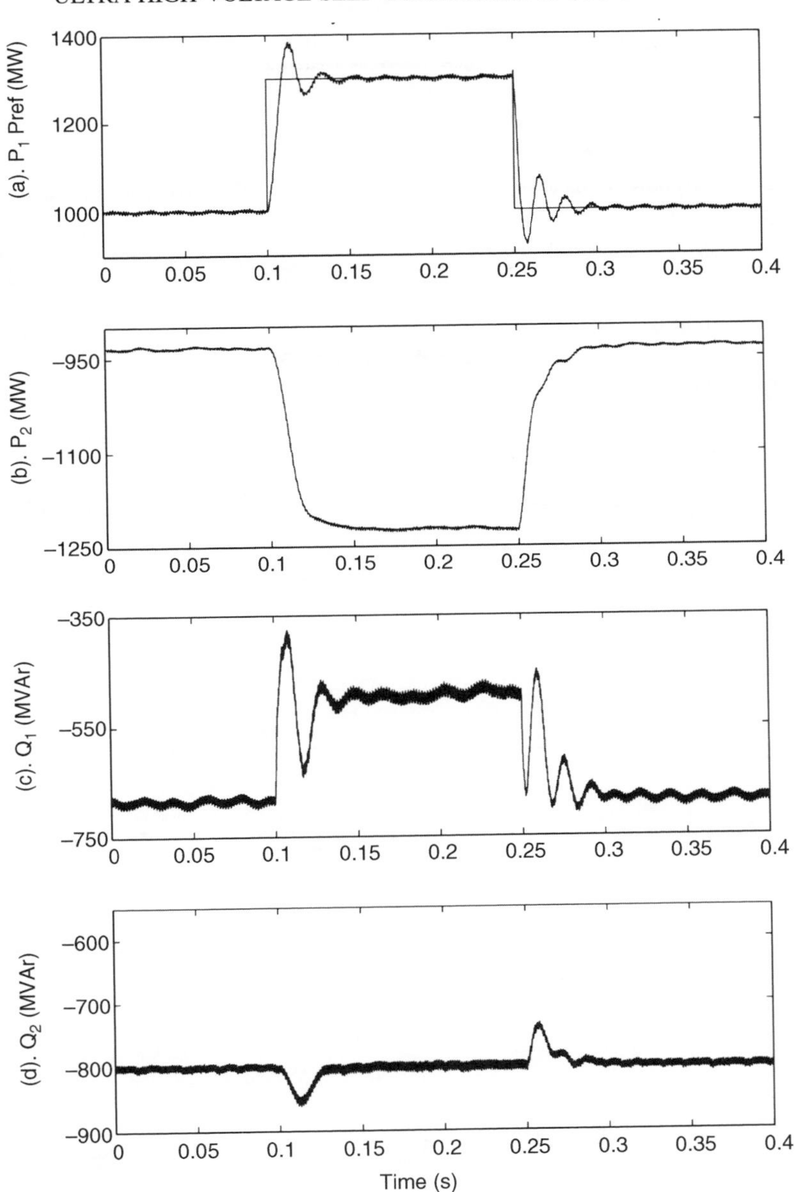

Figure 8.6 Real and reactive power response to a real power order.

power setting from the initial −800 to −600 MVAr as before, but without returning it to its original value. Following the step change at 100 ms, Figure 8.8(f) shows that the terminal voltage at the sending end (which controls the active power) drops down to $531kV$. However, in this case the transformer tap setting is changed from 1 to 0.925 after 250 ms (see Figure 8.8(e)). It is clear from Figure 8.8(f) that the terminal voltage is restored to its original value.

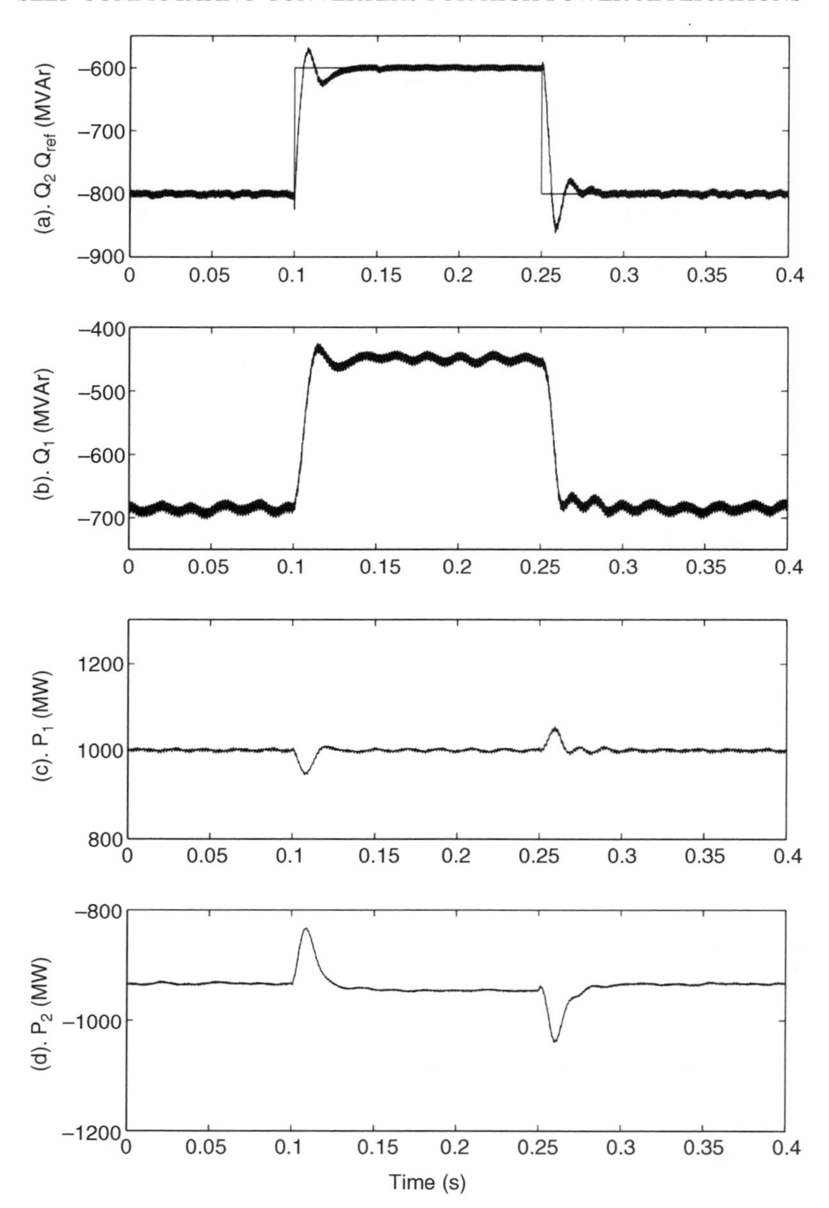

Figure 8.7 Real and reactive power response to a reactive power order.

8.4 Simulated performance following disturbances

8.4.1 Response to an AC system fault

Figure 8.9 illustrates the response of the MLCR-HVDC system to a three-phase to ground short circuit applied directly at the terminals of the receiving converter terminal, while the link is

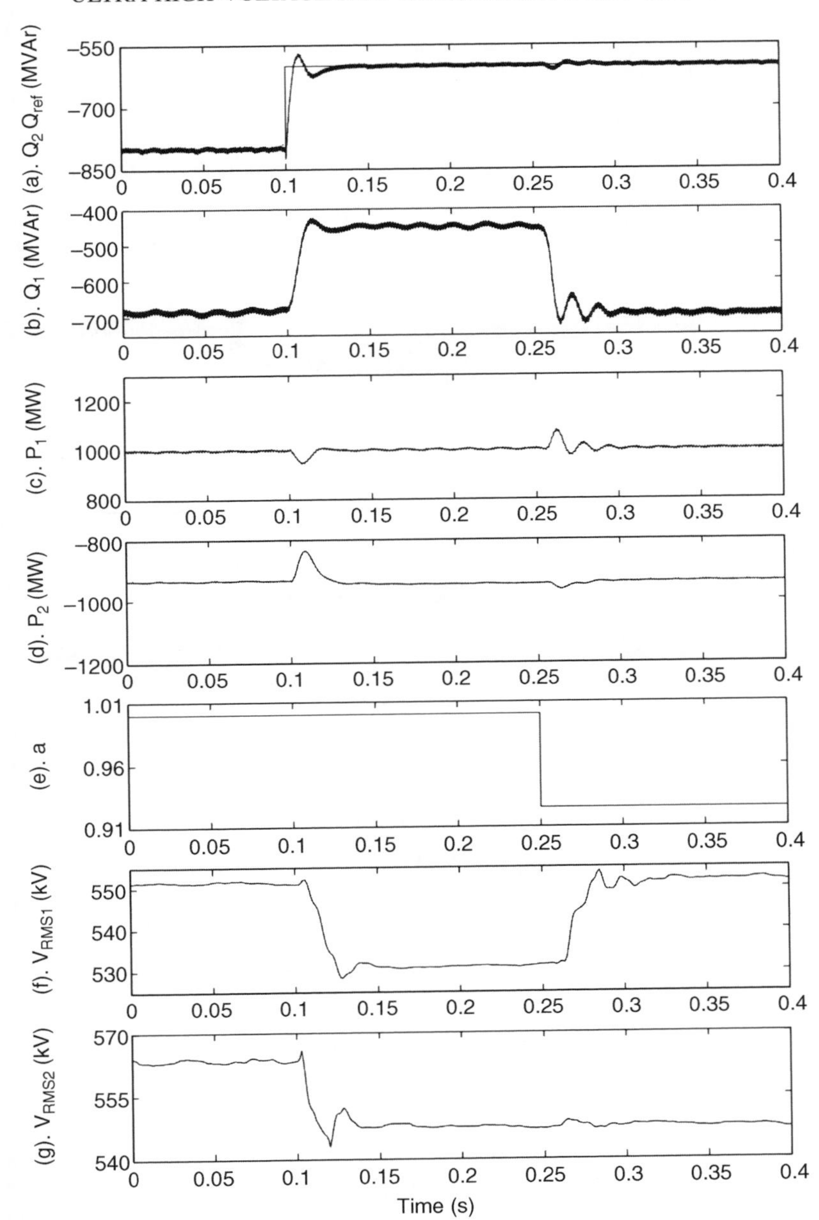

Figure 8.8 The effect of OLTC on the terminal voltage for a change in the reactive power order.

initially operating as specified in section 8.3. In the simulation the fault is assumed to be detected at the instant 100 ms and cleared (by the AC circuit breakers) after another 100 ms. The collapse of the receiving end voltage initially causes an increase in the DC current, which soon reaches the maximum setting and, therefore, changes the sending end control from

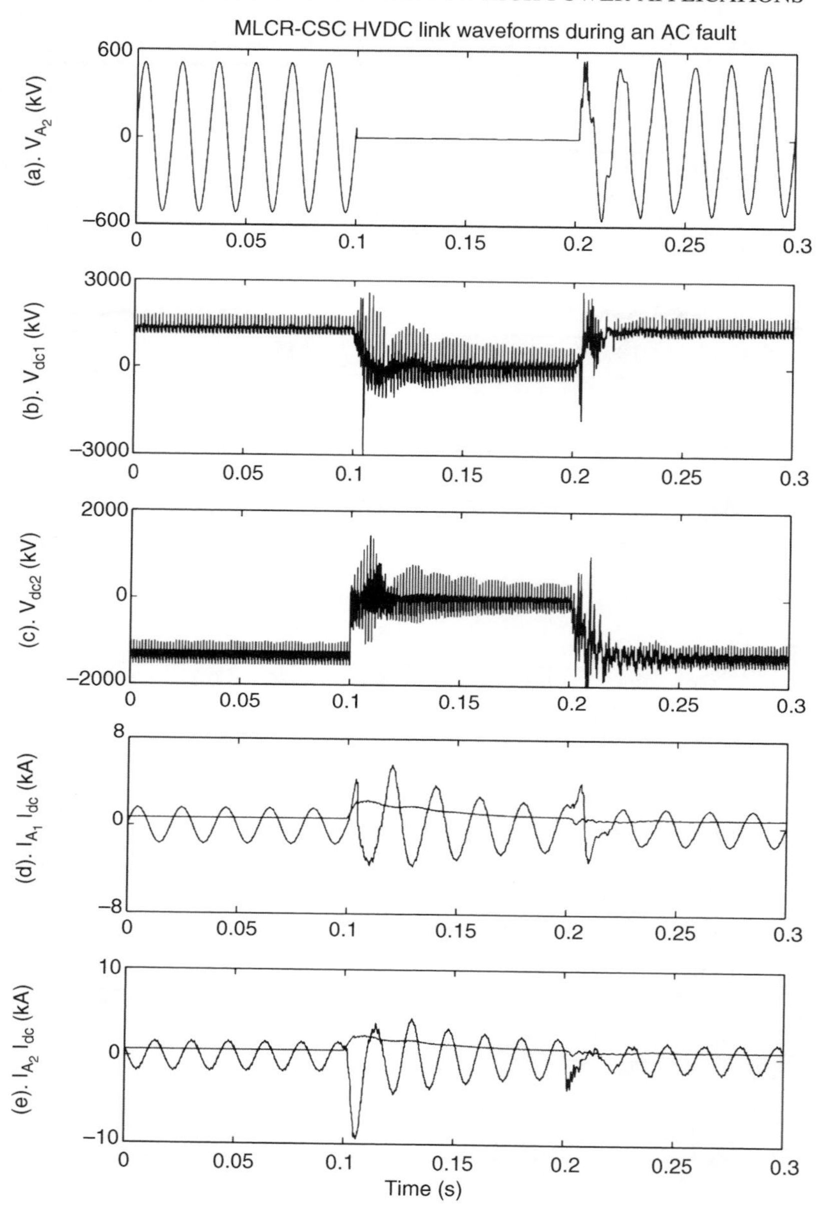

Figure 8.9 Response of MLCR HVDC to an AC side fault.

constant power to constant current. No further corrective action is required, as the link continues to carry practically normal current during the fault and will be ready to resume normal operation soon after the fault is cleared.

Figure 8.9(c) and (d) show that there is a 200% overshoot in the DC current and corresponding increases in the AC current for about 20 ms. Following fault clearance, at

200 ms, the system takes approximately a further 50 ms to regain the pre-fault operating conditions.

8.4.2 Response to a DC system fault

A bipolar short circuit is placed in the mid-point of the DC line when the system is operating with a real power order of 1000 MW controlled at the sending end and with a reactive power order of −800 MVAr at the receiving end. The fault initially increases the DC current to try and maintain the specified power flow. However, in the process, the maximum current limit is reached and the converter goes into constant current control. Similarly the inverter end current reduces and is kept at its minimum current setting. Therefore a small current (the difference between the rectifier and inverter current settings) will continue to flow at the fault point. Under those conditions the DC fault will not be self-clearing.

Instead, upon detection of the fault assumed to occur at the 100 ms point, the sending end converter is temporarily ordered to go into inversion. This action clears the energy stored in the DC system faster and, thus, after a time allowed for the arc deionization (arbitrarily set at 50 ms in the simulation), normal control operation resumes. The result of the simulation, shown in Figure 8.10, indicates that normal operating conditions are re-established after 100 ms from the instant of fault detection.

8.5 Provision of independent reactive power control [5]

Self-commutating thyristor-based HVDC transmission can be made more attractive if the converter power factor can be controlled independently from the DC voltage or current level. Such addition should provide the thyristor-based technology with the level of flexibility of PWM conversion. A multigroup firing-shift concept that performs that task is described below.

The exchange of reactive power between a current source converter and the AC system is determined by the sine of the firing angle (α). Altering α has an immediate effect on the DC voltage level and, thus, to maintain the specified DC power transfer through the link, a corresponding change of firing angle must be made at the other end, which in turn affects its reactive power exchange with the AC system. Therefore, under conventional converter control, the reactive powers injected at the two ends of a multilevel CSC link are interdependent.

The output current waveform produced by a current source converter consisting of two or more 12-pulse self-commutating groups is not altered when a shift is introduced between the firings of the constituent groups of the converter station. In multilevel CSC HVDC interconnections with at least two 12-pulse groups per terminal (such as shown in Figure 8.11) the same current waveform is produced by each of the 12-pulse converter groups, and thus the total output current waveform remains the same if a phase shift is introduced between the firings of the two groups constituting the converter station.

When a change of operating conditions at the receiving end demands more reactive power from the converter, and thus reduces the DC voltage, shifting the firings of the two sending end converter groups in opposite directions provides the required DC voltage reduction, while maintaining the reactive power constant (due to the opposite polarity of the two firing angle corrections). A relatively small change of active power will be caused by the variation of the

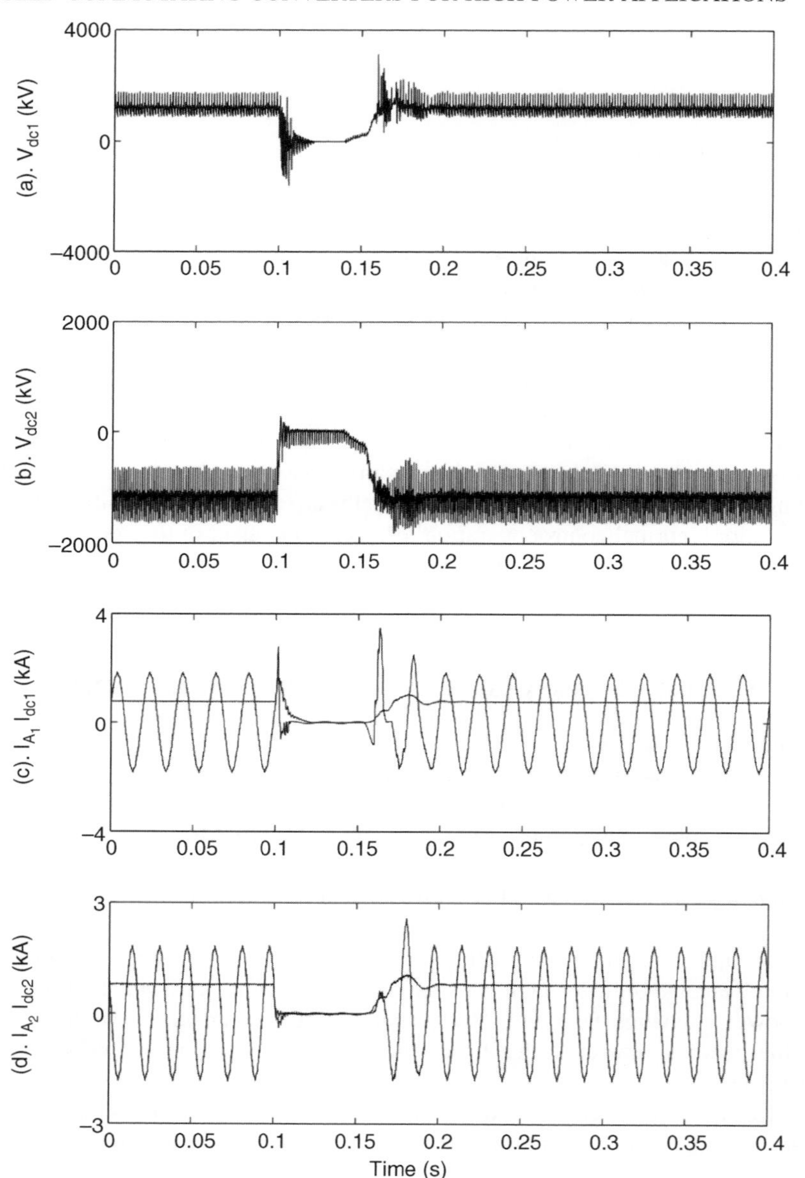

Figure 8.10 Response of MLCR HVDC to a DC side fault.

fundamental current produced by the shift, but this change can be compensated for by a small extra correction of the two firing angles.

For a converter to operate in the firing-shift mode (which in the above example is the sending end converter), the firing instants of one group (say group A) is kept on the positive side (thus providing reactive power), while the second group (say group B) may act as a source or sink of reactive power (i.e. the firing angle may be positive or negative).

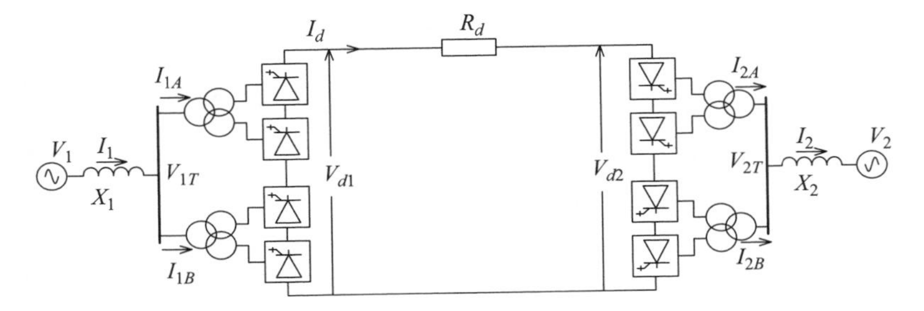

Figure 8.11 Simplified diagram of a DC link connecting two AC systems.

8.5.1 Steady state operation

To simplify the explanation of the steady state characteristics, the sending end operates under firing shift control, while the receiving end uses a common firing angle for the two converter groups. The generalized method with firing shift at both ends will be used in the dynamic simulation. Also, as shown in Figure 8.11, the interconnected AC systems are represented as simplified Thevenin equivalents, i.e. V_1, X_1 and V_2, X_2. At the receiving end the control specifications are the terminal AC voltage (V_1^{sp}) and the DC voltage (V_d^{sp}), while those at the sending end are the DC power transfer (P_{d1}^{sp}) and the reactive power (Q_1^{sp}).

8.5.1.1 Receiving end

If at this end no firing shift control is exercised, the firing angle will be the same for the two converter groups (i.e. $\alpha_{2A} = \alpha_{2B} = \alpha_2$). The dotted lines in the phasor diagram of Figure 8.12 represent the initial operating condition (with a Thevenin impedance of X_2) and the continuous line a new operating condition with a larger Thevenin impedance ($X_2^{(1)}$).

In both cases the system reactive power requirements are equally shared between the converter and the AC source and the firing angle is common to the two converter groups. The condition is represented by the following equations:

$$I_{2A} = I_{2B} = k_m I_d \tag{8.10}$$

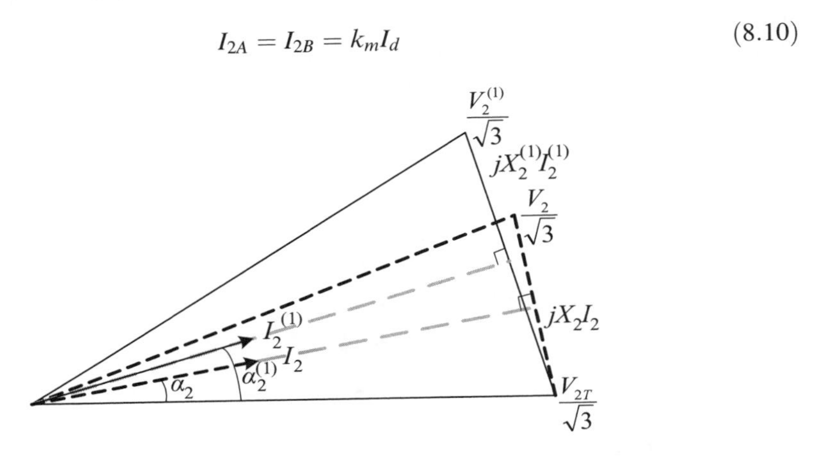

Figure 8.12 Operating conditions at the receiving end for two different system strengths.

or between the double converter AC and DC currents

$$I_2 = 2k_m I_d \tag{8.10a}$$

and

$$(V_2/\sqrt{3})^2 - (V_{T2}/\sqrt{3})^2 = X_2^2 I_2^2 - 2(V_{T2}/\sqrt{3})X_2 I_2 \sin(\alpha_2) \tag{8.11}$$

$$V_{d2} = 4(3\sqrt{2}/\pi)V_{T2}\cos(\alpha_2) \tag{8.12}$$

In terms of the specified power (which is normally controlled at the sending end) the DC voltages across the link are related by the expression

$$V_{d1} = V_{d2} + R_d \frac{P_{d1}^{sp}}{V_{d1}}$$

or

$$V_{d1} = \frac{V_{d2} + \sqrt{V_{d2}^2 + 4R_d P_{d1}^{sp}}}{2} \tag{8.13}$$

The solution of equations (8.10a) to (8.13) provides the initial values of $\alpha_2, V_{d2}, V_{d1}, I_2$.

8.5.1.2 Sending end

Figure 8.13 illustrates the two operating conditions in response to the change at the receiving end. To demonstrate the phase shift control principle, initially the sending end is set with one firing angle positive (α_{1A}) and one negative (α_{1B}). When a change in operating conditions occurs, represented by an increase in the Thevenin impedance and thus a change of firing angle at the receiving end, the sending end must compensate by an increase in firing angle ($\alpha_{1A}^{(1)}$ and $\alpha_{1B}^{(1)}$ in Figure 8.13) to maintain the specified active power transfer.

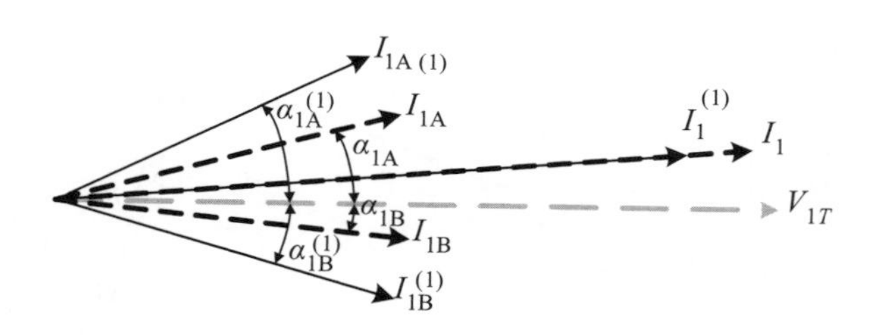

Figure 8.13 Firing shift control maintaining constant reactive power at the sending end for two different system strengths.

The following relationships apply to the sending end:

$$V_{d1} = 2\left(3\frac{\sqrt{2}}{\pi}\right)V_{1T}(\cos(\alpha_{1A}) + \cos(\alpha_{1B})) \tag{8.14}$$

$$I_{1A} = k_m I_d \tag{8.15}$$

and

$$I_1 = \sqrt{(I_{1A}\cos(\alpha_{1A}) + I_{1B}\cos(\alpha_{1B}))^2 + (I_{1A}\sin\alpha_{1A}) + I_{1B}\sin(\alpha_{1B}))^2}$$

or, because $|I_{1A}| = |I_{1B}|$ due to the series connection

$$I_1 = \sqrt{2}I_{1A}\sqrt{1+\cos(\alpha_{1A}-\alpha_{1B})} = 2I_{1A}\cos\left(\frac{(\alpha_{1A}-\alpha_{1B})}{2}\right) \tag{8.16}$$

The active power P_1 is equal to the specified DC power P_{d1}^{sp}

$$P_1 = V_{1T}I_1(\cos(\alpha_{1A}) + \cos(\alpha_{1B})) \tag{8.17}$$

and the reactive power

$$Q_1 = V_{1T}I_1(\sin(\alpha_{1A}) + \sin(\alpha_{1B})) \tag{8.18}$$

With V_1, V_{1T} and P_1 specified, the unknown variables are $V_{d1}, I_{1A}, I_1, \alpha_{1A}, \alpha_{1B}, Q_1$, which can be derived from the simultaneous solution of equations (8.13) to (8.18).

In the steady state, the values of α_{1A}, α_{1B}, and thus the internal reactive power circulation between the two converter groups, can be reduced by the use of transformer on-load tap change.

8.5.2 Control structure

For complete flexibility the sending end needs to control real and reactive power and the receiving end needs to keep the converter DC voltage constant (so as to minimize the DC current for a given real power setting) and control the reactive power. With reactive power control at both ends, the controllers can easily be configured for optimum power transfer at the system level depending on operating objectives, which usually involves providing constant power factor at the sending end and constant AC terminal voltage at the receiving end.

In order to control the real and reactive power over the complete operating range, the converter response needs to be linear. Standard PID controllers are unsuitable for this application as their gain is static, and although they may give suitable performance over a narrow band, the latter is not acceptable over the complete range. This is explained in more detail later.

Figure 8.14 illustrates the control ranges of the real and reactive power responses for α values of $\pm 90°$. It is clear that these controller surfaces are very non-linear, and it is not hard to understand why the use of a linear PID controller would be unsuitable.

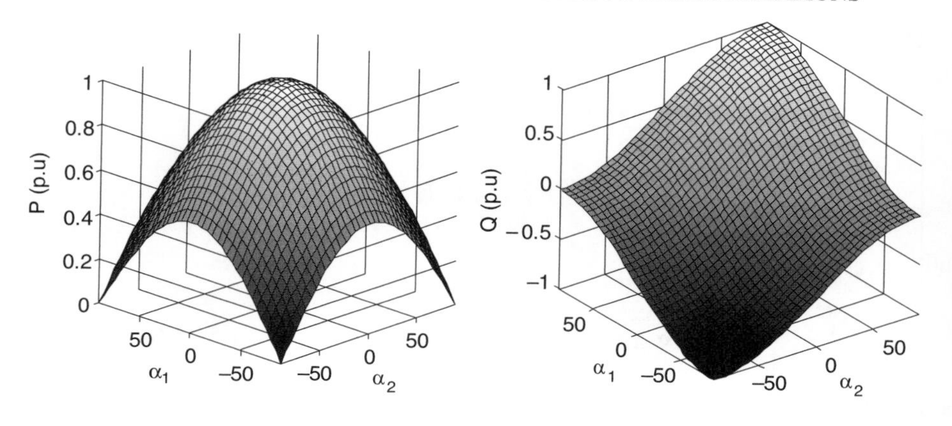

Figure 8.14 Calculated real and reactive power for varied firing angle (±90°).

Given the above controller surfaces, it is difficult to visualize how the controller must perform, especially since the controller firing angles are expected to operate equally well in the positive and negative regions. What is needed is a controller that operates for all combinations of P and Q without the need to manually switch controller gains and control actions. An example of four controller operating conditions is shown in Figure 8.15.

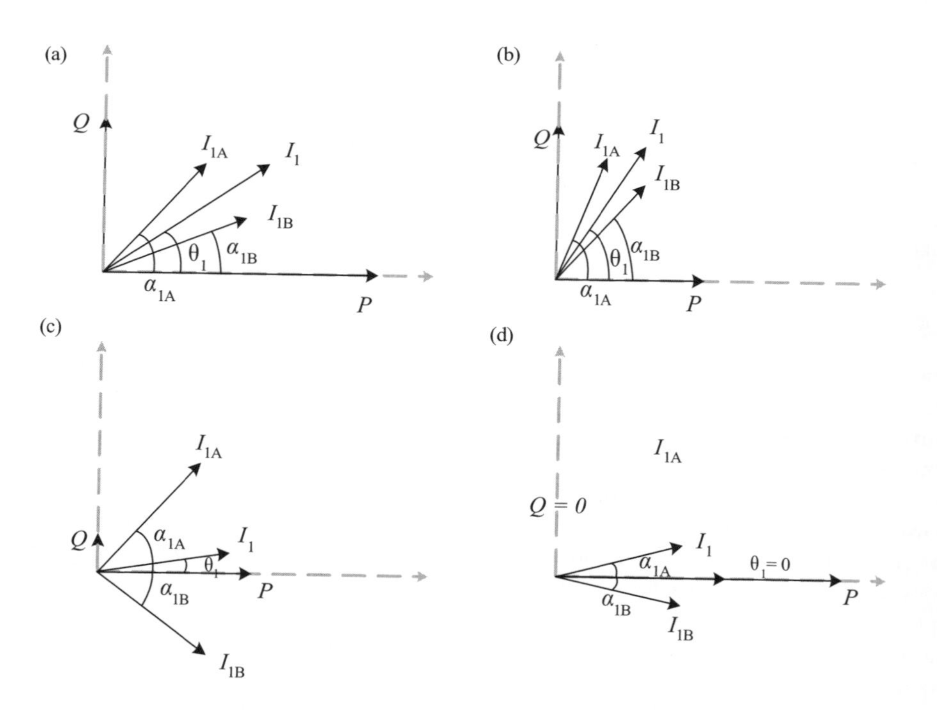

Figure 8.15 Firing shift control providing: (a) large P and *Q; (b) large Q and smaller P; (c) small P and Q; (d) large P, no Q.*

These diagrams show that the controller is expected to operate over a wide range of conditions and that the change in firing angle has the greatest influence on the real power near the Y axis and on the reactive power near the X axis. This is better explained by examining the real and reactive power contribution of one converter in isolation. The real power transferred by the converter depends on the cosine of the firing angle (α_1) while the reactive power depends on the sine of (α_1). What is of interest to control system designers is the rate of change of the controlled outputs P and Q, as this determines the level of gain (or sensitivity) in system response. Basic differentiation reveals that the rate of change is proportional to $-\sin(\alpha_1)$ for real power and $\cos(\alpha_1)$ for reactive power, making this a very strong non-linear system.

As mentioned earlier, conventional controller operation is confined to a relatively small range and functions with a fixed gain, thereby assuming that the system is linear over the small range. This control philosophy becomes even less suitable when we consider that an ideal independent and fully flexible controller should be able to provide a combination of α_1 and α_2 that satisfies the requirements of both P and Q simultaneously. Figure 8.16 illustrates a simplified block diagram of what the controller must achieve, the goal being a mapping function that translates P and Q into α_1 and α_2, to make the non-linear converter appear linear.

The only information representing the behaviour of the converter system is given by the steady state equations (8.8) and (8.9), which is sufficient initially, because they show the influence that $\Delta\alpha_1$ and $\Delta\alpha_2$ have on the output variables P and Q.

The rate of change of the output variables with respect to firing angles α_1 and α_2 can be expressed as:

$$\Delta P = \frac{\partial P}{\partial \alpha_1}\Delta\alpha_1 + \frac{\partial P}{\partial \alpha_2}\Delta\alpha_2 \tag{8.19}$$

$$\Delta Q = \frac{\partial Q}{\partial \alpha_1}\Delta\alpha_1 + \frac{\partial Q}{\partial \alpha_2}\Delta\alpha_2 \tag{8.20}$$

These are found by differentiating the steady state equations (8.17) and (8.18), i.e.

$$\frac{\partial \mathrm{P}}{\partial \alpha_1} = -3V_T I_1 \sin(\alpha_1) \tag{8.21}$$

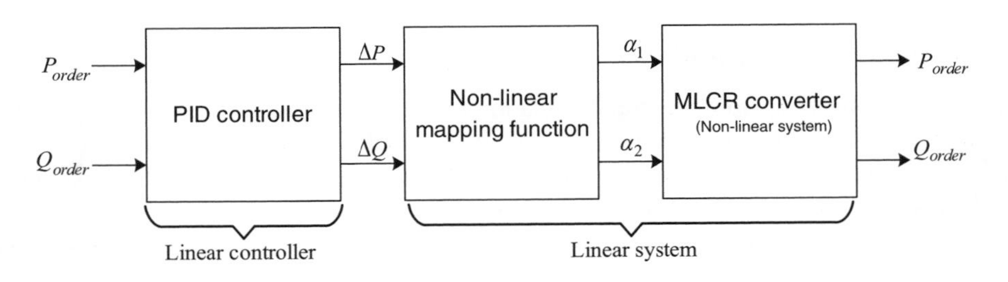

Figure 8.16 Block diagram of the non-linear system control objective.

$$\frac{\partial \mathrm{P}}{\partial \alpha_2} = -3V_T I_1 \sin(\alpha_2) \tag{8.22}$$

$$\frac{\partial \mathrm{Q}}{\partial \alpha_1} = 3V_T I_1 \cos(\alpha_1) \tag{8.23}$$

$$\frac{\partial \mathrm{Q}}{\partial \alpha_2} = 3V_T I_1 \cos(\alpha_2) \tag{8.24}$$

Expressing the dynamic system in matrix form

$$\begin{bmatrix} \Delta P \\ \Delta Q \end{bmatrix} = \underbrace{\begin{bmatrix} \dfrac{\partial \mathrm{P}}{\partial \alpha_1} & \dfrac{\partial \mathrm{P}}{\partial \alpha_2} \\ \dfrac{\partial \mathrm{Q}}{\partial \alpha_1} & \dfrac{\partial Q}{\partial \alpha_2} \end{bmatrix}}_{A} \begin{bmatrix} \Delta \alpha_1 \\ \Delta \alpha_2 \end{bmatrix} \tag{8.25}$$

Equation (8.25) can be solved using basic matrix theory.

Using the partial differentials of the steady state equations to P and Q in matrix A, it is possible to model the converter system's transient response (but not the system state). If the matrix is non-singular, its inverse can be used to linearize the converter system behaviour. The inverse of matrix A, with the common gain component grouped on the left side, becomes:

$$A^{-1} = \frac{1}{3V_T I_1 \sin(\alpha_1 - \alpha_2)} \begin{bmatrix} -\cos(\alpha_2) & -\sin(\alpha_2) \\ \cos(\alpha_1) & \sin(\alpha_1) \end{bmatrix} \tag{8.26}$$

This equation indicates that the overall system gain depends on the difference between the two firing angles ($\sin(\alpha_1 - \alpha_2)$) and the contribution of (for P) real power on the other group's firing angle, and (for Q) reactive power contribution on the other group's firing angle. While making sense in theory, this needs to be realized in practice.

Examining the system on an incremental basis (i.e. from $\alpha_1 \rightarrow \alpha_1 + \delta\alpha_1$), as the difference ($\delta\alpha_1$) is reduced the accuracy is increased, becoming very close to the continuous integral equivalent. It could be argued that in each partial differential equation the effect of $\Delta\alpha_1$ on $\Delta\alpha_2$ and vice versa is not fully captured, but in a practical system this effect can be minimized with suitable feedback.

8.5.2.1 Practical implementation

Figures 8.17 and 8.18 show the implementation of the theory into a real system controller.

In Figure 8.17, the controller has two separate channels for the P and Q components. For each channel the error is calculated by subtracting the measured power from the power order, and this is fed into the PID controller. Each of the increments, ΔP and ΔQ, become the inputs into a non-linear mapping function, which resolves these increments into their ($\Delta\alpha_{1P}$ and $\Delta\alpha_{2P}$) and ($\Delta\alpha_{1Q}$ and $\Delta\alpha_{2Q}$) components. The non-linear errors are then combined and $\Delta\alpha_1$ and $\Delta\alpha_2$ are integrated to provide the required outputs (α_1) and (α_2) as inputs into the converter firing

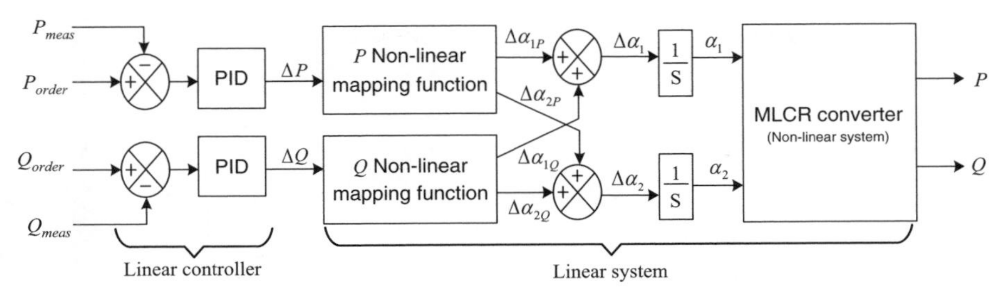

Figure 8.17 Diagram of the P,Q controller.

logic. The non-linear mapping functions in Figure 8.17 are represented byA_{11}^{-1}, A_{21}^{-1} (for P) and A_{12}^{-1}, A_{22}^{-1} (for Q)in Equation (8.26).

Figure 8.18 shows how the system is realized in a practical controller. The controller layout follows almost exactly the analytical development from equations (8.19) to (8.26), with only additional low-pass filters added to prevent ringing when the error is almost zero. It is important to note that the common component of the converter control is calculated separately (Figure 8.18(b)), as this determines the overall gain of the system. Hard limits on the calculation are provided so as to prevent wind up and instability which can occur if $\alpha_1 = \alpha_2$. To ensure that firing angle α_1 is always greater than α_2, limits are also placed on the integrators.

The receiving end controller topology is much the same as that of the sending end, but as it must control Vdc_r and Q_r, the layout is different. Using steady state equations (8.14) and (8.18), the inverse transfer function becomes:

$$A^{-1} = \frac{1}{3V_T \sin(\alpha_1 - \alpha_2)} \begin{bmatrix} \frac{-\sqrt{2}\pi\cos(\alpha_2)}{4} & \frac{-\sin(\alpha_2)}{I_1} \\ \frac{\sqrt{2}\pi\cos(\alpha_1)}{4} & \frac{\sin(\alpha_1)}{I_1} \end{bmatrix} \tag{8.27}$$

Given the steady state equations, and taking into consideration equation (8.15), it becomes apparent that although full control is justified by the theory, the range of Q control depends on the magnitude of I_{dc}. Optimum DC power transmission occurs when the DC current is minimized, as this also minimizes the DC link power losses; however this affects the range of Q controllability at both the sending and receiving ends. As the reactive power circulation is confined to the AC system side, the magnitude of the AC current in each converter group determines the level of reactive power controllability in the AC system. The real power, which is also a function of the AC current magnitude, is determined by the combination of V_{dc} and I_{dc} on the DC link. To understand the reactive power controllability limits, it must be realized that the same amount of real power can be transferred with a combination of high V_{dc}/low I_{dc}, or low V_{dc}/high I_{dc}.

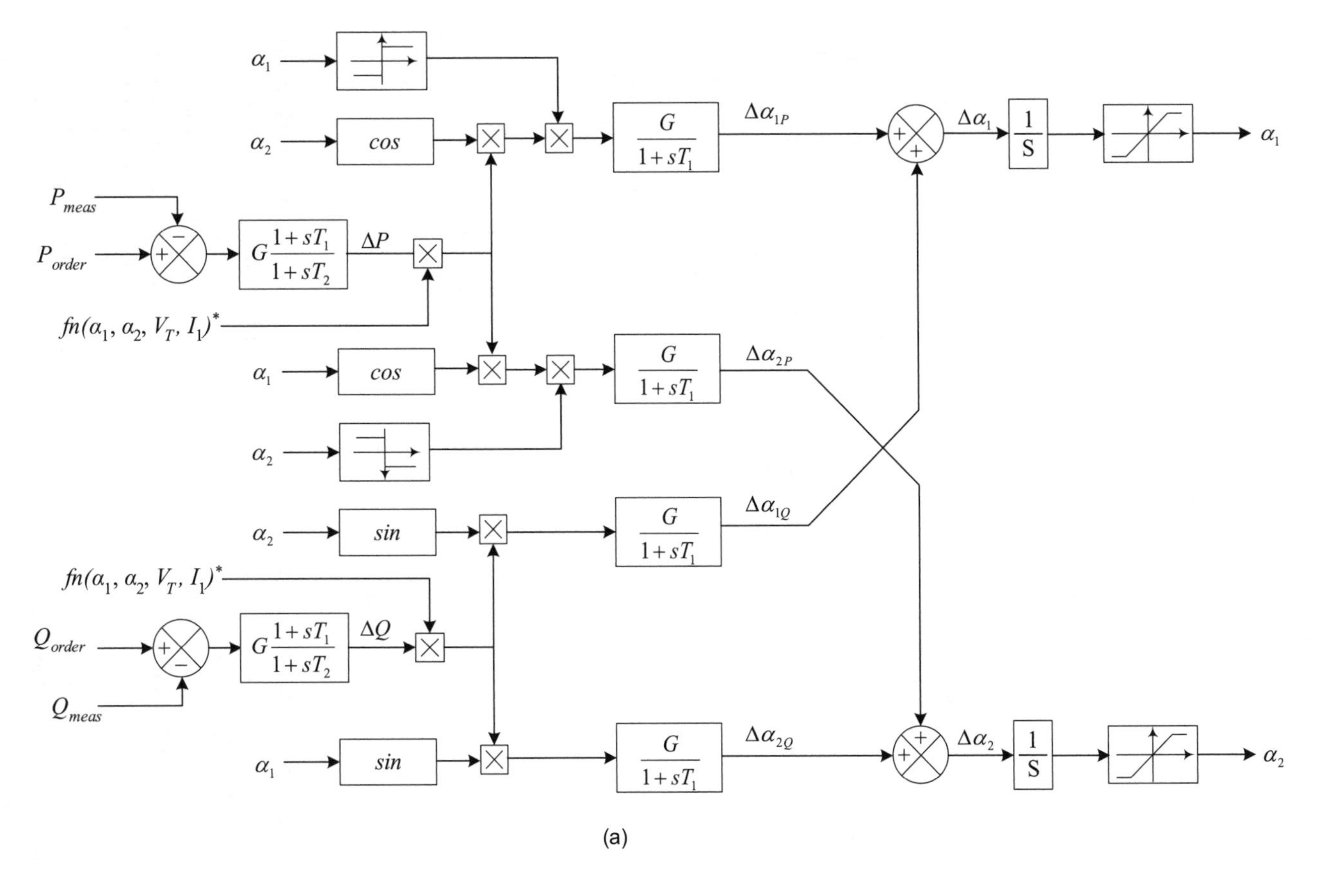

Figure 8.18 Sending end controller block diagram, with the main linearizing components in (a) and the common angle difference calculation in (b).

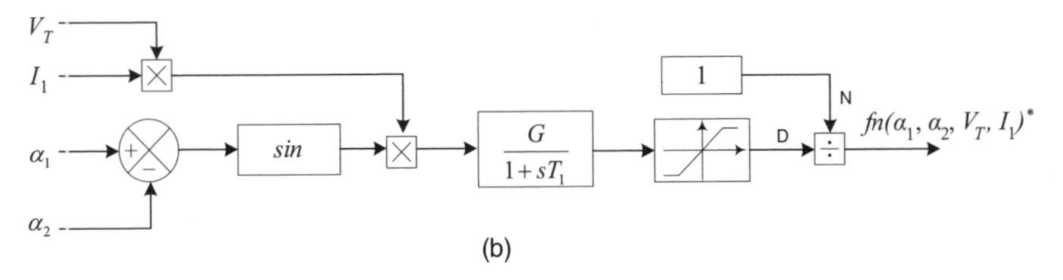

Figure 8.18 (Continued)

An example of this in a multigroup MLCR, in per unit terms is given below:

$$\text{High } V_{dc} : \; Pdc = 1pu = 5.5puV_{dc} \times 0.182\, puI_{dc} \tag{8.28}$$

$$\text{Low } V_{dc} : \; Pdc = 1pu = 2.25puV_{dc} \times 0.364\, puI_{dc} \tag{8.29}$$

and so with $I_1 = \sqrt{3}k_m I_{dc}$ and V_{ac} being the same in both cases, equation (8.18) shows that equation (8.28) would yield twice as much reactive power for a given firing angle as 8.29. So with conflicting objectives in real power efficiency and reactive power controllability, a compromise must be made between control range and overall efficiency during system design.

8.5.3 Dynamic simulation

The test circuit, shown in Figure 8.11, interconnects two AC systems represented by Thevenin equivalents. Each terminal consists of two five-level MLCR converter groups, for which the circuit configuration and steady state output current waveform are as shown in Figures 4.10 and 4.11 of Chapter 4.

Using the rated power (1000 MW) and voltage (220 kV) levels as base values, the source voltages are set at 1.02 and 1.06 pu at the receiving and sending ends respectively; the series impedance on both sides is initially equal to 0.254 pu (plus the transformer leakage reactance which is equal to 0.1 pu) and the DC line is represented by a resistance of 0.1 pu in series with a 2 H smoothing inductor. The active power transfer is controlled at the sending end and the terminal AC voltage at the receiving end.

A PSCAD model of the test system under the above operating conditions is used to start the EMTDC simulation. Figure 8.19 illustrates the response of the HVDC link to variations in the receiving end series impedance. Figure 8.19(a) shows the set value of the series reactance at the receiving end, which is initially equal to 0.254 pu. Upon reaching the initial steady condition, the value of the receiving end series reactance is increased by 50% (at 0.5 s) and by 100% (at 1.5 s) and then returned to the original values at 2.5 and 3.5 s respectively. The dynamic response at the receiving end is shown in graphs Figure 8.19(b) (active power), (c) (reactive power) and (d) (terminal voltage). The terminal voltage, active and reactive power responses at the sending end are shown in Figure 8.19(e), (f) and (g) respectively. In each case, after the oscillation that follows the change of condition disappears, the sending end parameters are maintained constant on the steady state, which clearly indicates the independence of the reactive power control on both sides of the link provided by firing-shift control of the double-group converter.

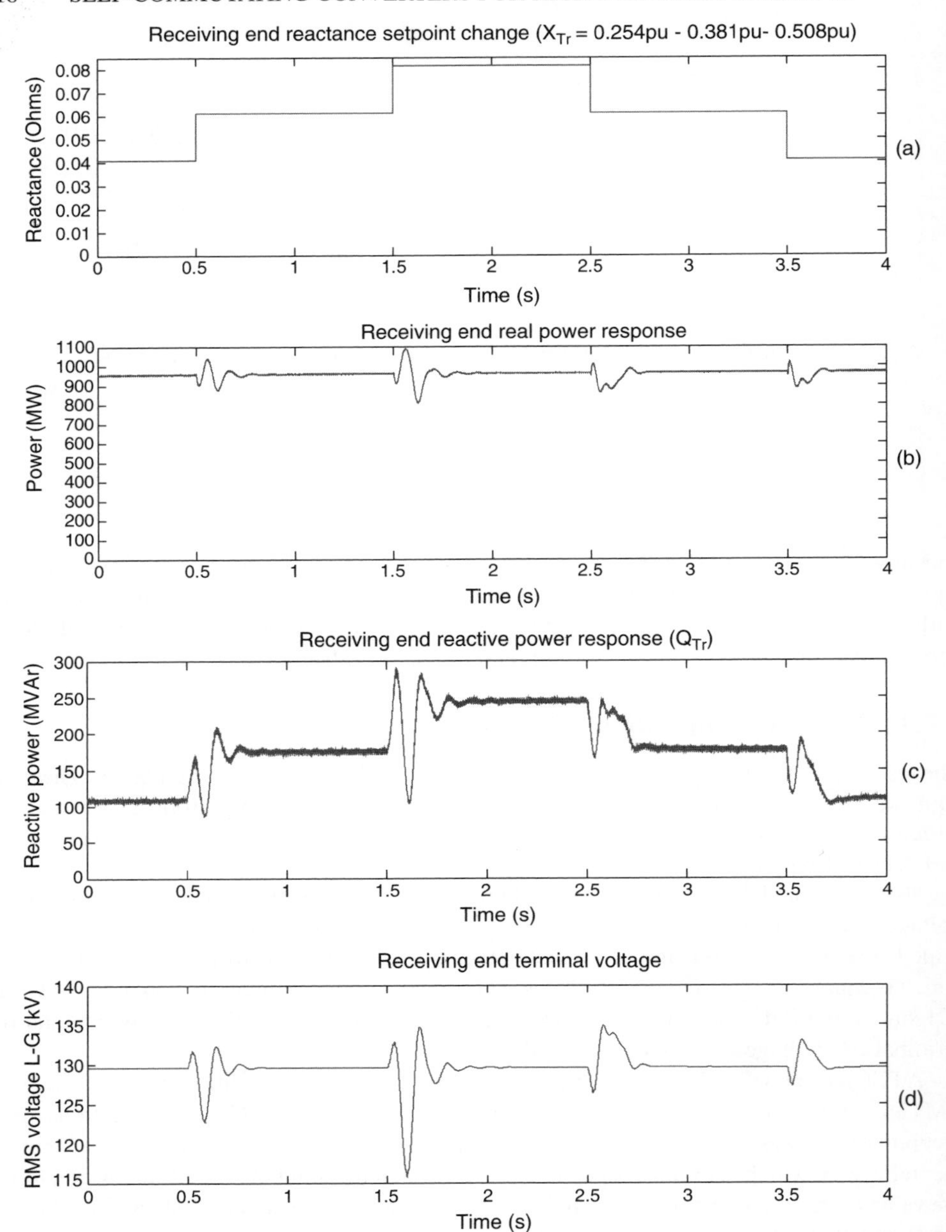

Figure 8.19 Receiving and sending end responses to changes in the reactive power conditions at the receiving end.

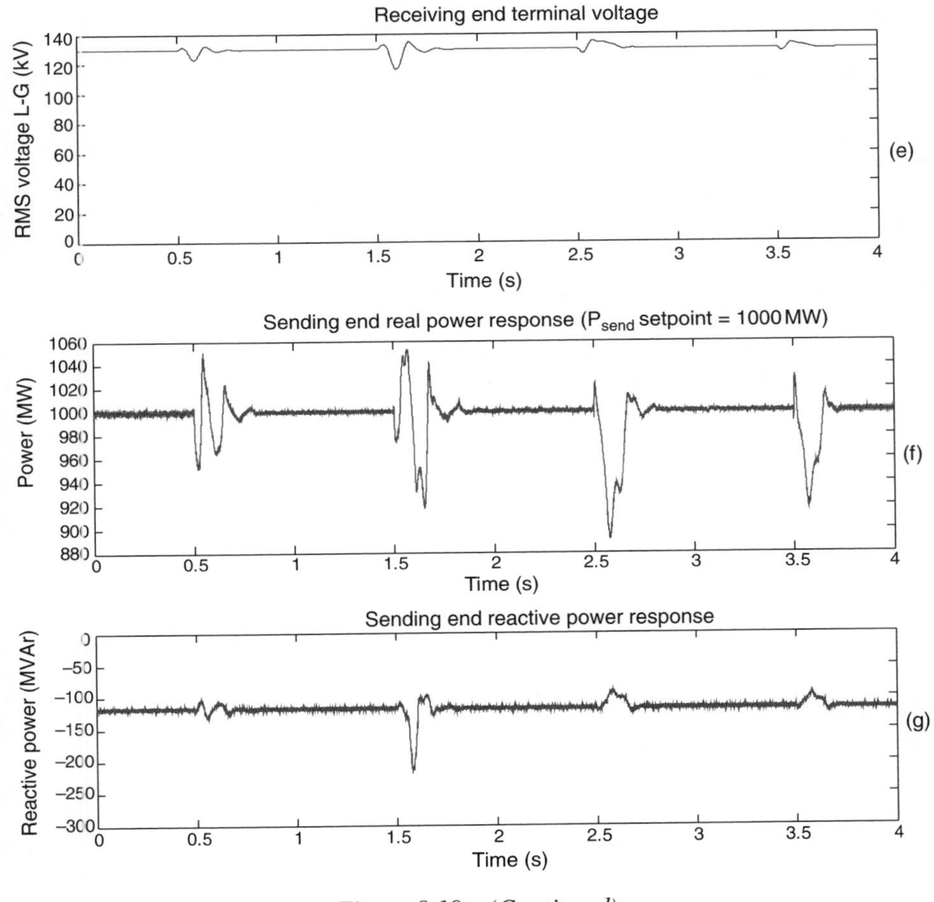

Figure 8.19 (Continued)

8.6 Summary

The power ratings of UHV transmission schemes will need two or more series-connected double converter groups at the multilevel based converter stations. These configurations are suitable for the type of converter control described in this chapter. It has been shown theoretically and verified by EMTDC simulation that the use of a controllable shift between the firings of the series-connected converter groups permits independent reactive power control at the two DC link terminals. This provides four-quadrant power controllability to multilevel current source HVDC transmission and, thus, makes this alternative equally flexible as PWM-controlled voltage source conversion, without the latter's limitations in terms of power and voltage ratings. It can be expected that MLCR combined with firing shift control should compete favourably with the conventional current source technology for very large power transmission schemes.

References

1. Adapa, R., Maruvada, S., Rashwan, M., Hingorani, N.G., Szechman, M. and Nayak, R. (2008) R & D needs for UHVDC at 800 kV and above. *CIGRE*, Paper BS-114.
2. Nayak, R.N., Schgal, Y.K. and Sen, S. (2008) Planning and design studies for ±800 kV, 6000 MW HVDC system. *CIGRE*, Paper B4-117.
3. Perera, L.B., Watson, N.R., Liu, Y.H. and Arrillaga, J. (2005) Multi-level current reinjection self-commutated HVDC converter. *Proceedings of IEE on Generation, Transmission and Distribution*, **152** (5), 1607–15.
4. Arrillaga, J., Liu, Y.H. and Watson, N.R. (2007) *Flexible Power Transmission – The HVDC Options*. John Wiley & Sons, London.
5. Murray, N.J., Arrillaga, J., Liu, Y.H. and Watson, N.R. (2008) Flexible reactive power control in multi-group current-sourced HVDC interconnections. *IEEE Transactions on Power Delivery*, **23** (4), 2160–67.

9

Back-to-Back Asynchronous Interconnection

9.1 Introduction

As the rating and acceptability of high power self-commutating switches improve, the boundaries between the HVDC and FACTS technologies are gradually becoming blurred. HVDC has already started using the new devices (thus improving its control flexibility) and FACTS controllers are increasing their power range (which may finally result in the control of the total power transfer of an asynchronous interconnection). Therefore, modern back-to-back (BTB) conversion could be considered as part of the HVDC and FACTS technologies.

Bidirectional BTB asynchronous interconnections with ratings of up to 1500 MW are already being used for the interconnection of networks of different frequencies, like the recent Brazil–Argentina scheme [1], and for trading reserves [2] or shifting peak energy loading times between networks with different time zones, such as between Finland and Russia [3].

The distinguishing feature of FACTS with respect to traditional HVDC has been its greater control flexibility. However, several fully flexible PWM-based HVDC voltage source conversion schemes in the 300 MW region are already operating successfully and the power rating capability of the PWM-VSC technology is on the increase.

Although the large power BTB interconnections under consideration are still based on conventional thyristor conversion, the self-commutating multilevel technology is likely to be a viable alternative in the near future. However, under conventional control the self-commutating multilevel BTB interconnection lacks independence of reactive power controllability between the two ends of the link, an important shortcoming.

The MLCR concept described in Chapter 4 and used in Chapter 8 for the series-connected double converter groups, is used here for the firing control of the parallel-connected BTB converter configuration to provide four-quadrant controllability. The parallel MLCR configuration has a simpler structure and is the preferred option for the large BTB asynchronous application, which will normally consist of groups of two paralleled bridges connected in series.

Self-Commutating Converters for High Power Applications J. Arrillaga, Y. H. Liu, N. R. Watson and N. J. Murray

9.2 Provision of independent reactive power control

Figure 9.1 presents a series/parallel-connected BTB link, interconnecting two separate AC systems represented by Thevenin equivalents V_s, Z_s and V_r, Z_r respectively. The converters are parallel-connected on their respective AC sides with a common terminal voltage (V_{Ts}) and series/parallel-connected on the DC side with a combined DC voltage (V_{ds}). The difference between V_{ds} and the receiving end DC voltage (V_{dr}) over the DC side resistance (Rd) determines the resulting DC current which is then inverted at the receiving end.

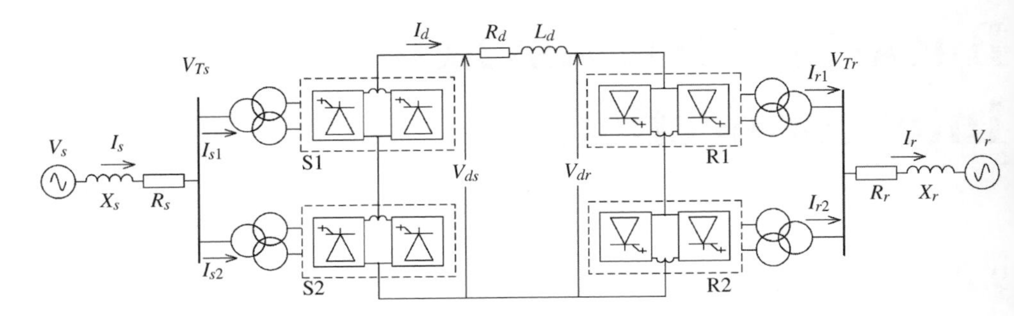

Figure 9.1 Block diagram of the back-to-back link configuration.

For the series-connected AC–DC rectifiers, the combined DC voltage at the sending end is

$$V_{dS} = k_1 V_{TS} \cos(\alpha_{S1}) + k_1 V_{TS} \cos(\alpha_{S2}) \tag{9.1}$$

where k_1 is a constant that depends on the converter topology and α_{S1} and α_{S2} are the firing angles which, under self-commutation, are capable of leading or lagging the terminal voltage phase angle.

The reactive power of rectifier S1 is given by

$$Q_{Ts1} = 3V_{Ts}k_2 I_d \sin(\alpha_{s1}) \tag{9.2}$$

where I_d is the DC current and k_2 is the AC to DC current ratio. For negative values of α_{S1} relative to the voltage V_{Ts}, reactive power is generated and supports this terminal voltage. The net reactive power generated or absorbed by the series-connected sending end rectifiers (at the common terminal) is therefore

$$Q_{Ts} = 3k_2 V_{Ts} I_d(\sin(\alpha_{s1}) + \sin(\alpha_{s2})) \tag{9.3}$$

By selecting suitable values of α_{s1} and α_{s2} in equation 9.3, the reactive power demands may be controlled. Moreover, by specifying firing angles that satisfy both equations 9.1 and 9.3, the DC voltage and reactive power at the sending end may be controlled simultaneously and independently, and on-load tap changers can be eliminated.

If the transformer real power losses are neglected, the sending end terminal real power P_T is given by

$$P_T = P_{ds} = V_d I_d \tag{9.4}$$

and substituting equation 9.1 into 9.4, the relationship becomes

$$P_t = 3k_2 V_{Ts} I_d(\cos(\alpha_{s1}) + \cos(\alpha_{s2})) \tag{9.5}$$

The supply current I_s is related to the DC current by

$$I_s = 2k_2 I_d \cos\left(\frac{\alpha_{s1} + \alpha_{s2}}{2}\right) \tag{9.6}$$

the magnitude and phase of supply current (I_s) through the system impedance (Z_s) determining the voltage V_{Ts}. By combining equation 9.6 with 9.3 and 9.5, the sending end terminal active and reactive powers (P_T and Q_T) may be rewritten as

$$P_{Ts} = 3k_1 V_{Ts} I_s \cos\left(\frac{\alpha_{s1} + \alpha_{s2}}{2}\right) \tag{9.7}$$

$$Q_{Ts} = 3k_1 V_{Ts} I_s \sin\left(\frac{\alpha_{s1} + \alpha_{s2}}{2}\right) \tag{9.8}$$

From equation 9.4, the DC power transfer may be achieved with any combination of V_d and I_d that satisfies a constant P_T. Unlike long-distance HVDC transmission, where V_{ds} is maximized to reduce transmission losses, the BTB link DC losses are negligible, and freedom exists to control the DC current across its full range. The level of reactive power controllability, at both the sending and receiving ends, depends on the magnitude of the DC current, and so, for a given real power transfer, increasing I_d is of real benefit when the converter interconnects two weak systems.

Equations 9.1 to 9.8 are modified for the receiving end by replacing the subscript (s) with (r). The receiving end DC voltage is thus described by:

$$V_{dr} = k_1 V_{Tr}(\cos(\alpha_{r1}) + \cos(\alpha_{r2})) \tag{9.9}$$

If the active power transfer across the BTB link is specified at the sending end, then the level of DC current may be modulated by adjusting the receiving end DC voltage; i.e. the DC current is increased by driving V_{dr} down relative to V_{ds} as given by

$$I_d = \frac{V_{ds} - V_{dr}}{R_d} \tag{9.10}$$

In practice the BTB link resistance is negligible, meaning that the sending and receiving end voltage are almost the same. The resistance has been introduced to illustrate the adjustment of DC current more easily. There will, however, be instantaneous differences in voltage across the smoothing reactor L_d.

Decreasing the receiving end DC voltage to increase the DC current has an immediate effect on the sending end power and, as a consequence, the sending end DC voltage must be reduced to maintain the specified power. Therefore in terms of specified DC real power (P_{dref}), receiving end DC voltage (V_{dr}) and DC resistance (R_d),

$$V_{ds} = \frac{V_{dr} + \sqrt{V_{dr}^2 + 4R_d P_{dref}}}{2} \tag{9.11}$$

The greater flexibility to choose the DC voltage and current in a BTB configuration also permits the use of large DC current at low real power settings, meaning that reactive power compensation is possible at both ends of the link, even when real power transfer is not required.

9.3 MLCR back-to-back link [4]

The MLCR configuration provides high pulse operation, and has the ability to operate with leading and lagging firing angles, whilst retaining conventional thyristors for the main bridges [5]. The series connection is needed for high-voltage transmission to minimize losses, but in the BTB configuration the use of a lower voltage allows a smaller station footprint. The parallel MLCR, shown in Figure 4.15, provides a better voltage and current balance for this application, with lower comparative losses and fewer switches, as compared to series multilevel reinjection, in the conduction path.

A small voltage phase shift across the transformer leakage reactance occurs under load, which depends on the converter firing angle. The phase shift, though normally very small, must be considered when the converter is required to operate near 0°. This effect is shown for the sending end in Figure 9.2, where V_{Ts1} and V_{Ts2} are the voltages on the converter side of the transformers.

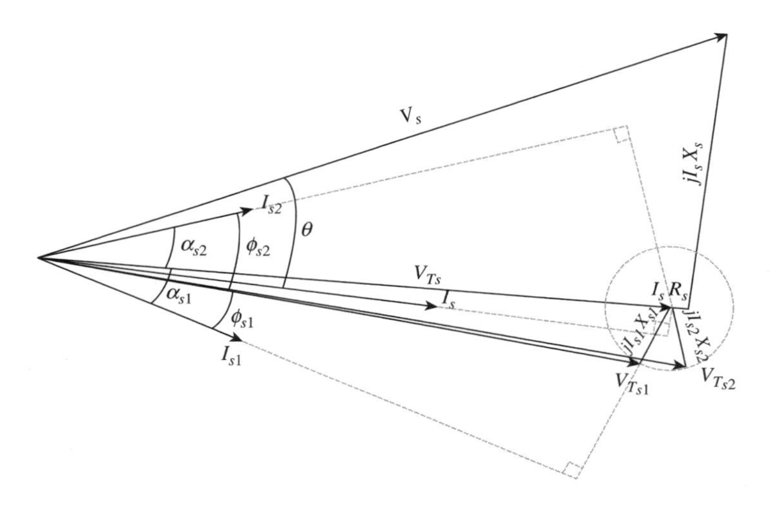

Figure 9.2 Phasor diagram of the phase shift across system impedance and transformer leakage reactances at the sending end of a series-connected parallel MLCR converter.

The locus centred about the peak of V_{Ts} describes the phase shift caused by the transformer leakage reactances $jI_{s1}X_{s1}$ and $jI_{s2}X_{s2}$ for a given DC current (as $|I_{s1}| = |I_{s2}| = kI_d$). The specified firing angles (α_{s1} and α_{s2} in the diagram) referenced to common terminal voltage V_T actually result in angles which differ slightly relative to their individual terminal voltages, as denoted byϕ_{s1} and ϕ_{s2}.

The common AC terminal voltage (V_{Ts} at the sending end) is often used in conventional HVDC as the reference for phase-locked loop (PLL) firing angle development, but considering that a firing angle specified with reference to this voltage may have a variable error of 1–2°, the star-connected winding on the secondary side is instead taken as the firing angle reference. The secondary side reference is exposed to the switching spikes which would ordinarily be attenuated by the converter transformer, and so the reference voltage must subsequently be low-pass filtered. When the magnitude and phase of the common terminal voltage (V_{Ts}) changes under load, a situation particularly pertinent to weak systems, an incorrect firing angle reference will change the reactive power absorbed or generated from the system, and thus have a direct effect on the terminal voltage phase and magnitude, which may compound and lead to voltage instability.

9.3.1 Determining the DC voltage operating limits

The firing angle limits at both converter ends are set by the active and reactive power requirements. The minimum and maximum values of the sending end system impedance define the limits of reactive power generation required by the rectifier to maintain the AC terminal voltage at rated active power transfer. In terms of the specified DC power (P_{Tsp}), equations 9.1 and 9.3 may be expressed as

$$\begin{aligned} Q_{Ts} &= 3V_{Ts}I_{s1A}(\sin(\alpha_{s1}) + \sin(\alpha_{s2})) \\ &= 3V_{Ts}\tfrac{I_d}{k_2}(\sin(\alpha_{s1}) + \sin(\alpha_{s2})) \\ &= 3V_{Ts}\tfrac{P_{Tsp}}{k_2 V_d}(\sin(\alpha_{s1}) + \sin(\alpha_{s2})) \end{aligned} \tag{9.12}$$

Rearranging to make α_{s2} the subject,

$$\alpha_{s2} = 2\arctan\left(\frac{Q_T k_1 k_2}{3P_{Tsp}}\right) - \alpha_{s1} \tag{9.13}$$

The valid range of DC voltages can be determined by substituting values of α_{s1} such that $-90 \leq (\alpha_{s1} \,\&\, \alpha_{s2}) \leq 90$.

An example of application is given in Figure 9.3 for an 1100 MW system with a short-circuit ratio of 2 at both the sending and receiving ends. The AC terminal voltage (V_{Ts}) at the sending end is maintained with 250 MVAr generated.

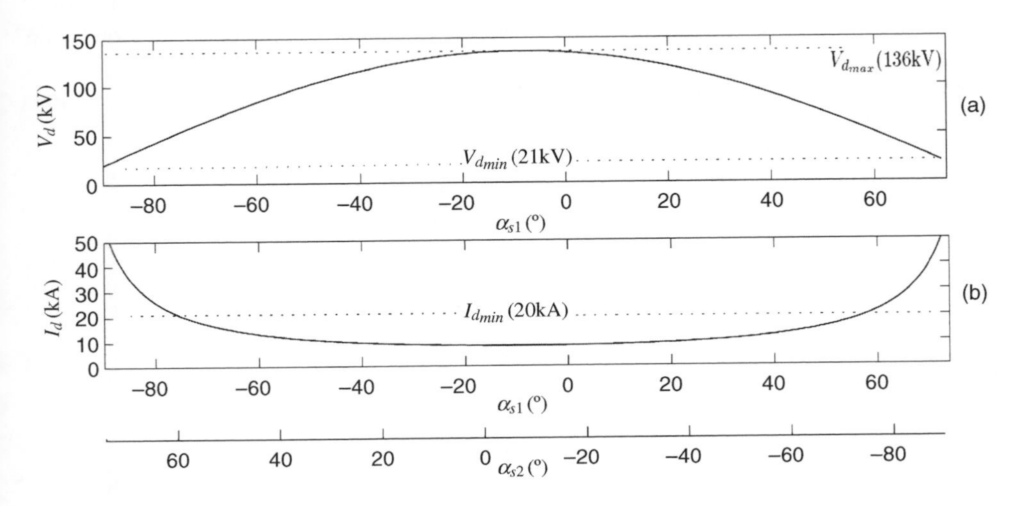

Figure 9.3 1100 MW back-to-back MLCR sending end DC characteristics while supplying 250 MVAr reactive power generation: (a) DC voltage range; (b) DC current range.

In the absence of losses and controller margin, the graph results show that the minimum and maximum absolute DC voltages are 21 kV and 136 kV. In practice the real power and reactive power controllers would be prevented from operating at these limits, instead allowing 5–15% at both ends to maintain control stability. The actual minimum DC voltage used in practice will also be dictated by the maximum DC current rating of the switches and smoothing reactor. As an example, a maximum DC current of 20 kA is used in Figure 9.3(b) which corresponds to a minimum DC voltage of 51 kV. Respecting these margins is especially important when both the rectifier and inverter are connected to weak systems, as reactive power generation keeps the AC terminal voltages (V_{Ts} and V_{Tr}) constant and maximum active power is reduced when either terminal voltage drops.

9.4 Control system design

In order to control the terminal active and reactive power independently, the converter responses are linearized by taking the partial derivatives of equations 9.3 and 9.5 with respect to α_{S1} and α_{S2}, the inverse of which, in matrix form, is given as:

$$A_s^{-1} = \frac{1}{3V_{Ts}k_2 I_d \sin(\alpha_{s1}-\alpha_{s2})}\begin{bmatrix} -\cos(\alpha_{s2}) & -\sin(\alpha_{s2}) \\ \cos(\alpha_{s1}) & \sin(\alpha_{s1}) \end{bmatrix} \tag{9.14}$$

and thus with

$$[A_s^{-1}]\begin{bmatrix} \Delta P_{Ts} \\ \Delta Q_{Ts} \end{bmatrix} = \begin{bmatrix} \Delta\alpha_{s1} \\ \Delta\alpha_{s2} \end{bmatrix} \tag{9.15}$$

and P_{Ts} and Q_{Ts} specified, the increment in both firing angles may be derived, without requiring the use of DC voltage as a control variable. In practice, when $\alpha_{S1} = \alpha_{S2}$ in equation 9.14, $\sin(\alpha_{S1}-\alpha_{S2})$ will be zero, with an undefined result, and so the denominator of equation 9.14 is limited to non-zero values to prevent control instability.

Similarly at the receiving end with the DC voltage and reactive power Q_{Tr} (and thus terminal voltage V_{Tr}) under control, the linearization is performed by

$$A_r^{-1} = \frac{1}{3V_{Tr}k_2 I_d \sin(\alpha_{r1}-\alpha_{r2})}\begin{bmatrix} \dfrac{-\pi\cos(\alpha_{r2})}{2\sqrt{2}} & \dfrac{-\sin(\alpha_{r2})}{k_2 I_d} \\ \dfrac{\pi\cos(\alpha_{r1})}{2\sqrt{2}} & \dfrac{\sin(\alpha_{r1})}{k_1 I_d} \end{bmatrix} \tag{9.16}$$

and

$$[A_r^{-1}]\begin{bmatrix} \Delta V_{Tr} \\ \Delta Q_{Tr} \end{bmatrix} = \begin{bmatrix} \Delta\alpha_{r1} \\ \Delta\alpha_{r2} \end{bmatrix} \tag{9.17}$$

The sending and receiving end controllers shown in Figures 9.4 and 9.5(a) are practically identical, aside from differences in lead–lag compensator parameters and the absolute values

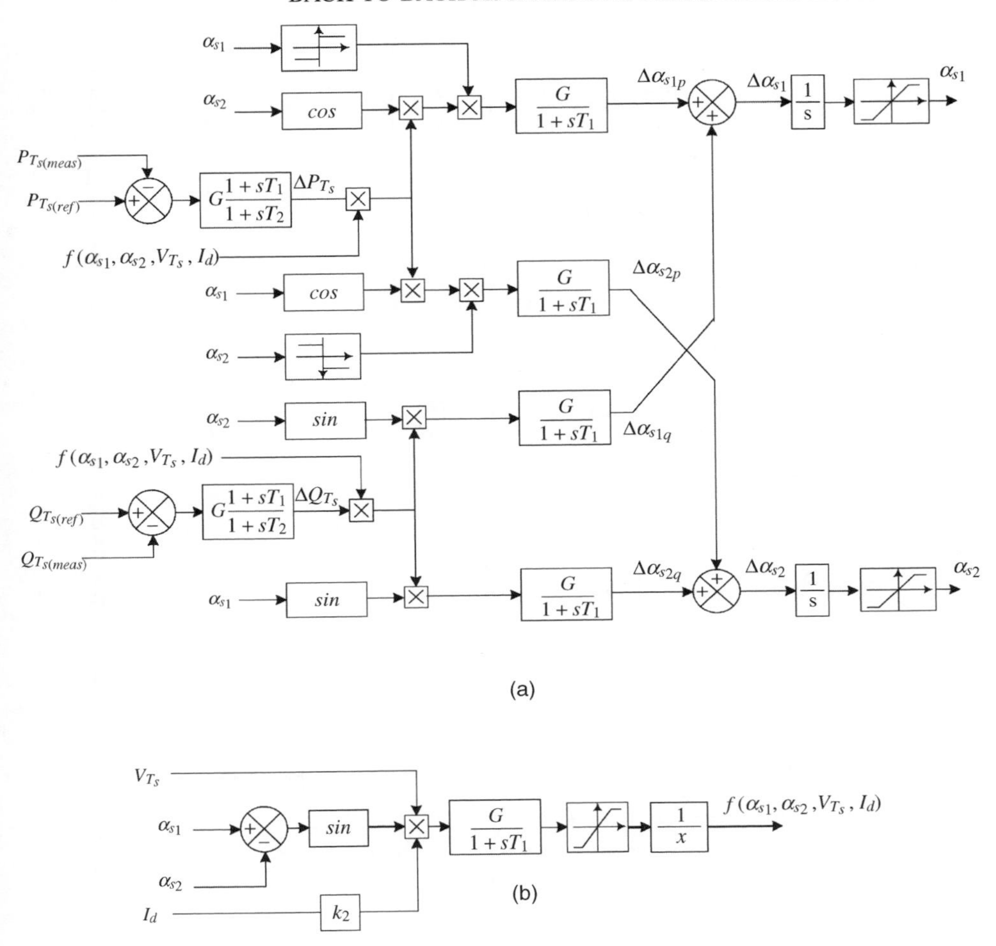

Figure 9.4 Sending end MLCR real and reactive power controller.

of the firing angles. Their main difference is the independence of current on the DC voltage controller at the receiving end.

The BTB link is bidirectional and when a power flow direction change is specified, the following process is initiated. The active power is decreased to zero, and the DC voltage order is raised to maximum to speed up active power reduction (by actively reducing the DC current). Upon the measured DC power and DC current reaching zero, the DC voltage order is set to zero, and once that is reached, both the sending and receiving end thyristor firing circuits are blocked. Then the integrators on each of the control channels are reset to their initial positions of $\pm 90°$ for the new sending and receiving ends. Active power control is then passed to the new sending end and the real power and DC voltage controls start up in the opposite direction.

The DC current is not maintained during the power direction reversal, in part due to the reactive power control required during the direction change, and because of the limitations

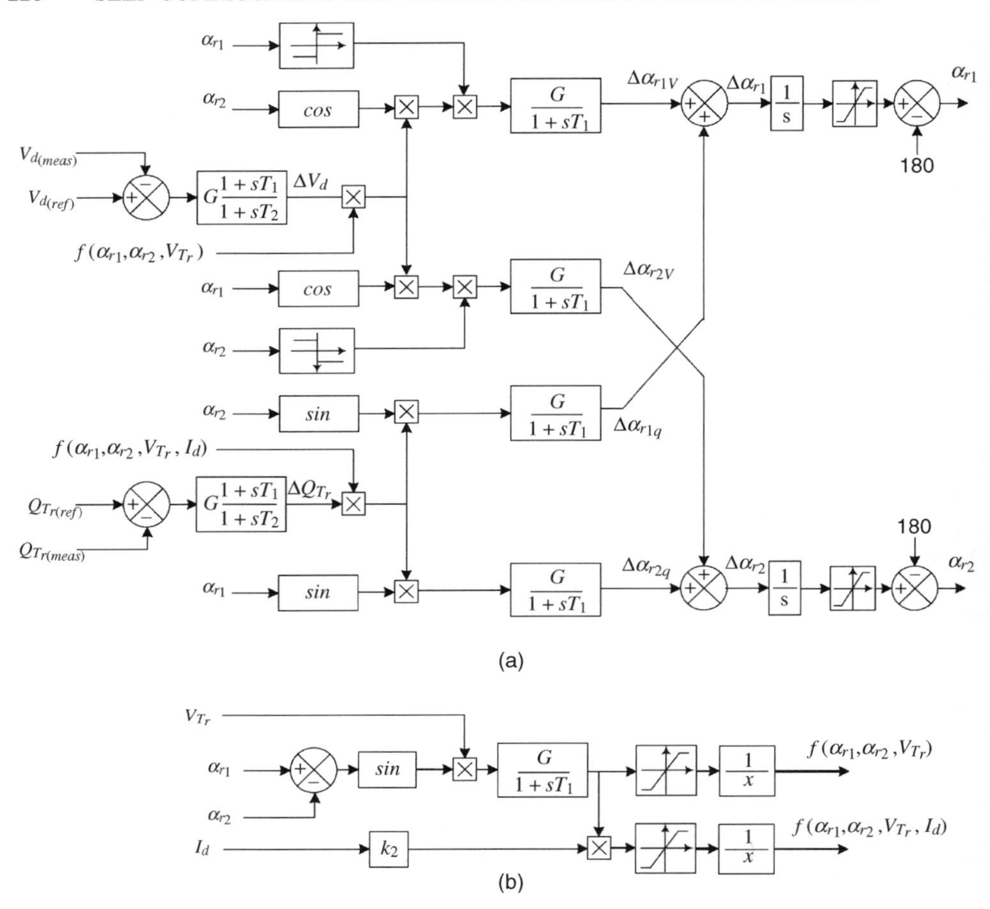

Figure 9.5 Receiving end MLCR real and reactive power controller.

with the PLL blocks in the PSCAD/EMTDC simulation environment. If during a real power direction reversal (where $\pm 180^\circ$ PLLs are used) a firing angle change from -179° to 179° can either pass through 180° or 0°; the consequences can be potentially catastrophic. When four firing angles must be coordinated simultaneously, the process becomes even more impractical.

With the linearization methods used in Figures 9.4 and 9.5, reactive power control is possible regardless of AC system strength. Terminal voltage control on the other hand is more difficult with a greater effect on voltage stability, and is therefore critical in weak systems, where a drop in terminal voltage at the sending end reduces the maximum available power transfer. To achieve stable terminal voltage control, a more suitable configuration for each end of the link is illustrated in Figure 9.6.

This diagram shows that the *rms* terminal voltage is calculated from each of the three-phase voltages (V_{Ta}, V_{Tb}, V_{Tc}), and is then compared to a nominal terminal voltage set-point

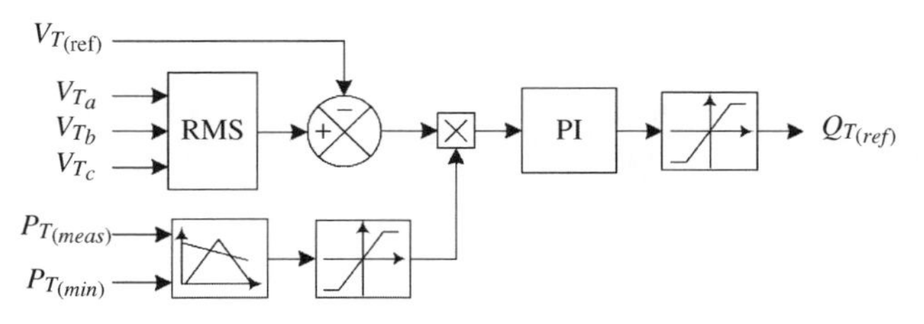

Figure 9.6 Terminal voltage control used at both ends of the link.

$V_{T(ref)}$. The error signal developed is fed through an enabling multiplier, which prevents terminal voltage control in the absence of real power transfer. The enabler is rate limited to provide a damped ramp up of the error signal. The PI controller outputs a reactive power order ($Q_{T(ref)}$) which is hard limited to prevent saturation of the controller output.

9.5 Dynamic performance

9.5.1 Test system

The test system is based on the series/parallel MLCR BTB link shown in Figure 9.7, with both ends supplied by Thevenin sources representing supply networks with short-circuit ratios of approximately 2.

Each Thevenin source supplies 500 kV through a series reactance of 0.3617 H to the common terminal bus, where two paralleled three-winding 700 MVA transformers provide 50 kV to two 48-pulse MLCR converters. The DC bus connecting the rectifier and inverter is represented by a 1.5 H smoothing reactor in series with a 100 mΩ resistor.

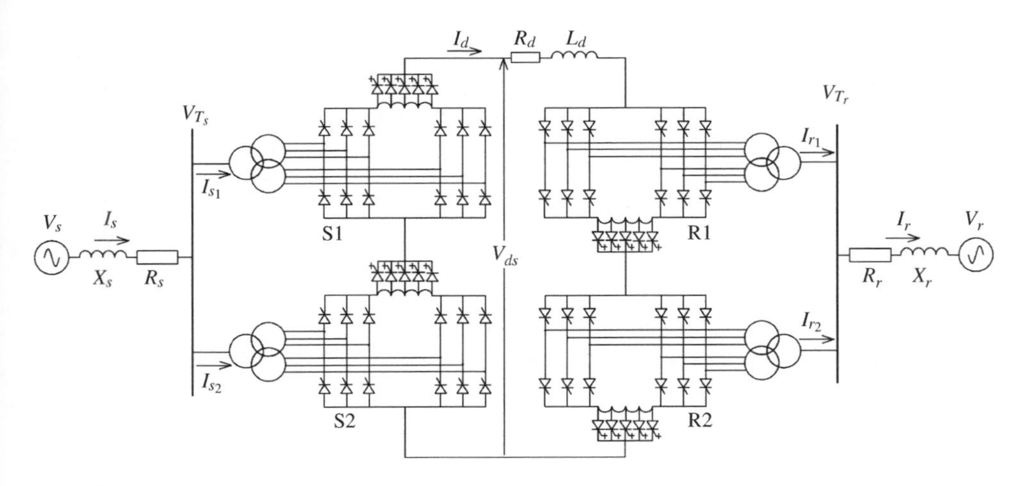

Figure 9.7 Back-to-back series/parallel MLCR test system.

9.5.2 Simulation verification

The test system is simulated using PSCAD/EMTDC over an 8 s period in order to verify the bidirectional transmission capability of 1100 MW, whilst controlling the AC terminal voltages for varying DC voltages (between 50 kV and 120 kV). The simulated results are shown in Figure 9.8.

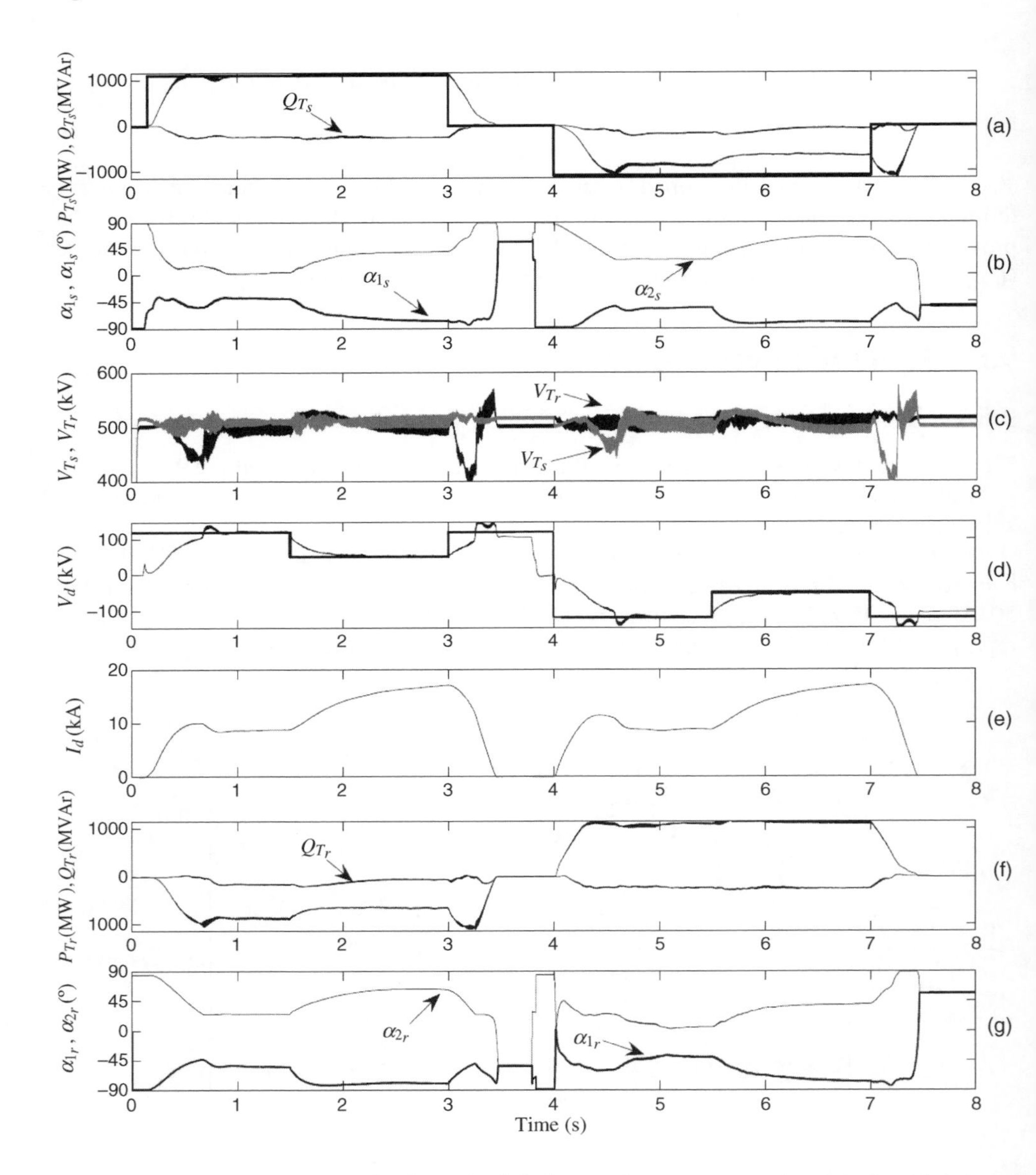

Figure 9.8 1100 MW MLCR back-to-back link simulation over an 8 s period, with (a) real and reactive power contributions of the sending end in, (b) firing angles, (c) terminal voltages, (d) DC voltage, (e) DC current, (f) receiving end real and reactive power in and (g) firing angles.

The link operation starts at t = 0.1 s. Figure 9.8(a) shows the sending end real power order (1100 MW), the measured power and the required reactive power generation (Q_{Ts}) at the common AC terminal controlled to maintain V_{Ts} constant at 500 kV (1 pu).The corresponding variables at the receiving end are shown in Figure 9.8(f). The reactive power controllers are automatically enabled when the active power increases over 5 MW, thus preventing attempted operation with negligible DC current.

The DC voltage order and the DC voltage response are shown in Figure 9.8(d). The initial set is 120 kV (between t = 0 and 1.5 s). The firing angles at the sending end, shown in Figure 9.8(b), reduce to 1.6° and −37.9° respectively; this indicates that one converter group is operating at maximum DC voltage, while the second group is supplying the reactive power compensation needed for the sending end to operate at its most efficient point.

Between t = 0 and t = 1.5 s the active power at the receiving end is about 9% lower than the power at the sending end, the difference being made by losses in the resistances, switches and transformers.

At t = 1.5 s the DC voltage is ordered to drop from 120 kV to 50 kV, which is implemented by decreasing α_{r1} and increasing α_{r2} at the receiving end (shown in Figure 9.8(g)), the actual values sent to the thyristor firing logic being ($180 - \alpha_{r1}$) and ($180 - \alpha_{r2}$) up to t = 3.4 s. These changes in firing angles at the receiving end cause the DC current to increase between t = 1.5 s and 3 s (with a corresponding rise in power losses to about 40%) and an increase in the two sending end firing angles (shown in Figure 9.8(b)). The effect on the active power at the sending end, however, is negligible and only a small change in terminal voltage is observed (shown in Figure 9.8(c)), which is quickly corrected by an increase in the Q_T order. A drop off in V_{Tr} of approximately 10% is observed between t = 0.4 and 0.8 s as the receiving end reactive power control is established; the control of Q_{Tr} is intentionally made slower than the DC voltage control, so that stable DC current is maintained, otherwise interaction between the sending and receiving end reactive power controllers would occur.

At t = 3.0 s a reversal in active power flow direction is ordered. The active power is ramped down, and the DC voltage control order raised to reduce the DC current. When the DC current falls to zero (at t = 3.5 s) V_{dorder} is decreased; at 3.85 s the main thyristor bridge firing circuits are disabled, and the duties of the active power and DC voltage controllers exchanged between the two ends.

At t = 3.85 s the firing angle origins are reset to ±90° and the controllers re-enabled. The 1100 MW order is applied to the new active power controller at t = 4.0 s and the DC voltage is then varied as in the first half of the simulation.

Thus the simulated test results validate the stated capability of the proposed control of providing the series/parallel double-group BTB MLCR link interconnecting two weak power systems, with independent reactive power control at the two ends of the link.

9.6 Waveform quality

The current waveforms and harmonic components of the sending and receiving ends for the test system described in the previous section are shown in Figure 9.9. The characteristic harmonic magnitudes at the sending end are 1.5% for the 12th, 1.9% for the 24th and 1.3% for the 48th, with a total harmonic distortion (THD) of 3.31%. At the receiving end, the characteristic harmonics are 3.1% for the 12th, 1.5% 24th and 1.2% for the 48th, with a THD of 4.34%.

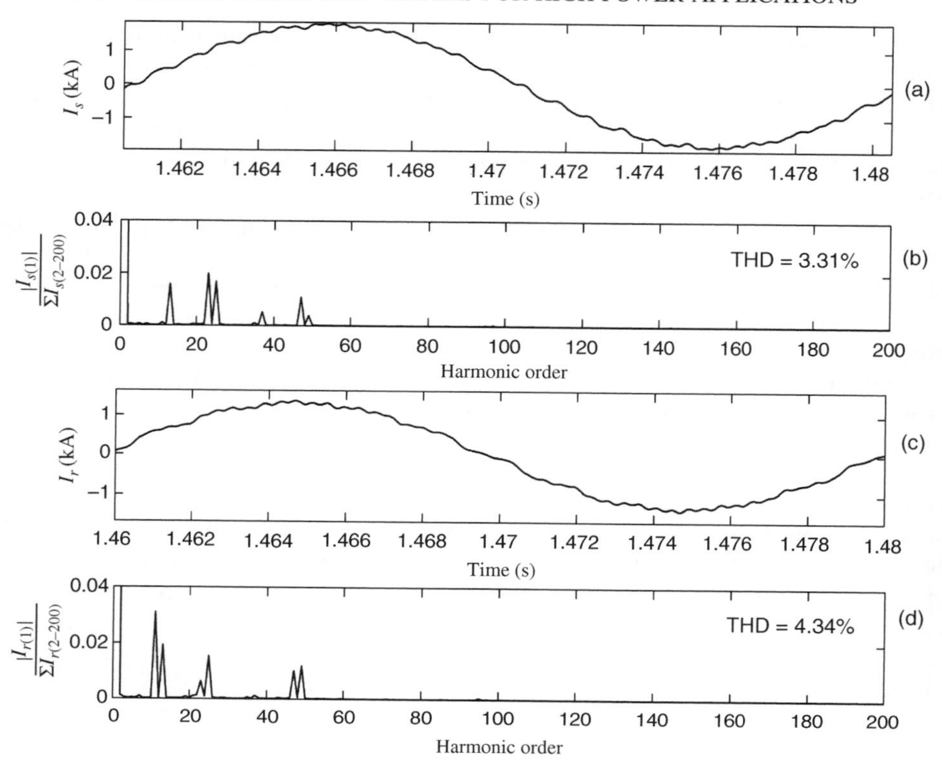

Figure 9.9 Harmonic performance of the sending (rectifying) and receiving (inverting) end supply waveforms for the first 200 harmonic orders.

9.7 Summary

The suitability of the MLCR-based BTB interconnection for bidirectional large power transfer between two weak AC systems has been demonstrated, as well as its capability to control DC voltage and current directly, whilst still satisfying the independent reactive power demands at both ends of the link. The total harmonic distortion of a hybrid series/parallel test system configuration has been shown to be 3.31% and 4.34% measured at the sending and receiving ends respectively.

These attributes make the BTB series/parallel MLCR perfectly adequate for high power interchange at high efficiency and with the same control flexibility provided by the PWM conversion technology.

References

1. Graham, J., Jonsson, B. and Moni, R.S. (2002) The Garabi 2000 MW interconnection back-to-back HVDC to connect weak AC systems. *Trends on Transmission Systems and Telecommunications*, library.abb.com.

2. Karim, A.M.H.A., Maskati, N.H.A. and Sud, S. (2004) Status of Gulf co-operation council (GCC) electricity grid system interconnection. *Power Engineering Society General Meeting, IEEE*, 6–10 June 2004.
3. Voropai, N.I. and Kucherov, Yu.N. (1999) Some aspects of Russian electricity policy taking into account the EU Electricity Directive. *Power Engineering Society Summer Meeting IEEE*, 1999.
4. Murray, N.J., Liu, Y.H., Arrillaga, J. and Watson, N.R. Paper accepted for publication by the *IET Proceedings on Power Electronics*.
5. Arrillaga, J., Liu, Y.H., Perera, B. and Watson, N.R. (2006) A current reinjection scheme that adds self-commutation and pulse multiplication to the thyristor converter. *IEEE Transactions on Power Delivery*, **21** (3), 1593–9.

10

Low Voltage High DC Current AC–DC Conversion

10.1 Introduction

Large-scale DC supplies are used throughout the industry with current ratings between 10 and 500 kA for the metal processing and refining of copper, zinc, manganese, steel and aluminium; they are also used in the production of chemicals such as chlorine. The rectified current is typically derived from saturable reactor controlled diodes or silicon controlled rectifiers. High current ratings require the use of parallel-connected converter structures, normally fed from phase-shifted transformers.

The control of power in present line-commutated rectification is restricted to a single quadrant and the provision of reactive power requires a combination of passive compensation and phase-shifted transformers provided with on-load tap changers (OLTCs), all of which results in costly converter stations and considerable maintenance. In practice, the supply voltage has always some unbalance and distortion that upsets the nominal phase shift and, thus, the phase-shifting configuration is normally restricted to 48 pulses.

Although the current ratings of thyristor switches with turn-off capability, such as the IGCT, have already reached levels that make them suitable for high current rectification, the conventional phase-controlled concept still prevents independent reactive power control (i.e. the rectification process is still restricted to a single quadrant). The reason is that the active and reactive powers in the phase-controlled rectification process vary in opposite directions, because adjusting the DC current is achieved by varying the DC voltage, which is itself determined by the cosine of the firing angle, while the reactive power is a function of the sine of the firing angle.

This chapter presents some self-commutating alternatives to conventional line-commutated phase angle control, capable of controlling the reactive power as well as reducing the structural complexity and increasing the operating flexibility of the existing rectifier stations used in high DC current industry applications. Much of the description is made with reference to the aluminium smelter technology, which currently uses the highest power ratings.

Self-Commutating Converters for High Power Applications J. Arrillaga, Y. H. Liu, N. R. Watson and N. J. Murray

In the following description, a basic rectifier unit refers to a six-pulse bridge, the group rectifier unit represents several parallel-connected six-pulse bridges operating under common firing-angle control and the double-group two-quadrant rectifier refers to two parallel-connected group rectifier units (of any pulse number) operating in two different quadrants and with different firing angles. The group configurations are paralleled on each of the AC and DC sides. These common connection points are referred to as the AC and DC buses respectively.

10.2 Present high current rectification technology

The largest user of high current rectification is the aluminium smelting industry, with typical applied DC voltages of 1 kV to 2 kV, and DC currents in hundreds of kA.

The applied DC voltage is controlled using a combination of OLTCs for large steps in voltage and saturable reactors for fine control. With DC currents in hundreds of kA, several OLTC/saturable-reactor DC supplies need to be parallel-connected, as shown in Figure 10.1.

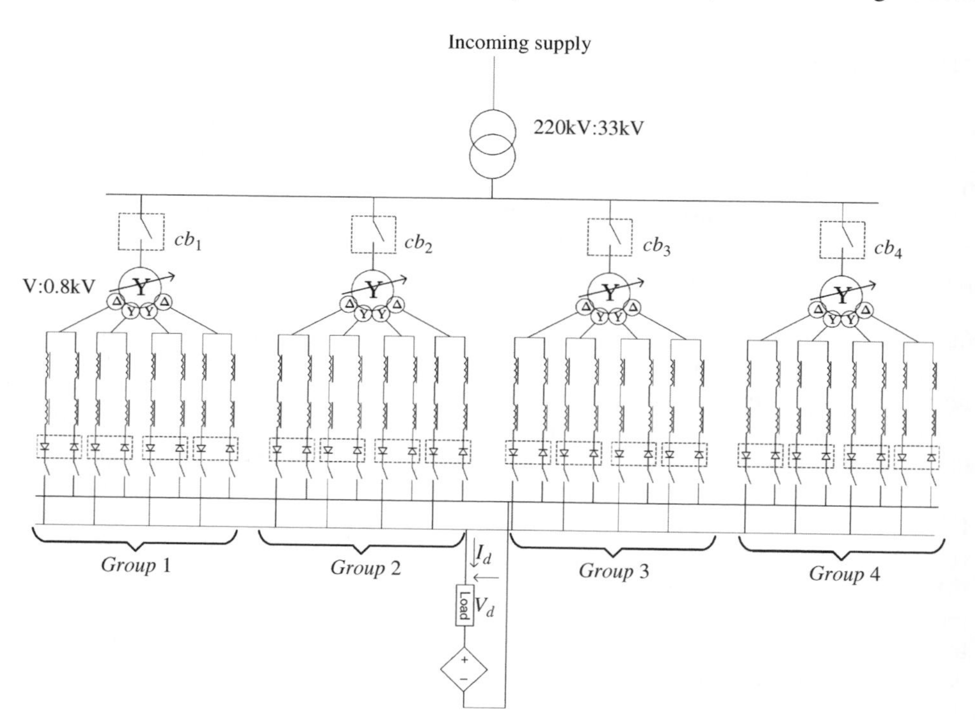

Figure 10.1 Typical smelter power system configuration.

Four zig-zag wound phase-shifted transformers with multiple secondaries are parallel connected on both the AC and DC sides of the converter, providing a common DC bus voltage to the load. Fine DC voltage, and therefore current, adjustment is made by controlling the DC excitation current of the saturable reactors, and transformer tap changes are made when the reactor voltage is out of range.

This electro-mechanical process has several deficiencies, the first being that tap positions are slow and change very frequently, requiring regular out of service maintenance. Secondly, since tap changing and saturable reactor adjustment are part of the normal operation process, given the small differences in manufacturing tolerance between each rectifier, differences in reactor bias current and taps positions can occur. Realization of ideal phase-shift, and thus harmonic performance, rely on equal transformer leakage reactances and balanced DC current, a situation not possible with this method of control. Also, minimum tap-change position of each zig-zag transformer is limited to around 10%, and thus to fully remove power, the circuit breakers (cb_1 to cb_4 in Figure 10.1) must be opened. Likewise, establishing power supply, either from cold or resuming from a power interruption, must begin at 10% full DC voltage, with a corresponding disturbance to the AC terminal voltage when the breakers close. Smelters with weak transmission systems are more susceptible to terminal voltage variation with changes in operating conditions, a case common when smelters are constructed in coastal locations far from generation, for access to deep sea ports for export. Voltage support is installed, [1, 2] with power-factor correction and harmonic filtering to improve voltage quality.

Thyristor-based rectification has been used commercially to augment existing smelter DC current capability since the 1980s, and more recently ABB has introduced thyristor rectifiers of similar ratings to existing diode equivalents. Similarly, Fuji now offers S-Former (integrated thyristor–transformer) modules, with ratings that also make them attractive alternatives.

While the use of thyristors gives an improvement in dynamic voltage control, it comes at the expense of increased reactive power demand, as reactive power absorption increases in proportion to the sine of the firing angle. Adding reactive power compensation is a logical but costly solution, and must be rated for full voltage and current. Alternatively, firing angle can be minimized by adding transformer OLTCs, but this reintroduces all of the electro-mechanical problems identified earlier, and makes the solution less attractive.

10.2.1 Smelter potlines

Smelting requires extremely large DC currents, often in excess of 320 kA, to facilitate the production of commercial quantities of aluminium using the reduction process. The DC load consists of one or more potlines, which in turn are made up of many series-connected reduction cells, often in groups of 250 or more. Each cell has a nominal operating voltage which is controlled to around 4.0–4.5 V. The reduction cell is constructed with a movable carbon anode, electrolyte, aluminium cathode, cathode carbon and steel collector as shown in Figure 10.2.

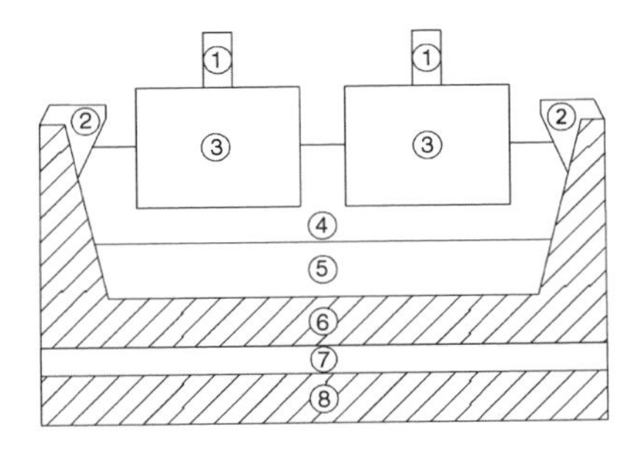

Figure 10.2 Simplified schematic of an aluminium reduction cell.

A normal reduction cell production cycle consists of: normal operation, operation for metal tapping (where aluminium is formed near the cathode and syphoned off periodically), alumina feeding and anode effect quenching. The cell anode is slowly consumed during the process, as are the cell walls, and must be periodically repaired or replaced.

10.2.2 Load profile

Under normal operating conditions the smelter load may be thought of as constant, with voltage and current fluctuations minimized by offsetting each cell's operating cycle within the potline. However, two additional operating conditions, that require increased control flexibility, must also be considered. The first, when a new smelter is commissioned, and the second following a power outage where smelter potlines must be restarted and the cells heated back up to operating temperature.

When a smelter is first commissioned, a small number of new cells is connected and brought up to operating temperature so that their voltages stabilize, with the positions of the other cells in the potline short circuited. More cells are connected, at a rate of one to two per day, and as they reach temperature, further cells are added; the process continues until the full complement is installed. As the series-connected cells have a common DC current, current magnitude must be maintained despite the increases in load, otherwise cells may become unstable. A smelter rectifier must therefore be able to supply full-rated current from very low voltage to full-rated voltage, with continuous control in between. It must also be able to compensate quickly when a new cell is installed, as it may initially be of high resistance, and require a much higher applied voltage across it to maintain the potline current flow.

A generic smelter voltage and current profile during commissioning is given in Figure 10.3. The DC current is ramped up in Figure 10.3(b), with the minimum number of cells connected, until full-rated current is reached, at which point further cells are connected and allowed to reach operating temperature. The small steps in voltage (Figure 10.3(a)) depict the connection of additional cells (from $t = 0.22$ to $t = 1.0$); 10 steps are shown for clarity, but in practice the number of steps will reflect the number of installed cells. The process may be coordinated across several potlines if a common DC supply is used, as applied voltage will also be common.

Once all cells are in service, and full aluminium production has begun, the DC current may again be considered constant, with the voltage varied depending on the number of series-connected cells in service. Periodically, some cells are removed for maintenance (to have the

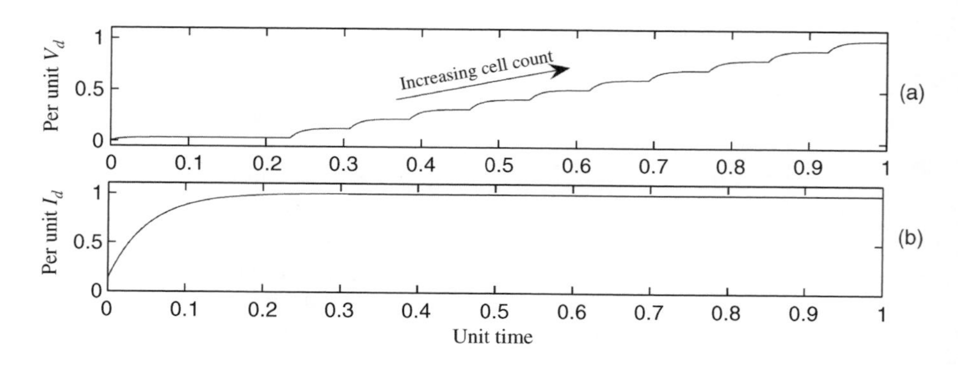

Figure 10.3 Idealized DC voltage and current during smelter commissioning.

anode carbon replaced or the cell walls repaired) and their position in the potline is short circuited. For normal operation, at least 90% of the cells are in service, but seldom are all in service at once.

If power is interrupted when a smelter is in full production, the smelter must be promptly restarted before the cells solidify and are made worthless. When the smelter is re-energized, the DC current is brought back up, the rate of increase depending on how long the cells have been out of service. The current may be increased in several stages to ensure that all cells maintain their voltage balance, although conventional smelters employing tap changers may not have this flexibility and may instead rely on controllers on each of the individual cells. An example of the smelter response is given in Figure 10.4, where the period and step sizes are arbitrary, and are used to illustrate that the current rise may be dependent on cell temperature.

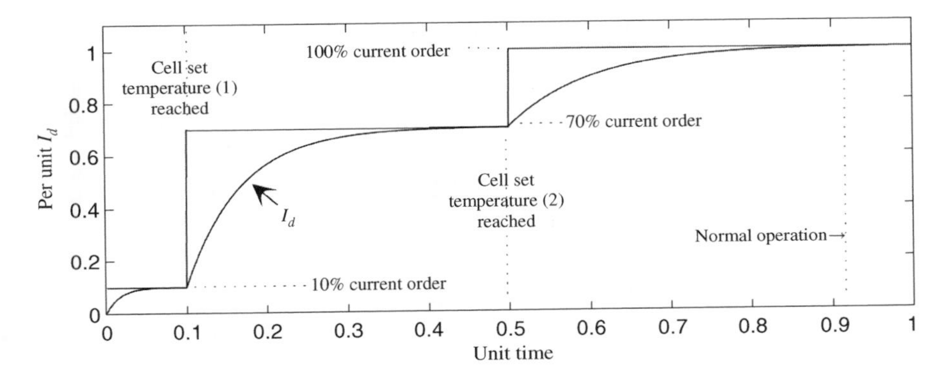

Figure 10.4 Ideal control response during smelter restart.

Figure 10.4 shows that the smelter is restarted (at t = 0) with a 10% current order and the current rise is continually monitored. When (at t = 0.075) the current level stabilizes, the cell temperature is compared to the expected level and when all the cells are within temperature tolerance (shown as label (1) in Figure 10.4) the next stage begins, which is initiated (at t = 0.1) with a 70% current order. Once 70% of the current is reached (at t = 0.5) and all the cells are balanced, the 100% current order is given corresponding to full applied DC voltage. At t = 0.92, full-rated DC current is obtained and, with all cells stabilized at operating temperature, production resumes.

10.3 Hybrid double-group configuration

The hybrid configuration, shown in Figure 10.5, consists of an IGCT group with all the rectifiers under common leading firing angle control and a thyristor group with all the rectifiers under common lagging firing angle. This arrangement does not alter the conventional high-pulse operation of the smelter. Likewise, by issuing the same DC current order to the leading and lagging group controllers, the sought-after high power-factor is implicitly assured.

At the smelter connection point, the power system is normally weak and this has a major influence on the maximum power transfer and reactive power requirements of the double-group configuration. In the leading group bridges, maximum power is reached when a further increase in current order requires a reduced firing angle, which reduces the reactive power generated and, thus, the AC terminal and DC side voltages; however, the reduction of DC

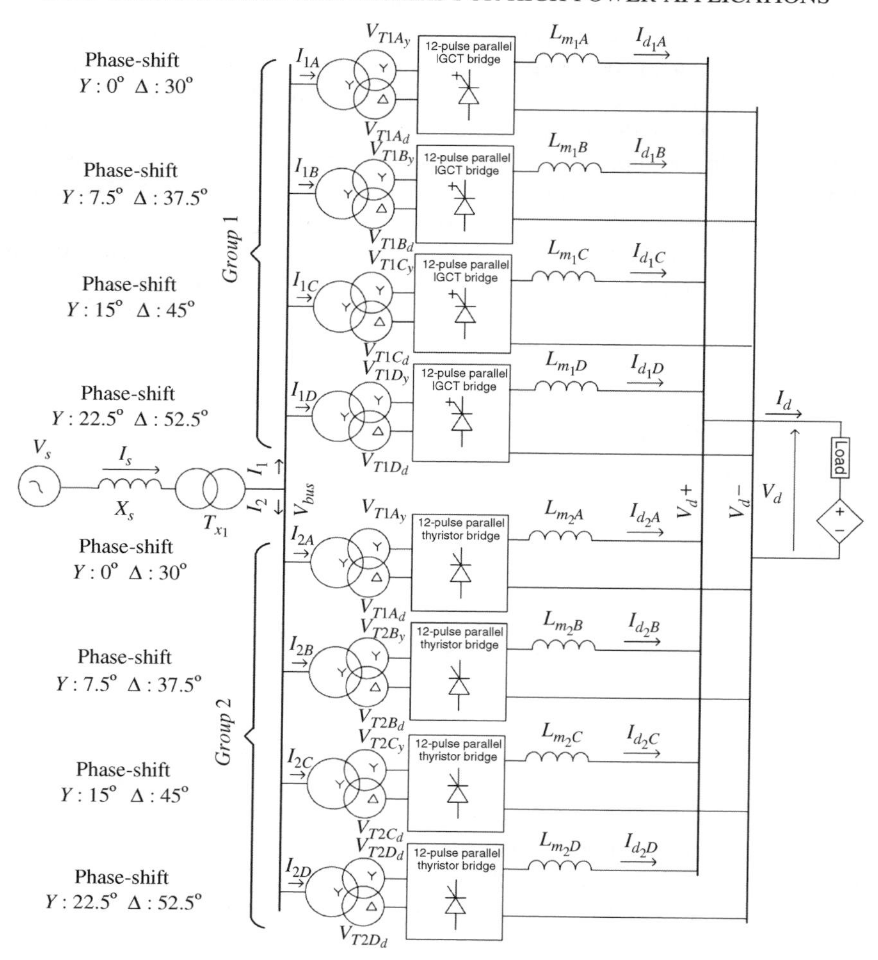

Figure 10.5 Configuration of the hybrid double-group 48-pulse test system.

voltage is compensated by a further reduction of firing angle, which in turn reduces the reactive power generated and lowers the terminal voltage.

Conversely in the lagging group bridges, a small increase in current order near the rated current (achieved by a reduction of the firing angle) reduces the demand for reactive power and this allows an increase in maximum power transfer. Thus, for some values of firing angle (of the order of 10–20°) it should be possible to improve the maximum power output of the smelter by increasing the current order of the lagging (thyristor-based) group and reducing the current order of the leading (IGCT-based) group.

10.3.1 The control concept

Figure 10.6 presents a single line diagram of the double-group two-quadrant configuration. The DC smelter load is represented in the diagram by an inductive (L_d) resistive (R_d) circuit in

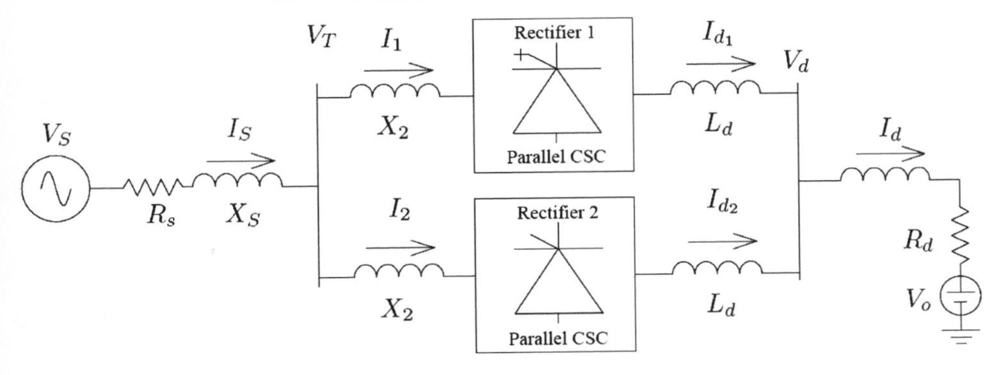

Figure 10.6 Generalized single line diagram of two paralleled rectifiers.

series with a back emf (V_o) and the total DC current (I_d) is distributed between the two group rectifier units. The magnitudes of the group AC currents (I_1) and (I_2) are both a function of their DC currents and their phase is dependent on the firing angles of their respective groups. Finally, the supply current (I_s) is the vector addition of the two individual AC group currents.

If the group firing angles, denoted α_1 and α_2, are selected such that one is leading and the other lagging the common AC terminal voltage (V_T), the reactive power injected into the AC system is $Q_T = Q_{T1} + Q_{T2} = V_T I_1 \sin(\alpha_1) - V_T I_2 \sin(\alpha_2)$ and can be controlled by varying $I_1, \alpha_1, I_2, \alpha_2$ without affecting the magnitude of the DC power supplied to the load. The magnitudes of the AC group currents (I_1 and I_2) are related to their respective interfacing transformers turns ratios (k_{n1} and k_{n2}) and their DC currents, i.e. $I_1 = k_1 k_{n1} I_{d1}$ and $I_2 = k_1 k_{n2} I_{d2}$.

As the two group rectifiers are connected in parallel, $V_{dc} = k_{n1} V_T \cos(\alpha_1) = k_{n2} V_T \cos(\alpha_2)$, and the reactive power injected into the AC system is

$$Q_T = V_T k_1 \left[k_{n1} I_{d1} \sqrt{1 - k_{n1}^{-2} k_{n2}^2 \cos^2(\alpha_2)} - k_{n2} I_{d2} \sin(\alpha_2) \right]$$

The transformer turn ratios are set at the design stage to achieve the best rating for each group during the operation when the group DC currents (I_{d1} and I_{d2}) do not reach their rated level.

If $k_{n1} = k_{n2}$ and the DC current is equally distributed between the groups ($I_{d1} = I_{d2}$), their imaginary components ($I_1 \sin \alpha_1$) and ($-I_2 \sin \alpha_2$) will cancel out and I_s will be in phase with V_T. The entire controlled rectification process will then appear to the AC system (at V_T) as a resistive load. This operating condition is illustrated in Figure 10.7.

10.3.2 Steady state analysis and waveforms

In the diagram of Figure 10.8, each group rectifier is represented by a single block, though in practice it will be made up of a number of parallel-connected six-pulse rectifier units. One of the groups (the one with lagging current) can use conventional thyristors, while the other (the one with leading current) must be self-commutating (IGCT-based). The following two subsections analyse the behaviour of this system without (the IGCT case) and with (the thyristor case) commutation overlap.

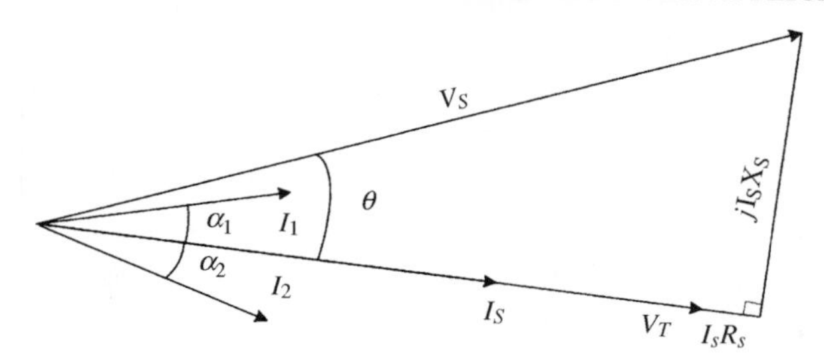

Figure 10.7 Phasor diagram illustrating the case of unity power-factor in a double-group rectifier.

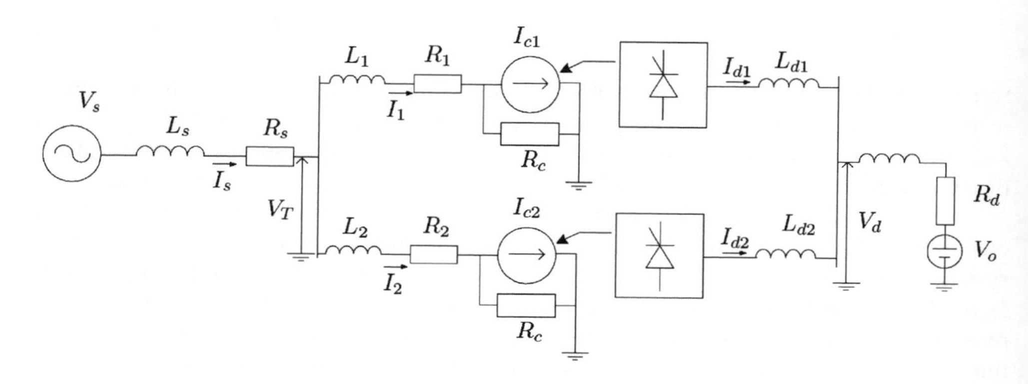

Figure 10.8 Simplified diagram of the hybrid double-group rectifier.

10.3.2.1 The self-commutating group

The circuit equations are derived for one half of the double group under ideal switching conditions. At the common AC bus (V_T) the equations for real and reactive power for one group rectifier unit are:

$$P_{T_1} = k_1 V_T I_1 \cos(\alpha_1) \tag{10.1}$$

$$Q_{T_1} = k_1 V_T I_1 \sin(\alpha_1) \tag{10.2}$$

where k_1 is a constant that depends on bridge rectifier configuration and α the firing angle.

Equating powers in terms of DC power (P_d)

$$P_{T_1} - P_{o_1} - P_{l_1} = P_{d1} \tag{10.3}$$

where P_{o_1} is the power loss, and P_{l_1} is the group contribution to the load power.

Given that

$$P_{d1} = V_{d1} I_{d1} \tag{10.4}$$

where

$$V_{d1} = L_1 \frac{dI_{d1}}{dt} \tag{10.5}$$

the combination of equations 10.3, 10.4 and 10.5 yields

$$L_{d1} I_{d1} \frac{dI_{d1}}{dt} = P_{T_1} - P_{o_1} - P_{l_1} \tag{10.6}$$

and

$$P_{o_1} = (R_{dc} + R_{ac}) I_d^2 \tag{10.7}$$

represents the total group power loss (AC and DC).

Equation 10.6 can be written as

$$\frac{L_{d1}}{2} \frac{d}{dt} I_{d1}^2 + (R_{ac} + R_{dc}) I_{d1}^2 = P_{T_1} - P_{L_1} \tag{10.8}$$

Rearranging to make I_{d1}^2 the subject, the equation in the (s) domain becomes:

$$I_{d1}^2(s) = \frac{1}{\frac{L_{d1}}{2} s + (R_{dc} + R_{ac})} \cdot (P_{T_1} - P_{L_1}) \tag{10.9}$$

Equation 10.9 gives the desired output I_{d1} in terms of measured inputs and rectifier parameters. Although the output is in terms of I_{d1}^2 this makes no difference to the control method, and allows the relationship to be considered as linear. No error is introduced as the group rectifier power flow direction and, therefore, the DC current are unidirectional. Thus the above relationships justify the implementation of a linear control system design.

The Thevenin source parameters could be represented in the equations by a suitable D-Q transform [3], from which the terminal voltage could be calculated; however, as parameters L_s and R_s are variable in practice, it is better to use the measured terminal voltage directly in the control calculations.

If ideal switches are used and the DC currents between the groups are balanced, the firing angles required to achieve unity power-factor operation will be equal and of opposite polarities and the DC output load voltage will be the average of the contribution from each group rectifier. Their individual contributions for a firing angle of $|15°|$ are given in Figure 10.9(a) and (b) for a leading and lagging angle respectively. The resulting waveform at the common DC bus, shown in Figure 10.9(c), is adapted from that of reference [4] for a 6-pulse basic rectifier unit.

From Figure 10.9(c) the upper peaks occur at $-30°, 30°$ and $90°$, with a maximum value of $\sqrt{2}\, V_T$, and lower peaks at 0° and 60° with a value of $\sqrt{\frac{3}{2}} V_T$; valid from $0 \leq |\alpha| \leq 30°$.

Above $|a| = 30°$ the upper peak disappears and $\sqrt{\frac{3}{2}} V_T$ becomes the new peak. The peak values and their valid ranges are given in Table 10.1.

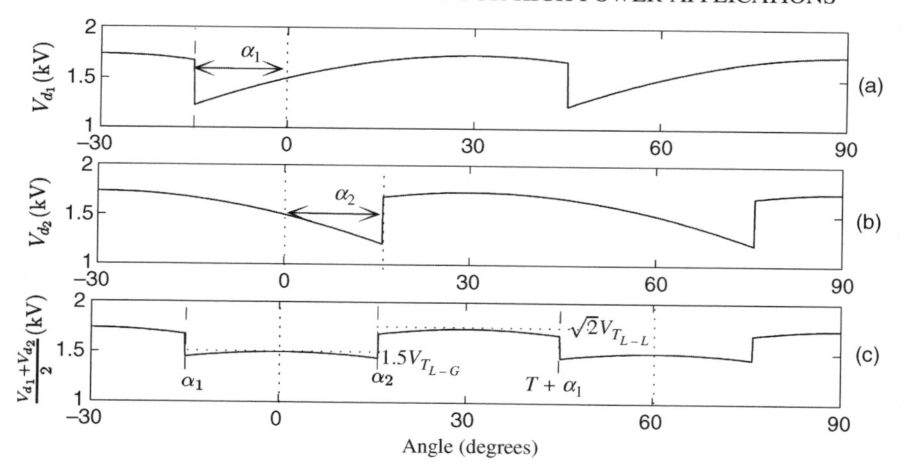

Figure 10.9 Idealized combined DC voltage waveform for six-pulse paralleled IGCT bridges for a firing angle of 15°.

10.3.2.2 The thyristor control group

To include the effect of the commutation overlap, equations 10.1 and 10.2 are replaced by:

$$P_{th} = 3k_{th}\left(\frac{3\sqrt{2}}{\pi}V_T\cos(\alpha_{th}) - \frac{3I_{d_{th}}X_c}{\pi}\right)I_{d_{th}} \tag{10.10}$$

$$Q_{th} = 3k_{th}\left(\frac{3\sqrt{2}}{\pi}V_T\sin(\alpha_{th}) - \frac{3I_{d_{th}}X_c}{\pi}\right)I_{d_{th}} \tag{10.11}$$

where the subscript$_{th}$ indicates applicability to the thyristor rectifier.

The value of the commutation reactance X_c in equations 10.10 and 10.11 varies with the amount of load connected to the supply system.

When comparing the DC output voltage of self-commutated and line-commutated rectifiers, the differences are most pronounced at small firing angles, where rated current is supplied. In a smelter where both thyristor types are used, the firing angle required for a specified DC voltage will differ between group rectifiers, the exact amount dependent on X_c and I_d.

Table 10.1 Waveform peaks for varied firing angle (α)

α	Upper peak	Lower peak
0–30°	$\sqrt{2}V_T$	$\sqrt{\frac{3}{2}}V_T$
30–60°	$\sqrt{\frac{3}{2}}V_T$	$\frac{V_T}{\sqrt{2}}$
60–90°	$\frac{V_T}{\sqrt{2}}$	0

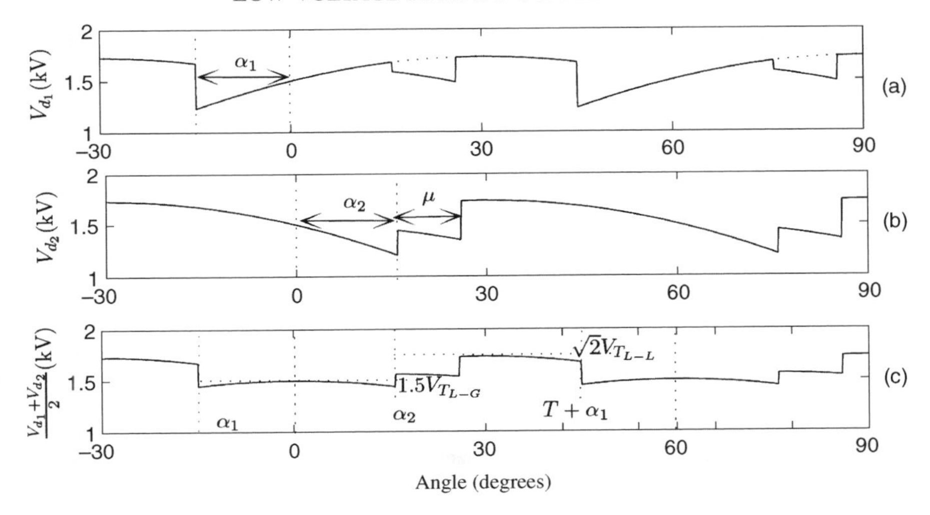

Figure 10.10 Idealized combined DC voltage waveform for six-pulse paralleled IGCT and thyristor bridges for a firing angle of 15° and commutation angle of 10°.

During the commutation two supply phases are effectively shorted together via their phase impedances, and any other circuit connected to the same terminal will experience the same voltage waveform. This is also the case if an IGCT bridge is connected in parallel with a thyristor bridge; in this case the AC supply voltage will be reduced by the voltage notches caused by the thyristor bridge commutations. The constituent and output DC voltage waveforms of the hybrid thyristor/IGCT combination are shown in Figure 10.10; where Figure 10.10(b) shows the DC output voltage of the thyristor bridge and Figure 10.10(a) the IGCT bridge voltage. In the latter waveform, the notches caused by the commutation in Figure 10.10(b) are evident from 15° to 25°, the dotted line giving the ideal trace. The average DC voltage as seen by the load is given in Figure 10.10(c), the peak levels of the two sections of the waveforms being $\sqrt{2}\,V_{T(L-L)}$ and $1.5V_{T(L-G)}$ respectively. Obviously once the commutation angle (μ) is greater than $(T+\alpha_1)$ the peak values will change, the new peak depending on the magnitude of I_d and X_c.

For equal current sharing to occur, the average DC voltages of the IGCT and thyristor bridges in Figure 10.10(a) and (b) must be the same under normal balanced operating conditions. Given that in this example the average DC voltage is 1.5250 $V_{T(L\text{–}L)}$ for the thyristor bridge, and higher at 1.5697 $V_{T(L\text{–}L)}$ for the IGCT bridge, equal current sharing would not occur, and so to be feasible, the bridges must have different firing angles. The exact difference depends on the average voltage drop due to the effect of thyristor commutation on each of the bridges.

Although attempts have been made to analyse and quantify these interactions analytically, the models used have either assumed infinite sources [4] or ideal switchings [5], both questioning the validity of the result. Moreover, even the simplified models require iterative solutions, which are more suited to computer simulation. No attempt is made here to develop an analytical model and concentrates, instead, on the complete solution that takes the interactions into account in the design of the control system.

10.3.2.3 Current and power ratings

Assuming that a ripple-free DC current is supplied equally, the fundamental *rms* AC current is

$$I_{(1)6p} = \frac{\sqrt{6}}{\pi} I_{d_{6p}} \tag{10.12}$$

for the IGCT rectifier and

$$I_{(1)6p} = k\frac{\sqrt{6}}{\pi} I_{d_{6p}} \tag{10.13}$$

for the line-commutated thyristor rectifiers [6], where

$k = \sqrt{}\{[\cos\ 2\alpha - \cos\ 2(\alpha + \mu)]^2 + [2\mu + \sin\ 2\alpha - \sin\ 2(\alpha + \mu)]^2\}/\{4[\cos\ \alpha - \cos\ (\alpha + \mu)]\}$,

$I_{d_{6p}} = \frac{I_d}{n_r n_g}$,

n_r is the number of 6-pulse rectifiers in a group,

n_g is the number of groups.

The *rms* thyristor and IGCT switch currents are given by

$$I_{sw} = \frac{1}{\sqrt{3}} I_{d_{6p}} \tag{10.14}$$

The rating for the star-delta connected transformer (0° phase shift) is

$$MVA_{0^o} = 2\frac{V_T}{k_T} I_{d_{6p}} \tag{10.15}$$

where V_T is the *rms* voltage of the common AC bus and k_T is the effective rectifier transformer turns ratio.

With multiple windings, the phase-shifting transformer's MVA rating [7] for each secondary (denoted by subscript$_s$) is

$$MVA_{s7.5-22.5} = 0.5 \times \Sigma(V_{w_n} \times I_{d_{6p}}) \tag{10.16}$$

where V_{w_n} is the equivalent sinusoidal *rms* voltage of each winding segment, and assuming that the two secondary winding sets are of equal rating.

The rating of the phase-shifted transformer primary is thus

$$MVA_{p7.5-22.5} = \Sigma(V_{w_n} \times I_{d_{6p}}) \tag{10.17}$$

The characteristic 12-pulse current waveform as seen from the primary side of each rectifier transformer is

$$i_{12p}(\omega t) = \frac{4\sqrt{3}}{\pi} I_{d_{6p}} \times \left(\cos(\omega t) - \frac{1}{11}\cos(11\omega t) + \frac{1}{13}\cos(13\omega t) - \frac{1}{23}\cos(23\omega t) + \frac{1}{25}\cos(25\omega t) - \ldots \right) \quad (10.18)$$

Combining all of the 12-pulse phase-shifted current waveforms, the current waveform as seen by the high-voltage transformer supplying one group from the AC bus becomes

$$i_{48p}(\omega t) = \frac{16\sqrt{3}}{\pi} I_{d_{6p}} \times \left(\cos(\omega t) - \frac{1}{47}\cos(37\omega t) + \frac{1}{49}\cos(49\omega t) - \frac{1}{95}\cos(95\omega t) + \frac{1}{97}\cos(97\omega t) - \ldots \right) \quad (10.19)$$

The *rms* fundamental current for one 48-pulse group is thus

$$I_{48p} = \frac{8\sqrt{6}}{\pi} I_{d_{6p}} \quad (10.20)$$

and when two groups are combined together

$$I_{s_{48p}} = \frac{16\sqrt{6}}{\pi} I_{d_{6p}}$$

which, written in terms of the total dc load current (I_d), is

$$I_{s_{48p}} = \frac{2\sqrt{6}}{\pi} I_d \quad (10.21)$$

The phase of each of the harmonic components in equation 10.19 will vary with firing angle and when two groups are combined their characteristic harmonics will either be attenuated or amplified for certain values of firing angle.

Finally, the MVA rating of the supply transformer (Tx_1 in Figure 10.5) with respect to the common AC bus is

$$MVA_{Tx_1} = V_T \times I_{s_{48p}} \quad (10.22)$$

10.3.3 Control system

The IGCT and thyristor controllers are derived from the circuit equations derived in section 10.3.2 and based on the cascaded control diagram of Figure 10.11. The subscript$_R$ is used to specify that the parameters apply to a generic group rectifier, and may be replaced by 1 or 2 for the IGCT and thyristor bridges respectively.

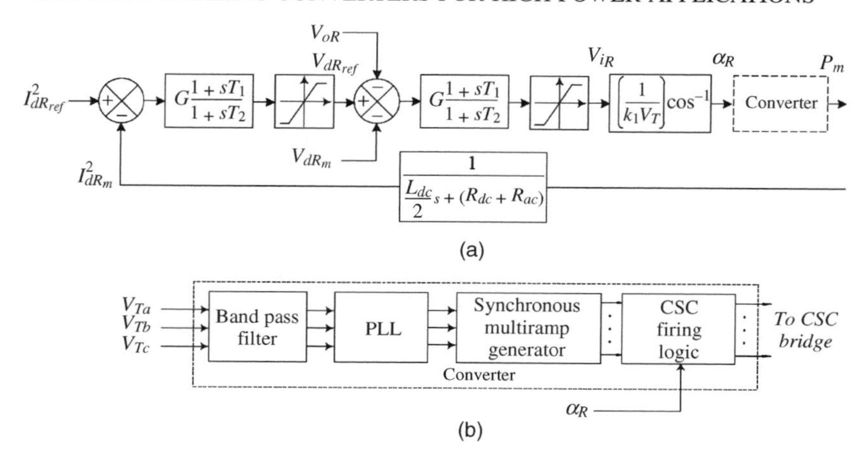

Figure 10.11 Block diagram of the generic smelter controller, with (a) current control and (b) firing angle control.

In Figure 10.11(a), the current order $I^2_{dR_{ref}}$ is compared to the calculated current $I^2_{dR_m}$ which is itself fed back through a block containing the DC parameters as calculated in equation 10.8. The resulting current error is passed to a lead–lag compensator whose output defines the voltage control order $V_{dR_{ref}}$. The signal is hard limited to prevent saturation and the difference between it and the measured DC voltage (V_{dR_m}), less the load back emf (V_{oR}), is passed to a second cascaded lead–lag compensator, the output of which defines the incremental voltage (V_{iR}) correction required. V_{iR} is divided by the conversion constant and *rms* terminal voltage ($k_1 V_T$) which yields $\cos(\alpha_R)$. The inverse cosine is taken to determine α_R, which is in turn passed to the converter firing logic, as given in Figure 10.11(b).

In the diagram a stable firing angle is developed with reference to the three-phase terminal voltage, which is band passed and used as the input to the three-phase phase-locked loop (PLL), which in turn is used to generate synchronized ramping control signals for each of the rectifier switches. These ramps are compared to the value of α_R to determine a precise switching pattern.

The same DC voltage must be produced by each rectifier for a given firing angle command, so that the DC load current is equally shared between all rectifiers in a group. When the phase is intentionally shifted, as is the case with phase-shifting transformers, the firing angle reference must also be shifted, either with a static increase in angle with respect to the common AC bus, or by physically moving the voltage reference (V_{Ta}, V_{Tb}, V_{Tc} in Figure 10.11(b)) to the low-voltage side of the transformer. The latter has the advantage of compensating for changes from the ideal shift, but the low-voltage waveform is typically more distorted and must be filtered; the filters potentially add other variable phase shifts which must be taken into account.

10.3.4 Simulated performance

10.3.4.1 Test system

The test system is a 300 MW smelter consisting of two group rectifiers, each made up of four 12-pulse equivalent rectifiers in the configuration shown in Figure 10.5. Only one potline is used, with a nominal DC voltage of 1 kV with the applied voltage and back emf of the individual cells being 4 V and 1.7 V respectively. Therefore to emulate the full smelter, 250 cells need to be placed in series. As shown in Figure 10.8, the test system also includes a load resistance (calculated to provide the full 320 kA load capacity at full voltage) and a lumped inductance (representing all the DC side inductances).

The incoming supply is represented by a Thevenin equivalent with V_s and X_s set to 220 kV and 0.17 H respectively, providing a short-circuit ratio of approximately 3. Transformer T_{X1} supplies 33 kV to the common bus that feeds the rectifier transformers and has a leakage reactance of 10%.

Each rectifier within a group is supplied by a phase-shifted transformer, providing conversion from 33 kV to 0.8 kV as input to the rectifier bridges, with leakage reactances for each specified as 5%.

The DC outputs are paralleled through small intergroup reactors ($L_{m1A}-L_{m1D}$ and $L_{m2A}-L_{m2D}$) which allow for instantaneous differences between each of the DC connections. The rectified DC voltage outputs from these groups range from 0 to a maximum (when unloaded) of 1100 V_{dc}.

The DC load is represented by a resistance and back emf of 1.5 mΩ and 420 V respectively, with V_d and I_d representing the load DC voltage and current measurements.

The DC side circuit parameters are in practice reasonably constant during normal operation, but they vary greatly during the cell warm-up phase. The temperature characteristics of the cells are well known and, thus, by monitoring the cell temperature during this operating phase the nominal controller gains can be adjusted, or alternatively the gain can be scheduled to ensure stable voltage and current control.

10.3.4.2 Dynamic response

The test system has been modelled in PSCAD/EMTDC and the response to a series of current step changes recorded over a 24 s period, a time scale much shorter than would be the case in practice; this is done to enable the computer solution in a realistic time; consequently the DC inductances are reduced by a factor of ten to produce a realistic di/dt. The solution period includes commissioning, normal operation, cell maintenance and shut-down phases, as well as a restart under current control. The shut-down phase constitutes a controlled reduction in DC load current, as opposed to the interruption of power when the AC circuit breakers are opened.

The smelter response is given in Figure 10.12. At time $t = 0.1$ s, commissioning begins, with initially 20% of the cells connected and the remaining positions in the potline short circuited. The smelter is energized with a 10% current order as given in Figure 10.12(b), which is then increased to 100% at t = 1 s; 90% current is reached at t = 5 s where additional cells are connected. Subsequent cells are installed at 0.5 s intervals until all are in service at $t = 8$ s. The DC voltage and real power traces increase as expected and negligible change is observed in I_d and Q_T, with high power-factor maintained throughout.

At t = 9 s, 10% of the installed cells are shorted to simulate their removal for maintenance; the DC voltage, as observed in Figure 10.12(c), decreases to maintain I_d constant as does the real power in Figure 10.12(a). The cell removal requires a reduction of the applied DC voltage in order to maintain the current constant. The DC voltage reduction requires an increase of the absolute firing angle and, thus, an increase in reactive power absorption in the thyristor group and an increase in reactive power generation in the IGCT group, such that the net reactive

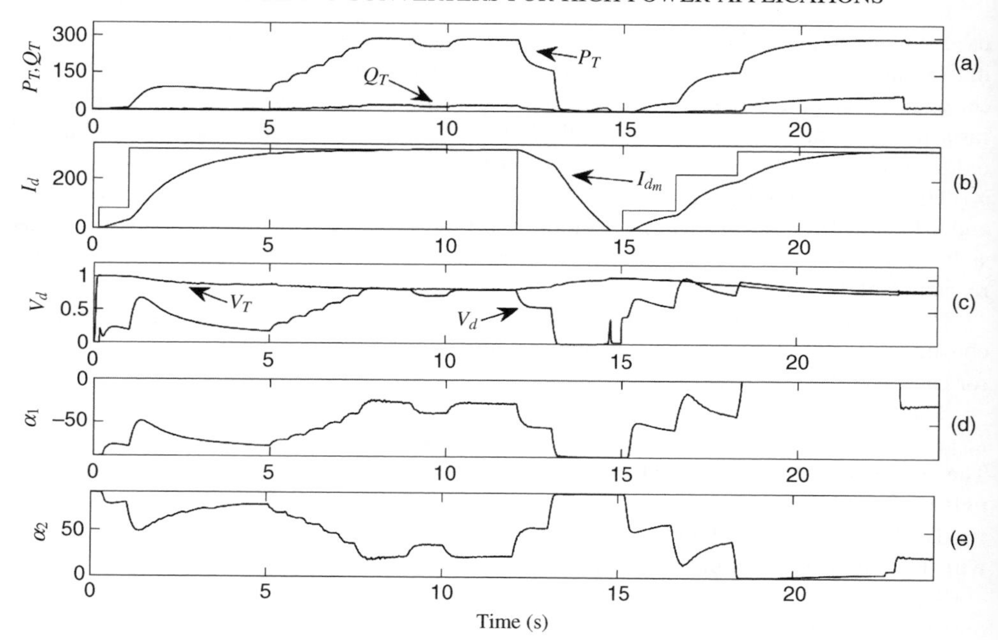

Figure 10.12 Simulated smelter performance, with (a) real and reactive powers, (b) ordered and measured DC currents, (c) normalized DC voltage and terminal voltage, (d) leading firing angles, (e) lagging firing angles.

power at the smelter terminals remains practically unchanged. The removed cells are reinstated at t = 10 s.

At t = 12 s the shut-down command is given and the controlled reduction of I_d begins. The current is first ramped down at a rate limited to 60 kA per second to prevent excessive reactive power circulation. When half power is reached, the ramp rate is increased to 120 kA per second, until the real power flow ceases (at 14 s) and the remaining current flow decays to zero (at t = 14.7 s). The HV supply circuit breakers are then opened.

Smelter restart begins at t = 15 s where a 64 kA (20%) current order is given, followed by a 225 kA (70%) order at 16.25 s, and finally a 320 kA (100%) current order at 18.25 s, with normal operation recommencing at approximately 23 s.

The active and reactive power responses on the high-voltage side of the supply transformer (T_{x1}) are shown in Figure 10.12(a). The lagging and leading reactive power contributions are shown in Figure 10.13(a) and (b), respectively. High power-factor operation is seen to be maintained (Figure 10.13(c)), with an average of 0.999 observed during normal operation and dropping briefly to 0.976 during restart (at about t = 22.8 s); this is due to the reactive power demands of all transformers when the rectifiers operate at zero firing angle.

10.3.4.3 Waveform quality

The high quality of the full load current waveform (I_s) at the smelter supply, shown in Figure 10.14(a), is clearly evident. Due to the lack of commutation overlap, the stepped 48-pulse waveform is more obvious in the case of the IGCT (shown in Figure 10.14(b)). The

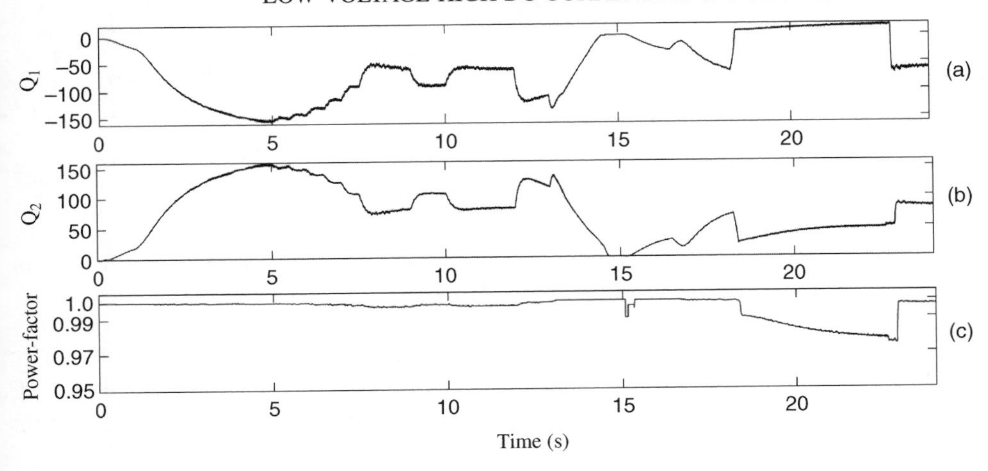

Figure 10.13 Smelter reactive power circulation with (a) generation in group 1, (b) absorption in group 2 and (c) power-factor as seen from the smelter terminal.

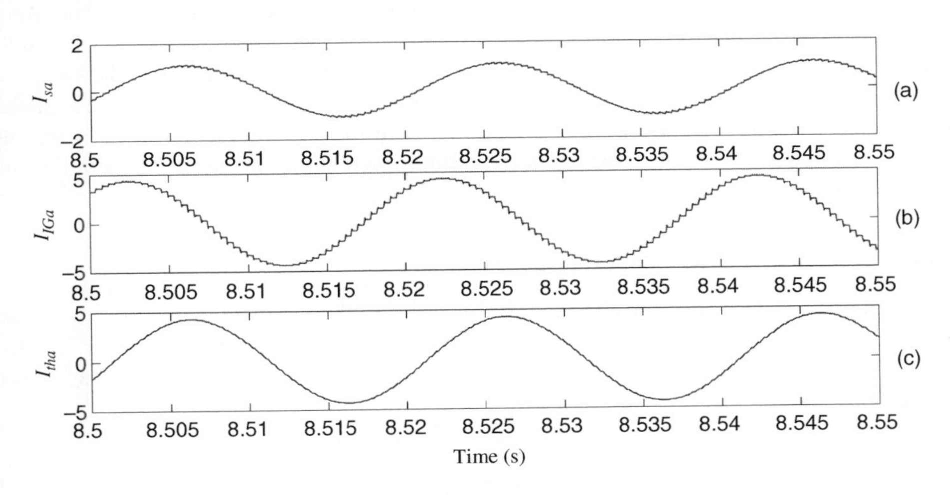

Figure 10.14 AC currents for: (a) the entire smelter; (b) combined IGCT bridges; (c) thyristor bridges.

corresponding harmonic spectra of the supply current, shown in Figure 10.15, indicate levels of 1.6% for the 47th, 49th, 0.99 for the 95th, 97th and 0.6% for the 143rd, 145th characteristic harmonics, the total harmonic distortion being 2.63%.

10.4 Centre-tapped rectifier option

A potential improvement to the IGCT-thyristor rectifier detailed in section 10.3 is the replacement of the transformers and rectifier bridge circuits for each six-pulse rectifier with a centre-tapped configuration. The immediate advantage of this is the reduction in switching

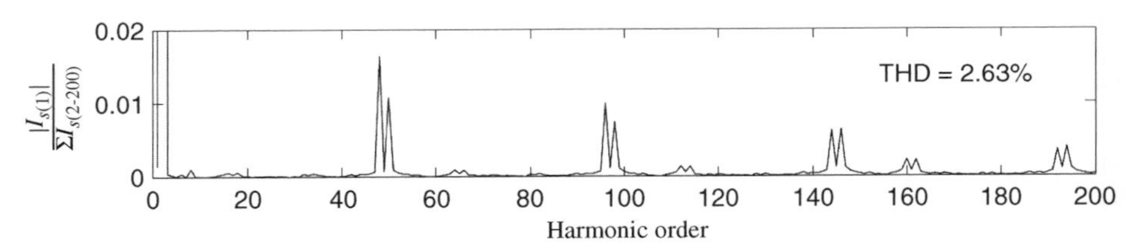

Figure 10.15 Harmonic performance of supply current (Is) for the first 200 harmonic orders, as taken from one cycle of Figure 10.14(a).

losses as only one thyristor conducts at any given time. Each thyristor only conducts for 60° of a cycle, as opposed to 120° for the bridge configuration, meaning the average thyristor DC current is half for the same rectifier power rating. The traditional argument for the use of the bridge configuration is a better utilization of the rectifier transformer, whereby the secondary windings conduct in both directions. This is justified by the need to connect switches in series to provide the voltage ratings required in medium- and high-voltage applications. However, the latter argument does not apply to the smelter case, where the DC voltage (below 2 kV) can be obtained with a single switch. Moreover, while the secondary transformer windings are simpler in the bridge configuration, especially when phase shifting is considered, it is at the expense of doubling their average current rating, which is a limiting factor in the design of the smelter transformers. The number of switches remains the same (six), each rated at double voltage and half average current. A consideration with the six-phase connection is that the primary winding must be delta connected to permit triplen harmonic current circulation, and must be sufficiently rated.

10.4.1 Current and power ratings

The rectifier thyristors conduct individually for 60° and their ratings in relation to the DC current of one six-pulse rectifier ($I_{d_{6p}}$) is given below, following on from equation 10.13.

The *rms* AC current in each phase and therefore each of the thyristor and IGCT switches in the six-phase rectifier is

$$I_{6phase} = \frac{1}{\sqrt{6}} I_{d_{6p}} \tag{10.23}$$

and the *rms* of the fundamental is

$$I_{(1)6phase} = \frac{\sqrt{3}}{\pi} I_{d_{6p}} \tag{10.24}$$

for the IGCT case, and

$$I_{(1)6phase_{th}} = \frac{\sqrt{3}}{\pi} I_{d_{6p}} \left[\frac{(\cos(2\alpha) - \cos 2(\alpha+\mu))^2 + (2\mu + \sin(2\alpha) - 2\sin(\alpha+\mu))^2}{4[\cos(\alpha) - \cos(\alpha+\mu)]} \right]^{\frac{1}{2}} \tag{10.25}$$

for the line-commutated thyristor rectifiers.

With series-connected phase-shifted windings, the transformer MVA rating for each secondary (denoted by subscript$_s$) is

$$MVA_{s7.5-22.5} = 0.5 \times \Sigma(V_{w_n} \times I_{d_{6p}}) \tag{10.26}$$

where V_{w_n} is the equivalent sinusoidal *rms* voltage of each winding segment, and assuming the two secondary winding sets are of equal rating.

The rating of the 12-pulse primary connection with two six-phase secondary windings is a combination of the *rms* values of each of the secondary AC currents.

$$MVA_{0^o} = 2\frac{V_T}{k_T} I_{d_{6p}} \tag{10.27}$$

where V_T is the *rms* voltage of the common AC bus and k_T is the effective rectifier transformer turns ratio. This is, as expected, the same as equation 10.15 for the bridge rectifier case. The characteristic harmonics, group current rating and supply transformer MVA rating are thus the same as in equations 10.18–10.22.

10.5 Two-quadrant MLCR rectification

Although the single MLCR rectifier is capable of developing the required pulse number, the current capacities of potential applications may require several MLCR rectifiers connected in parallel. By pairing up individual rectifiers, one with a leading and the other with a lagging firing angle, high power-factor operation is possible, and the DC current capability is increased. Potential applications include chlorine production plants, where DC currents are in excess of 40 kA and high pulse number is desired, but is uneconomical or impractical with conventional phase-shifted rectifier topologies. Likewise copper electro-winning is performed at currents of 10–100 kA and a number of MLCR high power-factor lead/lag pairs can be parallel-connected to achieve the required current levels. Other industrial processes including the production of manganese, magnesium, zinc and steel, require similar magnitude DC currents, but the largest DC current requirements come from the aluminium smelting industry where 500 kA is possible [8], with several pairs of rectifier units in parallel.

The MLCR rectifiers have more switches in the conduction path at a given instant, compared to conventional bridge rectifiers, but they use much simpler and more efficient transformer connections. With MLCR 48-pulse operation is possible using a single star–delta/star transformer as opposed to eight conventional phase-shifted secondaries if suitable transformer secondary ratings are available.

Figure 10.16 shows the block diagram of two MLCR units (A and B) that are connected in parallel on both the DC and AC sides. The AC system beyond the common AC bus (V_{bus}) is represented by a Thevenin equivalent (V_S and X_S). On the low voltage side of the transformers the AC voltages are V_{T1A} and V_{T1B} with further subscripts$_d$ and $_y$ used to indicate the delta and star connections respectively. On the DC side, small intra-group reactors (L_{m_1A} and L_{m_1B}) are used to allow the parallel connection despite the instantaneous voltage mismatches.

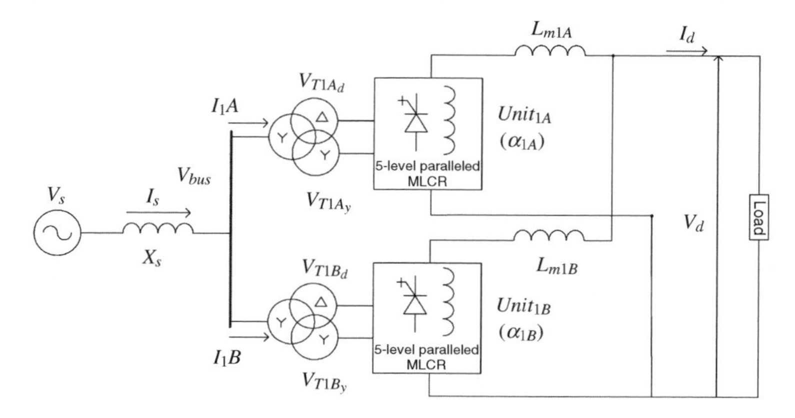

Figure 10.16 Rectifier group consisting of two paralleled five-level 48-pulse MLCR rectifiers.

Unlike traditional thyristor rectifiers, the MLCR switch commutation is performed under zero current, and the conventional reduction in DC voltage due to thyristor commutation is absent. The leading–lagging combination of Figure 10.16 should therefore produce the same average DC voltage for a given firing angle regardless of polarity.

The MLCR current waveforms in each of the supply transformer secondary windings are time varying functions. The quasi-sinusoid developed on the transformer primary side has the effect of phase-shifting the secondary transformer terminal voltage relative to the primary winding voltage across the complex impedance of the transformer. The magnitude and phase of the shift are dependent on the AC current magnitude, transformer leakage reactance and firing angle. The phasor representation for leading and lagging firing angle is shown in Figure 10.17. Figure 10.17(a), the leading case, shows the increased terminal voltage relative to the common bus voltage; conversely a decrease in Figure 10.17(b) is observed for a lagging firing angle.

When connected in parallel, the relationship between leading and lagging rectifier terminal voltages ($V_{T_{1A}}$ and $V_{T_{1B}}$ respectively) and the common DC voltage is

$$V_d = k_m V_{T_{1A}} \cos(\alpha_{1A}) = k_m V_{T_{1B}} \cos(\alpha_{1B}) \tag{10.28}$$

where k_m is the conversion constant which is dependent on topology, and the firing angles are referenced to each of the rectifier terminal voltages.

With reference to Figure 10.17, the relationship between the terminal voltage (V_T) and bus voltage (V_{bus}) is

$$V_{bus}^2 - V_T^2 = I^2 X^2 - 2IXV_T \sin(\alpha) \tag{10.29}$$

Rearranging equation 10.29 to isolate V_T yields

$$V_T = -IX\sin(\alpha) \pm \sqrt{V_{bus}^2 - I^2X^2\cos^2(\alpha)} \tag{10.30}$$

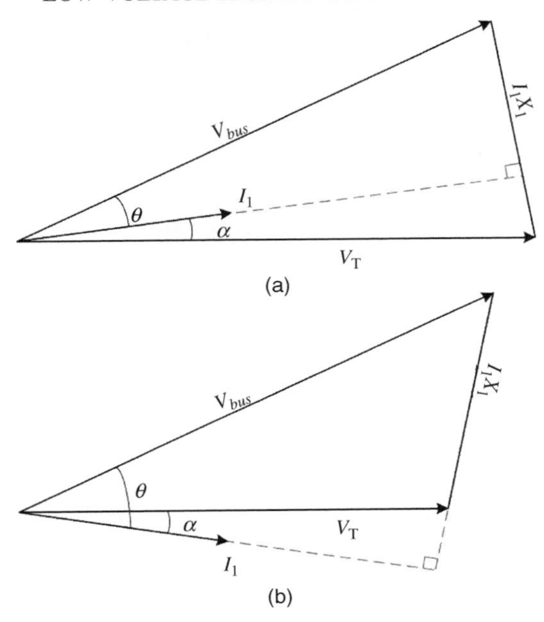

Figure 10.17 Operation of a single rectifier group for a leading (a) and lagging (b) firing angles.

which, when combined with equation 10.28 with respect to rectifier 1*A*, becomes

$$V_d = k_m \cos(\alpha_{1A})(\sqrt{V_{bus}^2 - I_{1A}^2 X_{1A}^2 \cos^2(\alpha_{1A})} - I_{1A} X_{1A} \sin(\alpha_{1A})) \quad (10.31)$$

The absolute values of firing angles α_{1A} and α_{1B} are dependent on the terminal voltages $V_{T_{1A}}$ and $V_{T_{1B}}$ respectively, which are in turn dependent on their AC current magnitude and angle and the resulting complex voltage across each of the rectifier transformer leakage reactances.

If the AC bus is supplied via a series impedance as in Figure 10.16, the AC bus voltage becomes a function of each rectifier's AC current and firing angle, making the system response very non-linear.

10.5.1 AC system analysis

The AC supply current waveform of the parallel MLCR (shown in Figure 4.15) is a combination of each of the thyristor rectifiers' and the reinjection circuit supply currents. A Fourier study is needed to find the expected peak, *rms* and THD of the combined current. The analysis is carried out under ideal conditions, with constant I_d, to simplify the solution. The solution for an m level MLCR bridge is summarized here, with a full proof available in reference [9].

Currents drawn by each of the star and delta bridges are found as follows.

The star-connected bridge contribution to the supply current is

$$I_{aYn} = \frac{2}{\pi}\int_0^{\pi} I_{aY}(\omega t)\sin(n\omega t)d(\omega t) = \frac{8[1-(-1)^n]I_d}{n\pi(m-1)}\sin\frac{n\pi}{12(m-1)}\cos\frac{n\pi}{6}S_{An} \tag{10.32}$$

for $m \geq 3, n = 1, 2, 3, \ldots$, and that of the delta connection

$$I_{ca\Delta n} = \frac{2}{\pi}\int_0^{\pi} I_{ca\Delta}(\omega t)\sin(n\omega t)d(\omega t) = \frac{8[1-(-1)^n]I_d}{3n\pi(m-1)}\sin\frac{n\pi}{12(m-1)}\cos\frac{n\pi}{6}S_{Bn} \tag{10.33}$$

for $m \geq 3, n = 1, 2, 3, \ldots$
where

$$S_{An} = (m-1)\sin\frac{n\pi}{6} + \sum_{i=1}^{m-2} i\sin\left(\frac{n\pi}{3} + \frac{in\pi}{6(m-1)}\right)$$

$$S_{Bn} = (m-1)\sin\frac{n\pi}{3} + 2\sum_{i=1}^{m-2} i\cos\frac{n\pi}{6}\sin\left(\frac{n\pi}{3} + \frac{in\pi}{6(m-1)}\right)$$

The total bridge supply current $I_{a_R}(\omega t)$ is found by combining equations 10.32 and 10.33, giving:

$$I_{a_R}(\omega t) = \frac{1}{k_n}\left[I_{aY}(\omega t) + \sqrt{3}I_{ca\Delta}(\omega t)\right] \tag{10.34}$$

where k_n is the interface transformer turns ratio.

The Fourier components of $I_{a_R}(\omega t)$ are

$$I_{aRn} = \frac{8[1-(-1)^n]I_d}{\sqrt{3}k_n n\pi(m-1)}S_{Cn}S_{Dn} \tag{10.35}$$

for $m \geq 3, n = 1, 2, 3, \ldots$
where

$$S_{Cn} = \sin\frac{n\pi}{12(m-1)}\cos\frac{n\pi}{6}$$

$$S_{Dn} = 2\left(\cos\frac{n\pi}{6} + \frac{\sqrt{3}}{2}\right)S_{An}$$

The fundamental peak value of the parallel MLCR bridge output current, derived from equation 10.35 is

$$I_{a_R1} = \frac{16\sqrt{3}I_d}{k_n\pi(m-1)}\sin\frac{\pi}{12(m-1)}\left[\frac{(m-1)}{2} + \sum_{i=1}^{m-2} i\cos\left(\frac{\pi}{6} - \frac{i\pi}{6(m-1)}\right)\right] \tag{10.36}$$

The output *rms* line current of the bridge is thus:

$$I_{a_Rrms} = \sqrt{\frac{1}{\pi}\int_0^{\pi} I_A(\omega t)^2 d(\omega t)} = \frac{\sqrt{4+\sqrt{3}}}{3k_n} I_d \sqrt{1+\frac{11-6\sqrt{3}}{13(m-1)^2}} \tag{10.37}$$

Given that a five-level reinjection, 48-pulse MLCR configuration is used throughout this section, equations 10.36 and 10.37 for $m=5$ become:

$$I_{a_R1} = \frac{1.7063}{k_n} I_d \tag{10.38}$$

$$I_{a_Rrms} = \frac{0.8}{k_n} I_d \tag{10.39}$$

The THD of the output current for the five-level configuration is

$$THD_{I_{aR}} = \sqrt{\frac{2I_{a_Rrms}}{I_{a_R1}^2} - 1} = 4.00\% \tag{10.40}$$

10.5.2 Component ratings

Approximate ratings are calculated based on the reinjection bridge being directly connected to an ideal source and under steady state operation. The rated supply current is specified as $I_{a_{rated}}$ and taken from the fundamental current $I_{a_R(1)}$. The following calculations apply to the five-level ($m=5$) MLCR.

Supply transformer:

The MVA rating of the supply transformer is

$$S = 3V_{T_{rated}} I_{a_{rated}} \tag{10.41}$$

where $V_{T_{rated}}$ and $I_{a_{rated}}$ are the rated fundamental *rms* phase voltage and line current of the supply transformer primary. The fundamental component of the output current I_{a_R} is the rectifier system rated current $I_{a_{rated}}$.

From equation 10.36, the rated current $I_{a_{rated}}$ is expressed as

$$I_{a_{rated}} = k_i \cdot I_d \tag{10.42}$$

where

$$k_i = \frac{\sqrt{6}(2-\sqrt{3})}{4k_n \pi \sin\frac{\pi}{48}} \tag{10.43}$$

The supply transformer primary voltage rating is taken as V_T, with the star and delta secondaries rated at $\frac{V_T}{k_R}$ and $\frac{\sqrt{3}V_T}{k_R}$ respectively.

Similarly the transformer phase current RMS ratings are tabulated below:

Y primary $$I_{a_{rms}} = \frac{\sqrt{4+\sqrt{3}}}{3k_n k_i}\sqrt{1+\frac{11-6\sqrt{3}}{52}}I_{a_R rated}$$

Y secondary $$I_{aY_{rms}} = \frac{\sqrt{2+(4)^{-2}}}{3k_i}I_{a_R rated}$$

Δ secondary $$I_{ca\Delta_{rms}} = \frac{\sqrt{2+(4)^{-2}}}{3\sqrt{3}k_i}I_{a_R rated}$$

The fundamental (*rms*) component of the phase currents are given as:

Y primary $$I_{a_{R(1)} rms} = I_{a_R rated}$$

Y secondary $$I_{aY_{R(1)} rms} = \frac{k_n I_{a_R rated}}{2}$$

Δ secondary $$I_{ca\Delta_{R(1)} rms} = \frac{k_n I_{a_R rated}}{2\sqrt{3}}$$

Multi-tapped reactor:

The operating frequency of the 48-pulse MLCR reactor is six times the supply frequency. The phase voltage *rms* rating is:

$$\sqrt{2}k_n^{-1}V_T \sin\frac{\pi}{12}\sqrt{2-\frac{6}{\pi}\cos 2\alpha_R} = \sqrt{6}k_n^{-1}\sin\frac{\pi}{12}\sqrt{2+\frac{6}{\pi}}V_{T_{rated}}$$

As the turns ratio of each of the reactor windings is equal, the proportion of the DC current is shared across them, with $\frac{3}{4}I_d$ across the first and fourth windings, and half I_d across windings two and three.

The *rms* current rating of a four-winding reactor is generalized as:

$$\sqrt{\frac{1}{4}\sum_{i=1}^{3}\left[\frac{iI_d}{4}\right]^2} = \frac{\sqrt{21}}{4}I_d = \frac{\sqrt{21}}{4k_i}I_{a_R rated}$$

Switching devices:

Voltage rating (forward/reverse) of the main bridge switches:

$$V_{sw} = \sqrt{6}k_n^{-1}V_{T_R rated}$$

rms current rating of each of the main bridge switches:

$$I_{sw_{rms}} = \frac{\sqrt{2+4^{-2}}}{3\sqrt{2}k_i}I_{a_R rated}$$

rms current rating for the reinjection switches S_{j1} and S_{jm}:

$$I_{jsw_{rms}} = \frac{I_{a_R rated}}{k_i 2\sqrt{2}}$$

and for the all other reinjection switches:

$$I_{jsw_{rms}} = \frac{I_{a_R rated}}{2k_i}$$

The maximum voltage (forward/reverse) to which a reinjection switch is subjected occurs when either switch 1 (S_{j1}) or switch 5 (S_{j5}) conducts. In an m-level rectifier, if S_{jm} conducts, the general expression for the voltage across the kth reinjection switch, S_{jk} is

$$V_{S_{jk}} = \frac{(5-k)}{4} V_M k = 1,2,3\ldots m$$

and

$$max\left[V_{S_{jk}}\right] = \frac{2\sqrt{6}k_n^{-1}(m-k)}{(m-1)} \sin\frac{\pi}{12} V_{SR}$$

Therefore the voltage rating of the reinjection switches in a five-level MLCR is

$$V_{jsw} = 2\sqrt{6}k_n^{-1} \sin\frac{\pi}{12} V_{SR}$$

10.5.3 Multigroup MLCR rectifier

In very high-current applications several MLCR rectifier groups are parallel-connected on the AC and DC sides to form a multigroup configuration. Figure 10.18 presents an 'n' group MLCR scheme, with each of the rectifiers connected to a common medium voltage AC bus (V_{bus}). The transformers, rectifiers and smoothing reactors of the various groups are of identical rating and construction.

As each group constitutes a five-level reinjection high power-factor rectifier, harmonic performance is assured regardless of the current contributions of each rectifier. Moreover, as long as the DC current is balanced between the leading and lagging rectifiers of the group, high power-factor is maintained. Therefore, unlike traditional phase-shifted high-pulse rectification, each rectifier DC current may be controlled directly and independently. As a consequence, the DC side parallel connections are greatly simplified, and are made without the need for interphase reactors, a limitation of traditional high-pulse phase-shifted transformer rectifiers.

10.5.3.1 Independent group current control

Independent DC current control also affords a rectifier group with increased flexibility. If power ratings allow, the current may be deliberately increased in the leading or lagging rectifier to change the phase of the group supply current (e.g. I_1 in Figure 10.18) relative to the bus voltage (V_{bus}), providing additional, but limited power-factor control. Generating a surplus of reactive power has the potential to correct for other inductive loads on the common connection (such as transformer reactances and induction machines), or to export back to the

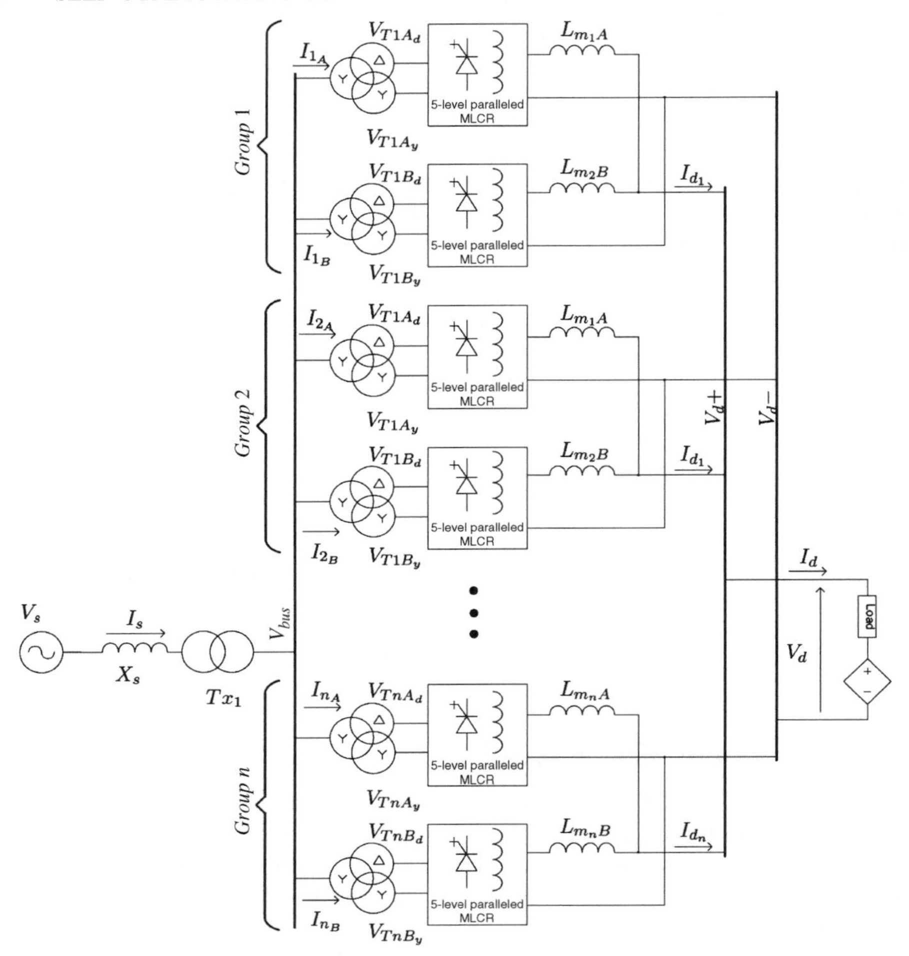

Figure 10.18 Paralleled test system comprised of 'n' groups.

system for terminal voltage support. Such power-factor control is very non-linear, since the magnitude and phase of each rectifier is intimately linked to firing angle and cannot be decoupled.

Generally, to maximize efficiency, the rating of the rectifiers and the load are matched, meaning that changing the proportioning of group DC current by decreasing the contribution of one rectifier and increasing that of the other would overload the latter. If on the other hand, the rectifiers were only partially loaded, altering the current bias within a group would be possible. Additionally, some groups could intentionally be run at rated DC current, with others lightly loaded or disconnected completely, to increase the multigroup efficiency during periods of partial loading.

In the case of aluminium smelters during commissioning, full current is reached with a minimum number of installed cells, requiring full-rated current from all rectifiers in the multigroup, but with low voltage and therefore large firing angles. Conversely, during a smelter restart with a full complement of installed cells, the current profile (similar to that given in

Figure 10.4) is such that load current could be supplied by some rectifier groups, with others brought online as demand increases.

The goal of such a strategy would be to minimize the number of active MLCRs for a given current order, thus providing the most efficient means of power supply. Two methods of managing this power delivery are explored below.

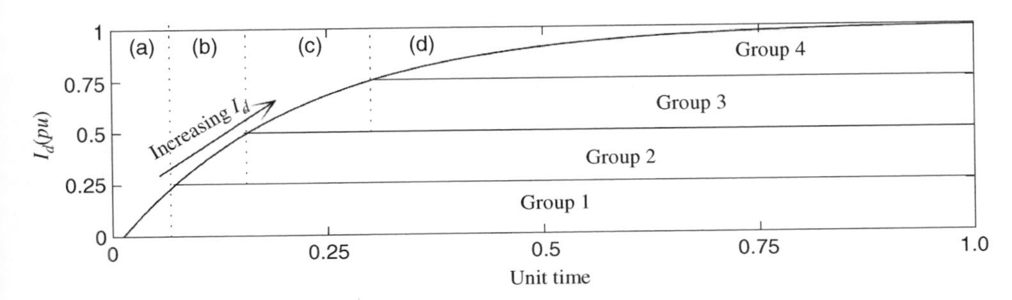

Figure 10.19 Ideal DC current response to minimize losses during increasing current order (in per unit).

In the first method, displayed graphically in Figure 10.19, rated current is maintained in all but one active rectifier, with it supplying the balance. In region (a) all the load current is supplied by Group 1 as DC current order is increased. Upon reaching rated current in Group 1, Group 2 is enabled (region (b)) and this provides the balance of current, until it too reaches rated current, where Group 3 is enabled and so on. In region (d), rated DC load current is shared equally between the full complement of rectifiers, so as to aggregate the losses.

The second method involves deriving a current order based on the number of active rectifier groups and adding new groups as the current order is increased, spreading the current delivery evenly between active groups. Some additional complexity is caused by the need to coordinate the rectifier group switch-on based on an overall set point and rectifier rating, and also by the controlled disconnection of groups when the current order decreases. To achieve this, a hysteresis band is introduced, so that a new MLCR rectifier is turned on when an increase in order will put the existing rectifier over 100% current rating, and taken off when removing one rectifier will result in all remaining rectifiers operating at a safe limit, say 75% of their rating.

The requirements are as follows:

Condition to trigger n + 1 rectifier group activation

$$\frac{I_d^*}{rating} > Grps\_active_n \tag{10.44}$$

Condition to trigger n rectifier group deactivation

$$\frac{I_d^*}{((Grps\_active - 1) - deadband^{\dagger}(\%))} \leq Grps\_active_n \tag{10.45}$$

The current order to each of the active groups is

$$I_{d_{grp}}^* = \frac{I_d^*}{(Grps\_active_n)} \tag{10.46}$$

where:

$I_d^* =$ load current order,

$I_{d_{grp}}^* = I_d$ group order,

$\text{Grps\_active}_n =$ number of groups active,

Rating = rating of rectifier,

deadband$^\dagger = 1 -$ minimum loading (%).

The methods of operation described above assume that full current is reached at full rated voltage, a condition that is only true if the full DC load is installed and the current order varied, as occurs during a restart (covered in section 10.3).

10.5.4 Controller design

The control system block diagram capable of implementing both methods of control is shown in Figure 10.20. Section (*I*) consists of a hard limit block and integrator which are configured as a simple integral controller through the use of negative feedback. When a change in current order is required, an error signal is generated from the difference between the present current order and the new order. The error magnitude is conditioned by user-defined bounds through the hard limit block and the resulting signal is integrated, again with user-defined values. By selecting suitable parameters, the overall signal ramp rate can be controlled and the response for small changes is unaffected.

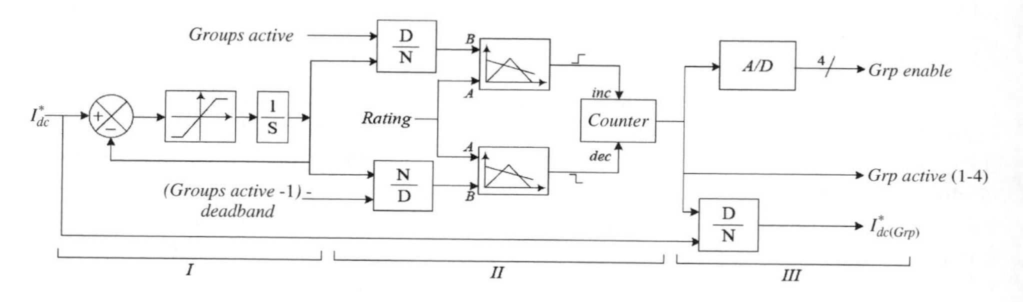

Figure 10.20 Multigroup rectifier scheduler.

The conditioned current order is fed as the input into section (*II*) where the two comparator blocks determine the number of groups to be activated, based on ratings and hysteresis. The upper comparator performs the increment function as in equation 10.44 and the lower performs the decrement function as in equation 10.45. The actual comparison block works as follows: when signal A increases to the point where it exceeds signal B, a pulse is output. In the case of the upper comparator, this corresponds to a counter increase (incrementing the number of groups active by 1), while the lower corresponds to a counter decrease.

Section (*III*) takes the counter output and provides the group I_d order and raw enable/disable signals to the rectifier groups. It also outputs a status level of how many groups are active.

Another important point is the possible $\frac{dI}{dt}$ at turn off if a rectifier group is deactivated with current still flowing in it, and the resulting negative V_d spike which can be several kV in magnitude. To ensure that abrupt removal of a parallel rectifier does not occur when a reduction in current order is requested, two things must happen. First the zero current order must wait until a 'deactivate' command is given, and second, the thyristors must only be turned off when current in the group falls to zero, thus allowing natural commutation to occur and preventing $\frac{dI}{dt}$ related voltage spikes. The logic required to correctly enable and disable the firing circuits is shown in Figure 10.21.

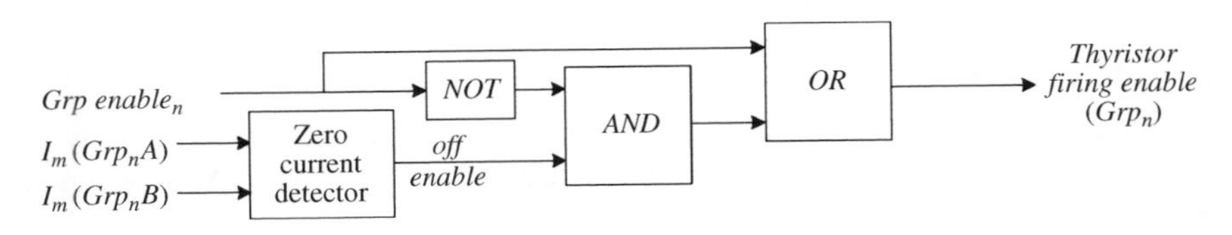

Figure 10.21 Group enabler logic.

The current order for each group is derived from simple PID feedback loops, as shown in the block diagram in Figure 10.22. With reference to the figure, the input to the group controller is rate limited to restrict $\frac{dI^*_{d(grp)}}{dt}$, particularly during the activation of an additional group, using the same order conditioning technique as in Figure 10.20(I) but with a much faster integration rate. The conditioned signal is split equally between the leading and lagging controllers (upper and lower in Figure 10.22) and becomes the current order for each unit. The error input to the lead–lag compensator (PID) block is determined by the difference between the measured group I_d and current order, whose output is rate limited and then passed as DC voltage order to the voltage controller. A second lead–lag compensator derives the firing angle order, which is scaled by the measured AC terminal voltage.

From there, the firing angles (α_{nA} and α_{nB}) are fed into the thyristor firing logic, derived from multiramp PLLs; each unit uses its filtered AC terminal voltage on the transformer low-voltage side as a reference in preference to the common AC bus.

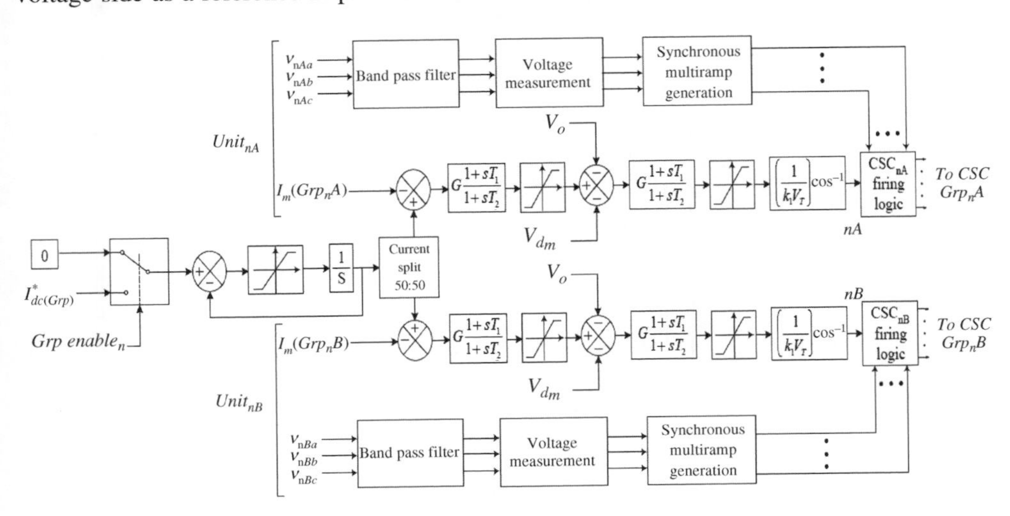

Figure 10.22 Rectifier group control and firing logic for group 'n'.

The control block diagrams in Figures 10.20–10.22 are also modular to enable virtually any number of groups to be paralleled. The controller gain parameters must be the same for all rectifiers, so that if a change in current order occurs when all rectifiers in the multigroup are active, all will respond at the same rate, thereby ensuring that current sharing between groups is maintained.

10.5.5 Simulated performance of an MLCR smelter

10.5.5.1 Test system

The test system of Figure 10.23 consists of eight identical 48-pulse MLCR rectifiers which are paired into four groups, forming a 300 MW smelter DC power supply. A Thevenin source represents the incoming transmission system, with the source voltage (V_s) and source impedance (X_s) specified as 220 kV and 0.17 H respectively. Transformer Tx_1 provides 33 kV

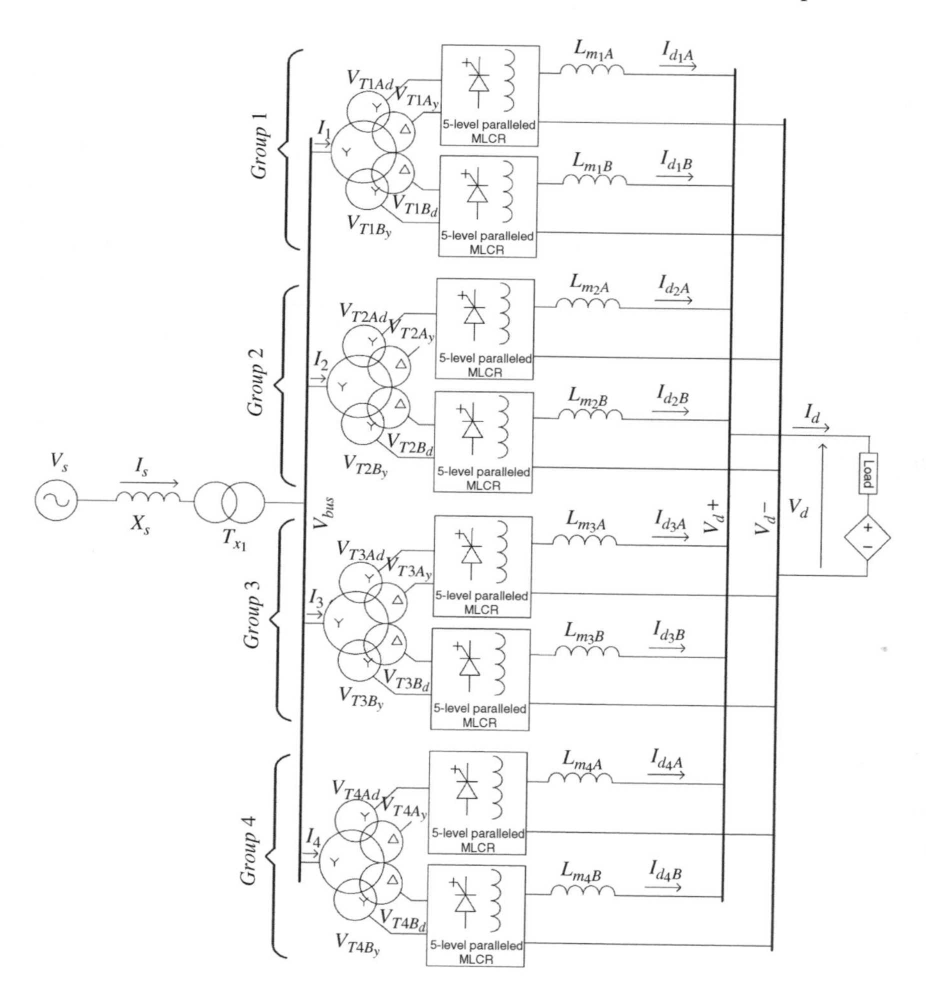

Figure 10.23 Paralleled test system comprised of four 48-pulse reinjection groups.

to a common AC bus (V_{bus}), which feeds each of the three-winding star–delta MLCR rectifier transformers.

In the figure, the $_A$ subscript denotes a rectifier with a leading, and $_B$ with a lagging firing angle. Each of the rectifier DC outputs are paralleled through small intergroup reactors ($L_{m1A} \ldots L_{m4B}$) which permit small instantaneous differences in DC voltage. The rectified DC voltage outputs from these groups range from 0 to 1100 V_{dc} unloaded.

The DC load is represented by a resistance and a small back emf of 1.5 mΩ and 420 V respectively, with V_d and I_d representing the load DC voltage and current measurements.

All transformer leakage reactances are specified as 10%, which when combined with X_s provide the system with a short-circuit ratio of approximately 3.

10.5.5.2 Dynamic Response

The test system is modelled using PSCAD/EMTDC package and a series of step changes is made over a 24 s period to emulate smelter commissioning, on-load maintenance, restart and normal operation, the results of which are presented in Figure 10.24.

The magnitude and timing of the step changes are identical to those of section 10.3.4.2, so that a direct comparison may be made. A summary of the steps in current order and number of installed cells is given in Table 10.2.

The simulation is started with a current order of 10% to energize all rectifiers, followed by a 100% current order at t = 1 s. Commissioning of the smelter potlines is initiated at t = 5 s where additional cells are connected in stages until 100% are in operation at t = 7.5 s. The rectifiers are all issued with an equal current command and the measured current in Figure 10.24(c) maintained evenly across all rectifiers (hence the singular trace).

At t = 12 s the shut-down command is given and the reduction in total DC current is observed in Figure 10.24(b). From t = 13.25 s to t = 14.65 s a slight difference in DC current response is noticeable between the lagging and leading rectifiers in Figure 10.24(c), with a slightly faster decrease rate from the latter.

When a decrease in DC current order is given, each rectifier reduces its absolute firing angle. For the leading rectifiers an immediate consequence of this is a lower AC terminal voltage to that prior to the decrease command. Conversely, the lagging rectifiers observe an increase in relative terminal voltage. As a result the lowering terminal voltage of the leading rectifiers reduces the maximum DC voltage proportionally, assisting in current decrease, while increased terminal voltage hinders the lagging type.

From the control diagram of Figure 10.22, the phenomenon is partly compensated in the voltage controllers by terminal voltage scaling prior to firing angle output, but since the terminal voltage is a calculation performed over time, and averaged to minimize noise, it exhibits a slight delay, which prevents ideal compensation. Since minimal real power is drawn from the power system during this period, the effect on the network is minimal.

Smelter restart begins at t = 15 s where a 64 kA (20%) current order is given, followed by a 225 kA (70%) order at 16.25 s, and finally a 320 kA (100%) current order at 18.25 s, with normal operation recommencing at approximately 19.2 s.

The active and reactive power responses on the high-voltage side of the supply transformer (T_{x1}) are given in Figure 10.24(a). The leading and lagging reactive power contributions of each of the rectifiers are shown in Figure 10.25(a) and (b) respectively. High power-factor operation is seen to be maintained (Figure 10.25(c)), with an average of 0.997 (4.45° lagging) observed during normal operation and dropping briefly to 0.99 (8.1° lagging) during restart (at

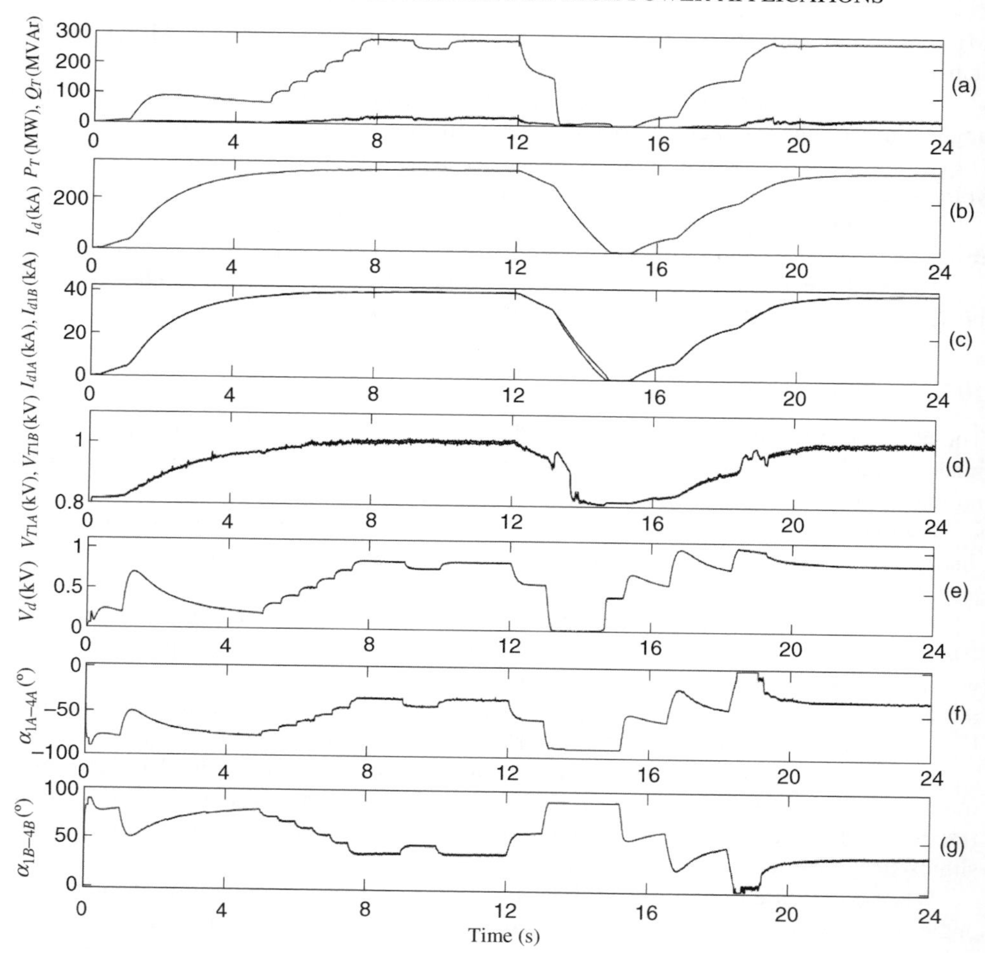

Figure 10.24 Control responses of four MLCR rectifier groups to a series of step changes with (a) total real and reactive power, (b) DC current, (c) group currents, (d) AC terminal voltages, (e) DC voltage, (f) leading firing angles and (g) lagging firing angles.

Table 10.2 Summary of step changes in the simulation

Time (s)	0.15	1	5	5.5	6	6.5	7
$I_{d_{order}}$	25%	100%	100%	100%	100%	100%	100%
Installed cells	14%	14%	28%	43%	57%	71%	86%
Time(s)	7.5	9	10	12	15	16.5	18.25
$I_{d_{order}}$	100%	90%	100%	0%	25%	70%	100%
Installed cells	100%	100%	100%	100%	100%	100%	100%

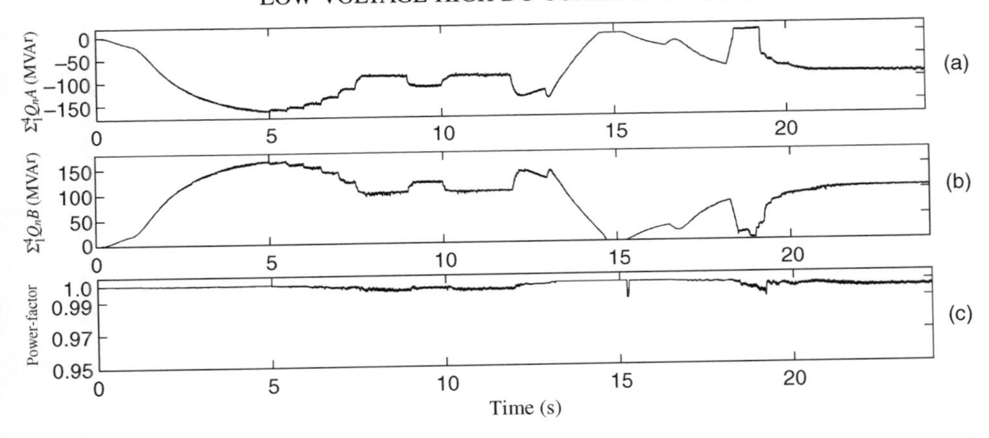

Figure 10.25 Reactive power contributions and overall smelter power-factor, with (a) total reactive power generation, (b) reactive power absorption and (c) power-factor as seen from transmission system.

about t = 19.0 s); this is due to the reactive power demands of all transformers when the rectifiers operate at zero firing angle.

Once again, the timescale used in this simulation is indicative only, as the cell commissioning period alone could take up to 6 months to complete. Similarly a restart would take much longer in practice, a period of 30–60 minutes being typical.

10.5.5.3 Waveform quality

The supply current, leading and lagging current waveforms of the multigroup MLCR rectifier are given in Figure 10.26. The waveform snapshot is taken at rated current, and the multilevel

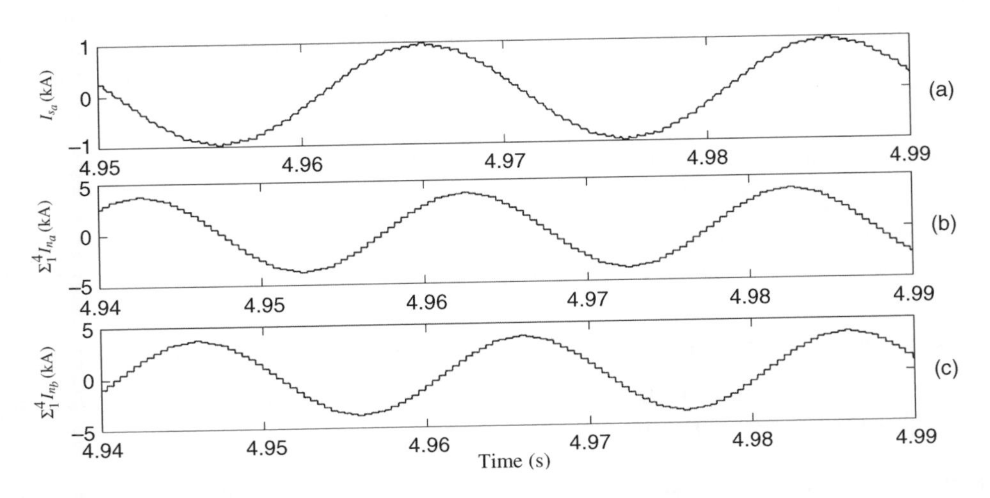

Figure 10.26 AC currents for the multigroup MLCR rectifier, with (a) the combined supply current, (b) leading rectifier current and (c) lagging rectifier current.

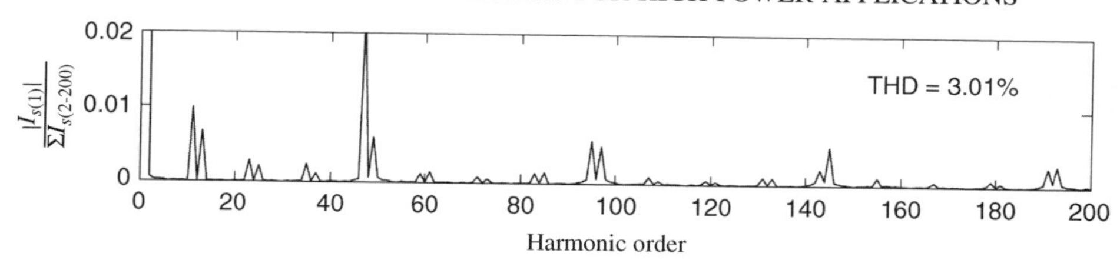

Figure 10.27 Multigroup MLCR AC supply current (Ia) harmonic performance for the first 200 harmonic orders.

waveform steps are clearly evident for the leading rectifier in Figure 10.26(b) and lagging rectifier in Figure 10.26(c).

The harmonic components of Figure 10.26(a) are analysed and the results shown in Figure 10.27. The predominant 47th, 49th harmonic has a magnitude of 2%, with the 11th, 13th at 0.9%, 45th, 97th at 0.57% and 143rd, 145th at 0.42%, which combine for a THD of 3.01%.

10.5.6 MLCR multigroup reactive power controllability

Given that the MLCR rectifiers are capable of controlling their DC currents directly, several options exist to alter the reactive power characteristics of the multigroup, to potentially improve operating efficiency, reduce restart time or to provide additional voltage support to the power system.

The reactive power contribution of rectifier$_n$ is given by

$$Q_n = \sqrt{3}k_1 V_{T_n} I_{d_n} \sin(\alpha_n) \tag{10.47}$$

Thus adjustment of the terminal voltage (V_{T_n}), DC current (I_{d_n}), conversion constant (k_1) and firing angle (α_n) all have a direct effect on the reactive power characteristics of the rectifier. While adjusting the terminal voltage and conversion constant require hardware changes, the DC current and firing angle may be manipulated purely by modifying the control method, and thus several potential control variations arise to improve on the multigroup reactive power characteristics. They are:

- The number of rectifiers active may be changed if the DC current demand permits (i.e. if the DC side is partially loaded).
- The balance of current within a group may be biased to either the leading or lagging rectifier (within a small amount if fully loaded) to make the group a net generator of reactive power.
- The lead–lag AC current balance across the entire multigroup may be altered.
- Reversal of the firing angle polarity of one or more of the rectifiers in the multigroup to alter reactive power generation.

Performance of each of the altered configurations is compared to the base case as given in Figure 10.24 and is shown as a dotted reference trace on each of the Figures 10.28 to 10.33. The measure of performance improvement is made in terms of operating efficiency in each case by

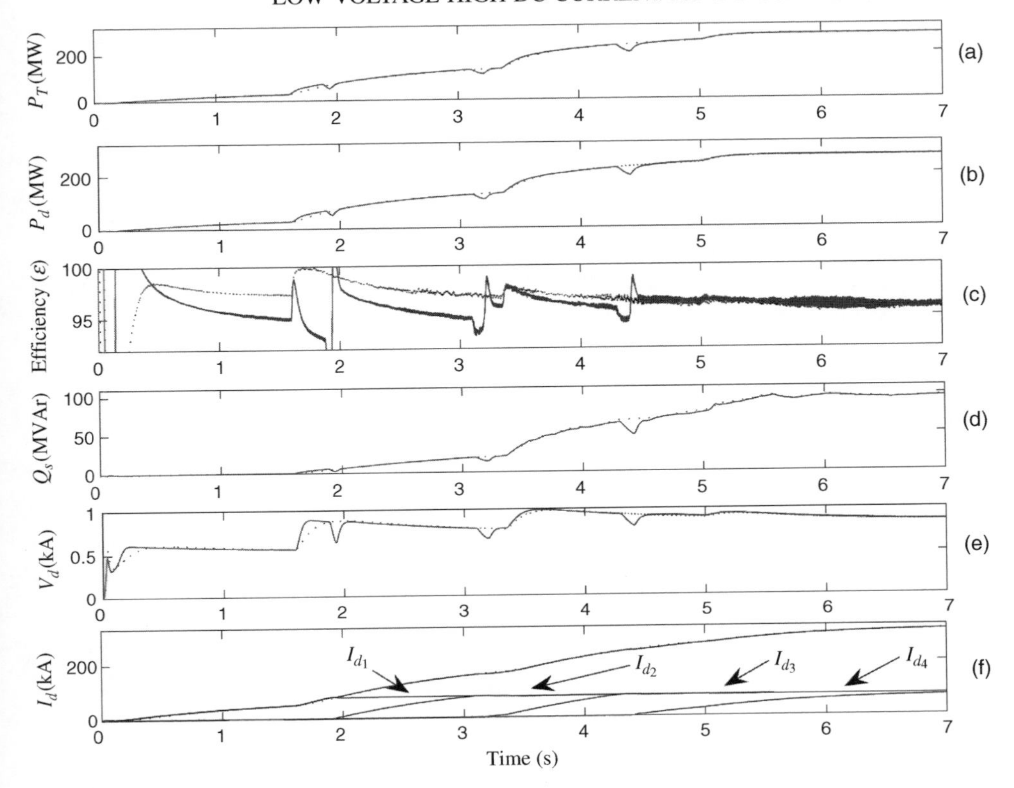

Figure 10.28 Simulated performance of a multigroup rectifier with staggered current share.

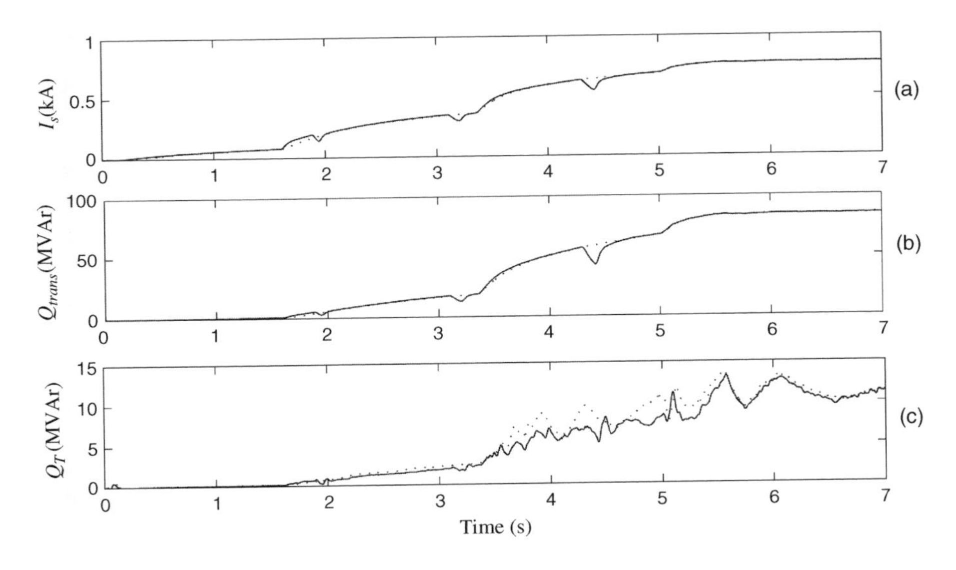

Figure 10.29 Multigroup reactive power performance with staggered current share.

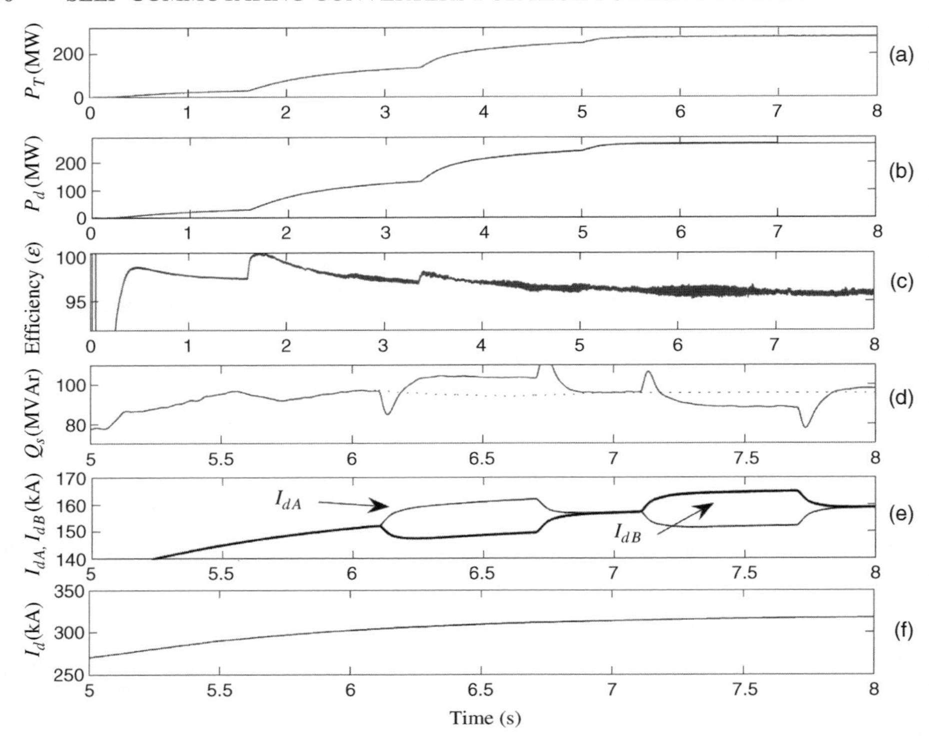

Figure 10.30 Multigroup reactive power performance with DC current biased to the leading and lagging rectifiers as compared to the base case (dotted trace in (d)).

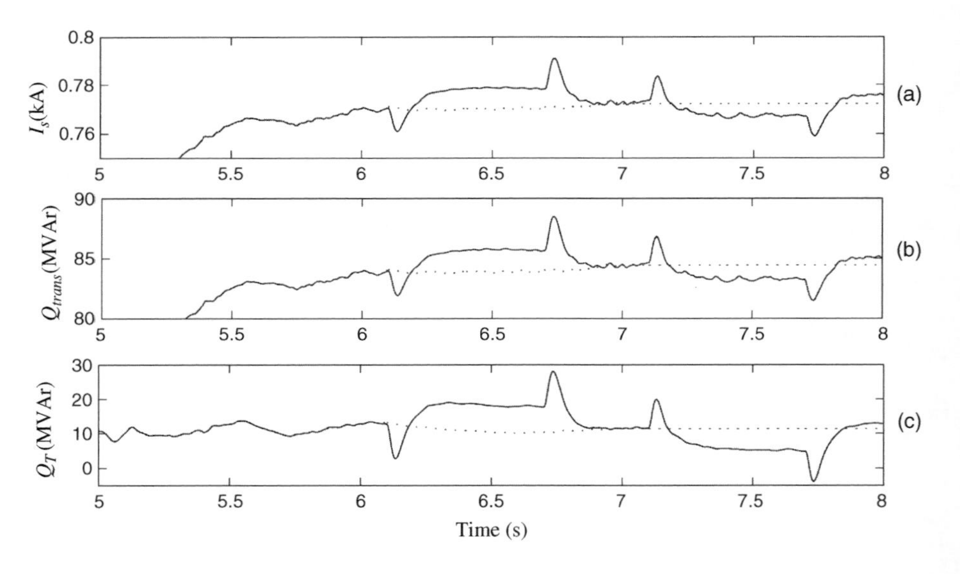

Figure 10.31 Multigroup reactive power performance with DC current biased to the leading and lagging rectifiers as compared to base case.

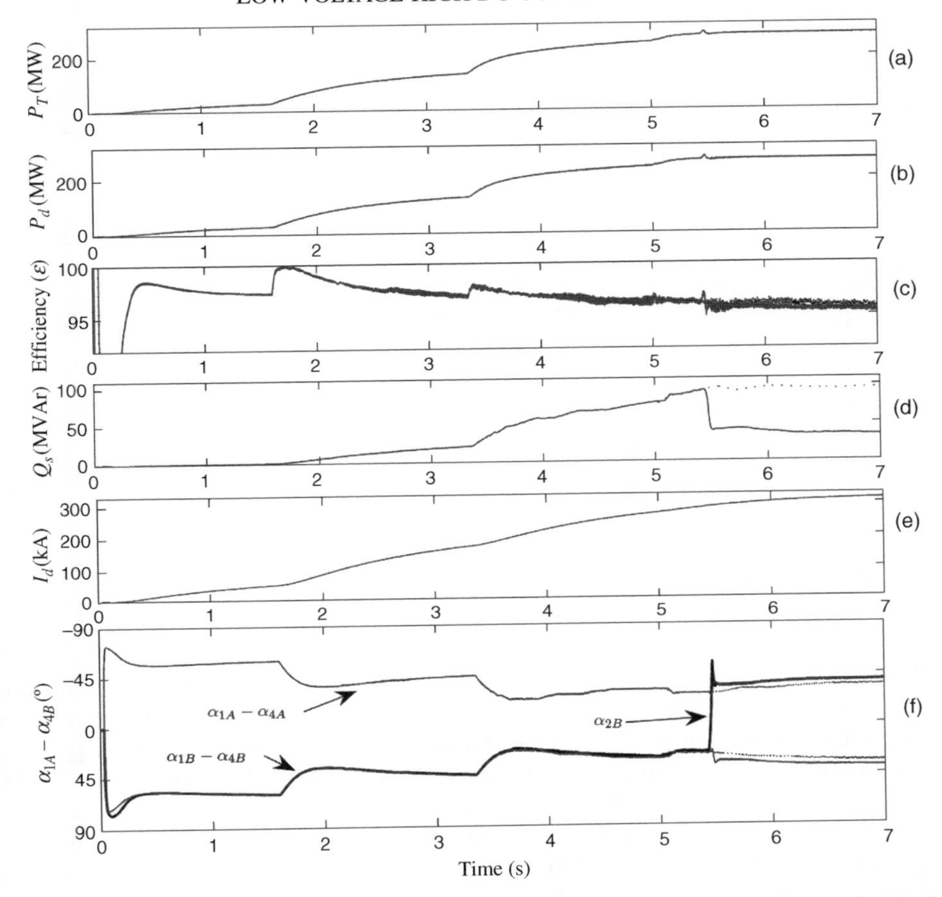

Figure 10.32 Multigroup performance with controlled polarity reversal of firing angle 2B as compared to base case.

comparing reactive power demands, and AC supply current magnitude. Incidentally, the multigroup efficiency for each situation is

$$efficiency(\varepsilon) = \frac{P_T}{V_d \times I_d}(\%) \tag{10.48}$$

with P_T representing the measured AC power at the HV terminal and $V_d I_d$ the measured DC power.

The following cases have been simulated on the same test system (with a short-circuit ratio of 3) and may not be applicable to a stronger system, where the terminal voltage variation with changing reactive power demand is lower.

10.5.6.1 Demand-based DC current staggering

The first configuration is made in accordance with the control theory in Figures 10.20–10.22, to analyse the performance of enabling the rectifier groups on demand. The results of this simulation over a 7 s period are given in Figure 10.28.

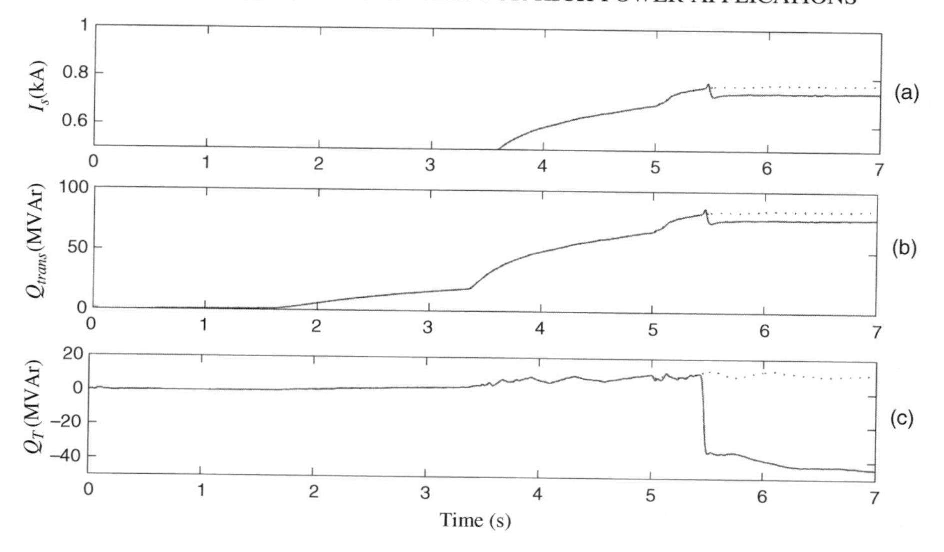

Figure 10.33 Multigroup reactive power performance with controlled polarity reversal of firing angle α_{2B}as compared to base case.

The first rectifier pair is activated and supplies all the DC current until demand increases above the rated (80 kA) value; other pairs are activated at t = 1.9, 3.1 and 4.3 s. Figure 10.28(a) shows the input power, Figure 10.28(c) the efficiency, Figure 10.28(d) the reactive power as observed from the source, Figure 10.28(e) the DC load voltage and Figure 10.28(f) the DC current contributions of each of the rectifier pairs and total DC current.

Figure 10.28(a) shows that the input power change is minimal, aside from small perturbations as each new group is activated, as evident in all traces at 1.75, 3.2 and 4.4 s. Comparing the traces of Figure 10.28(c), this configuration is actually less efficient in transitioning from zero to full load current as compared to the dotted reference trace. The lower efficiency is due to the amount of time that each new rectifier pair must have its firing angle at 0° (i.e. maximum DC voltage) to establish current flow. As the rate of current increase in a new group depends on the difference between load voltage and maximum DC voltage, the current rise takes longer as load voltage is raised. Previously with all rectifiers working in unison, that is, with the same DC current, there were few excursions to 0° firing angle during an increase in current, as all rectifiers worked together to increase a common DC voltage, so their individual increase rates were less and thus required less DC voltage for a given current rise.

The supply current in Figure 10.29(a) and the reactive power demands in 10.28(d) and 10.29(c) show negligible improvement over the whole simulation period.

10.5.6.2 DC current bias within the multigroup

Altering the balance of DC current spread between leading and lagging rectifiers has the potential to change the amount of reactive power generated or absorbed by the multigroup. The effects of altering the current share between a leading and lagging rectifier are the same whether a pair of rectifiers or a whole multigroup have their current split altered. For this reason the simulation is carried out with adjustment made across the whole multigroup.

The multigroup simulation in Figure 10.30 is initialized as in the base case and allowed to stabilize at full DC current, which occurs at approximately t = 6 s. At t = 6.1 s the DC current order of the leading group is increased by 4 kA (corresponding to 1 kA per 48-pulse MLCR rectifier) and the lagging group reduced by the same amount. The immediate effect as seen in Figure 10.30(d) is an increase in reactive power absorption by the multigroup, up 10 MVAr from 94 MVAr to 104 MVAr. At t = 6.7 s the DC current bias is returned to the original even split. At t = 7.0 s, the DC current of the lagging rectifiers is increased, with an immediate reduction in reactive power absorption in Figure 10.30(d) from 94 MVAr of the base case, to 88 MVAr. At t = 7.6 s the 50 : 50 DC current split is restored. The reactive power and supply current traces of Figure 10.31 exhibit a similar response as expected, with the reactive power absorbed by the rectifier and supply transformers in Figure 10.31(c) reduced significantly. The AC and DC powers of Figure 10.30(a) and (b) are unchanged as are the efficiency and DC current in Figure 10.30(c) and (e) respectively.

The reactive power absorbed by the multigroup in this example is thus affected more by changes in firing angle than by current order, or with reference to equation 10.47,

$$\frac{dQ_n}{d\alpha_n} > \frac{dQ_n}{dI_{d_n}}$$

as the sensitivity of the multigroup reactive power to a change in firing angle is greater than that of an increased DC current.

The reduced reactive power demand does come at the expense of an increased DC current rating in the lagging group, but in this case a 2.5% increase in current reduces the reactive power absorption by 9%, which may be an acceptable design consideration.

Obviously under different load conditions, or when connected to a transmission system of differing impedance, the firing angles for maximum power transfer will differ and this relationship and method of control may no longer be possible.

10.5.6.3 Firing angle polarity reversal within the group

The multigroup rectifiers modelled in these simulations are fed from a relatively weak transmission system and, as a result, the optimum firing angles for maximum power delivery are (for the high power-factor multi-group) −35° for the leading and 34° for the lagging groups, under steady-state conditions and when current is balanced. In the base case 91 MVAr are drawn from the ideal source through the transmission system.

At full load, the reactive power circulating in each of the MLCR groups is approximately 20 MVAr, generated by the leading rectifier and 20 MVAr absorbed by the lagging rectifier. If one lagging rectifier were to have its firing angle advanced so that it was leading, but with the same DC voltage, the DC contribution would be the same but the reactive power generated by the group would be 40 MVAr. This control opportunity is exploited in Figure 10.32(f) at t = 5.5 s, with the polarity of firing angle α_{2B} changed on-load, taking advantage of the long DC current time constant as compared to the reactive power polarity change time. Consequently, as shown in Figure 10.32(d), this increased reactive power contribution from the multigroup partially corrects for the reactive power absorbed by the transmission system (Figure 10.33(b)), reducing the complex voltage drop and the apparent power from the source, thus reducing the supply current, as seen in Figure 10.33(a), from 0.771 kA to 0.737 kA, and the associated transmission losses. The result of this is the possibility of increased maximum power transfer, increased transmission efficiency and a further increase in firing angle for a given DC load.

Consideration must be made for the other rectifiers supplying DC current, as advancing one rectifier's firing angle through zero will increase its DC voltage to maximum, potentially overloading that rectifier and reducing the apportioned current to the remaining rectifiers. If the time taken to change from lagging to leading firing angle is much shorter than the DC current time constant, the effect on the DC current share will be negligible, but the real power of the advancing group will increase momentarily. In Figure 10.32(a) and (b) there is a small increase in both AC and DC powers of approximately 5%, but its duration is only 50 ms. Likewise, there is a small difference in efficiency, but that settles quickly back to 96.1%.

Depending on the DC operating conditions, the ratio of leading to lagging firing angles could be set before the multigroup rectifiers are started, or polarity could be changed under load as the reactive power demands of the connected power system change.

With respect to aluminium smelter operation, this control method may not be suitable during commissioning, because the low DC voltage (and thus large firing angles) at rated current would cause a large imbalance in reactive power circulation and, potentially, an undesirably high terminal voltage. In this case the firing angle polarity would need to be switched under load at near full-rated DC voltage.

10.6 Parallel thyristor/MLCR rectification

Figure 10.34 presents a block diagram of a hybrid thyristor and MLCR groups configuration. For simplicity several phase-shifted thyristor rectifiers are shown as one power block so their applicability to any number of pulses can be retained. The relative sizes of the components are used to illustrate that the majority of power transfer is through the thyristor bridges.

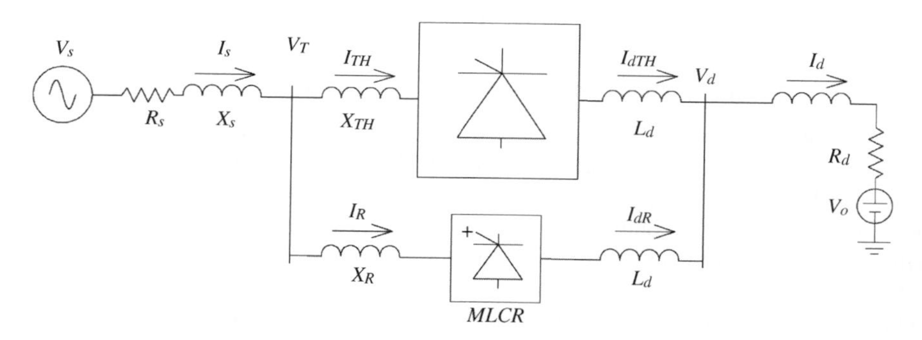

Figure 10.34 Simple block diagram of a hybrid thyristor MLCR-based smelter power supply.

The rectifiers are supplied from an AC system represented by a Thevenin equivalent circuit consisting of source voltage V_s and series impedance (X_s and R_s), which in turn feeds a common AC bus with voltage V_T. The transformer leakage reactances are represented by X_{th} and X_R for the thyristor and MLCR branches respectively. The bridges are also paralleled on the DC side, with small intergroup reactors (represented by L_d) included to allow for small instantaneous mismatches in DC voltage.

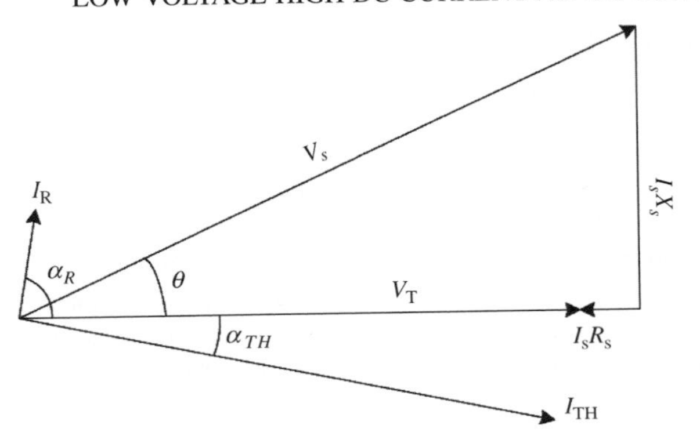

Figure 10.35 Phasor diagram for a hybrid thyristor MLCR-based smelter power supply.

An idealized phasor diagram of the configuration is shown in Figure 10.35. The source and terminal voltages, V_s and V_T respectively, are separated by the phase shift created by the supply current (I_s) and Thevenin impedance (X_s and R_s). Thyristor firing angle (α_{TH}) and AC current (I_{TH}) are referenced to the terminal voltage (V_T), as are the MLCR firing angle and AC current represented by α_R and I_R respectively. The magnitudes of each of the AC currents provide the same amount of reactive power, but their real power components differ greatly.

To enable the large reactive power correction needed for high power-factor operation whilst maintaining high efficiency, a reduced transformer ratio is used on the MLCR bridge. Increasing the secondary output voltage of the MLCR rectifier permits full DC load voltage to be developed at a large firing angle, and thus large reactive power compensation for a small real power output. This topology exploits the relationship between Q_R (MLCR reactive power) and α_R (MLCR firing angle) in the following equation:

$$Q_R = \sqrt{3}k_R I_{dR} V_T \sin(\alpha_R) \tag{10.49}$$

For a specified reactive power level (Q_R) and rated DC current (I_{d_R}), increasing constant k_R (the MLCR transformer turns ratio) must be coupled with a decrease in $\sin(\alpha_R)$. Thus with a suitable k_R and α_R near $-90°$, a small change in α_R will yield a large change in I_{d_R}, and therefore a large change in Q_R. Put simply, $\frac{dQ_R}{d\alpha_R}$ near $-90°$ is small and negative, whereas $\frac{dQ_R}{dI_{d_R}}$ is large and positive by comparison.

In a practical situation, the reactive power compensation level would be known or specified, as would the DC current rating of the MLCR switches and nominal DC load voltage. Considering the firing angle bounds of between $-90 \leq \alpha_R \leq -80°$ where $\cos(\alpha_R)$ is almost linear (regression residuals are all positive with a maximum error of 0.00034) minimum k_R may be calculated by:

$$k_R = \frac{Q_{R_{specified}}}{\sqrt{3}I_{R_{rated}} V_T \sin(\alpha_{R_{\max}})} \tag{10.50}$$

This enables predominantly reactive power generation rather than real power and keeps MLCR switch current ratings small, with a corresponding increase in voltage rating. Given the higher voltage rating and large reactive power demands, the MLCR transformer must be sized accordingly. With the MLCR, maximum DC output voltage will be k_R times that of the thyristor group and the voltage ratings of the MLCR switching components must be increased by the same level.

The AC side function of the MLCR rectifier is similar to conventional reactive power compensation with SVCs or STATCOMs, in that it draws controlled leading current to maintain high power-factor. However, SVCs and STATCOMs must be rated for the maximum reactive power required, which in the smelter case is only required during starting conditions, and do not contribute to the smelter's active power. Conversely, the MLCR rectifier of the hybrid system can provide practically its full DC current rating as active power under normal operating conditions when relatively little reactive power is required.

10.6.1 Circuit equations

In order to control the phase-shifted thyristor and MLCR rectifiers as a single system, the relationships between the real and reactive powers, terminal voltage and DC operating state must be defined. The real and reactive powers are related to the common terminal AC voltage (V_T) using the orthogonal d-q transform.

The transform identity matrix **M** is defined as:

$$\mathbf{M} = (\mathbf{M}^{-1})^T = \sqrt{\frac{2}{3}} \begin{bmatrix} \sin(\omega t) & \sin(\omega t - 120^\circ) & \sin(\omega t + 120^\circ) \\ \cos(\omega t) & \cos(\omega t - 120^\circ) & \cos(\omega t + 120^\circ) \\ \frac{1}{\sqrt{2}} & \frac{1}{\sqrt{2}} & \frac{1}{\sqrt{2}} \end{bmatrix} \tag{10.51}$$

If the three-phase system is balanced then:

$$\mathbf{I}_s = \sqrt{2} I_s [\sin(\omega t + \phi_1) \sin(\omega t + \phi_1 - 120^\circ) \sin(\omega t + \phi_1 + 120^\circ)]^T \tag{10.52}$$

$$\mathbf{V}_s = \sqrt{2} V_s [\sin(\omega t + \phi_s) \sin(\omega t + \phi_s - 120^\circ) \sin(\omega t + \phi_s + 120^\circ)]^T \tag{10.53}$$

which when transformed become:

$$\mathbf{I}_{S_{dq}} = \mathbf{M}\mathbf{I}_s = \sqrt{3} I_s [\cos(\phi_1) \sin(\phi_1) 0]^T = [I_{S_d} I_{S_q} 0]^T \tag{10.54}$$

$$\mathbf{V}_{S_{dq}} = \mathbf{M}\mathbf{V}_s = \sqrt{3} V_s [\cos(\phi_s) \sin(\phi_s) 0]^T = [V_{S_d} V_{S_q} 0]^T \tag{10.55}$$

To simplify the notation, matrix **N** is defined as:

$$\mathbf{N} = \begin{bmatrix} 0 & -1 \\ 1 & 0 \end{bmatrix}$$

Using the d-q reference frame the terminal real power (P_T) is calculated as:

$$P_T = \mathbf{V}_{\mathbf{T_{dq}}}^{\mathbf{T}} \mathbf{I}_{\mathbf{S_{dq}}} = P_S - P_L = \mathbf{V}_{\mathbf{S_{dq}}}^{\mathbf{T}} \mathbf{I}_{\mathbf{S_{dq}}} - R_S \mathbf{I}_{\mathbf{S_{dq}}}^{\mathbf{T}} \mathbf{I}_{\mathbf{S_{dq}}} \tag{10.56}$$

and reactive power as:

$$Q_T = \mathbf{V}^{\mathbf{T}}_{\mathbf{T_{dq}}}\mathbf{N}\mathbf{I}_{\mathbf{S_{dq}}} = Q_s - Q_L = \mathbf{V}^{\mathbf{T}}_{\mathbf{S_{dq}}}\mathbf{N}\mathbf{I}_{\mathbf{S_{dq}}} - \omega L_s \mathbf{I}^{\mathbf{T}}_{\mathbf{S_{dq}}}\mathbf{I}_{\mathbf{S_{dq}}} \tag{10.57}$$

where P_L and Q_L are the real and reactive power transmission losses respectively.

As P_T and Q_T are scalars they can be added.

$$\mathbf{I}^{\mathbf{T}}_{\mathbf{S_{dq}}}\mathbf{V}_{\mathbf{T_{dq}}} = \mathbf{I}^{\mathbf{T}}_{\mathbf{S_{dq}}}\mathbf{V}_{\mathbf{S_{dq}}} - R_s \mathbf{I}^{\mathbf{T}}_{\mathbf{S_{dq}}}\mathbf{I}_{\mathbf{S_{dq}}} \tag{10.58}$$

$$\mathbf{I}^{\mathbf{T}}_{\mathbf{S_{dq}}}\mathbf{N}\mathbf{V}_{\mathbf{T_{dq}}} = \mathbf{I}^{\mathbf{T}}_{\mathbf{S_{dq}}}\mathbf{N}\mathbf{V}_{\mathbf{S_{dq}}} - \omega L_s \mathbf{I}^{\mathbf{T}}_{\mathbf{S_{dq}}}\mathbf{I}_{\mathbf{S_{dq}}} \tag{10.59}$$

which combined allows the terminal voltage to be expressed as:

$$\mathbf{V}_{\mathbf{T_{dq}}} = \mathbf{V}_{\mathbf{S_{dq}}} - R_s \mathbf{I}_{\mathbf{S_{dq}}} - \omega L_s \mathbf{N}\mathbf{I}_{\mathbf{S_{dq}}} \tag{10.60}$$

The *rms* terminal voltage is then calculated by

$$V_{T_{rms}} = \mathbf{V}^{\mathbf{T}}_{\mathbf{T_{dq}}}\mathbf{V}_{\mathbf{T_{dq}}} = \mathbf{V}^{\mathbf{T}}_{\mathbf{T_{dq}}}\mathbf{V}_{\mathbf{S_{dq}}} - R_s \mathbf{V}^{\mathbf{T}}_{\mathbf{T}}\mathbf{I}_{\mathbf{S_{dq}}} - \omega L_s \mathbf{N}\mathbf{V}^{\mathbf{T}}_{\mathbf{T_{dq}}}\mathbf{I}_{\mathbf{S_{dq}}} = \mathbf{V}^{\mathbf{T}}_{\mathbf{T_{dq}}}\mathbf{V}_{\mathbf{S_{dq}}} - R_s P_T - \omega L_s Q_T \tag{10.61}$$

The terminal power consists of the individual power contributions of the thyristor and MLCR rectifiers. The thyristor real power is thus:

$$P_{th} = \sqrt{3}k_{th}\left(\frac{3\sqrt{2}}{\pi}V_T\cos(\alpha_{th}) - \frac{3I_{d_{th}}X_c}{\pi}\right)I_{d_{th}} \tag{10.62}$$

which accounts for the average DC voltage reduction due to the commutation period and commutation reactance (X_c). With this topology, the conversion constant k_{th} is approximately 1.4.

The MLCR active power contribution is defined as:

$$P_R = \sqrt{3}k_R I_R V_T \cos(\alpha_R) \tag{10.63}$$

where constant k_R is the MLCR conversion ratio as given by equation 10.50.

The reactive power demands of the thyristor rectifier are similarly calculated:

$$Q_{th} = \sqrt{3}k_{th}\left(\frac{3\sqrt{2}}{\pi}V_T\sin(\alpha_{th}) - \frac{3I_{d_{th}}X_c}{\pi}\right)I_{d_{th}} \tag{10.64}$$

The DC side equations given below are used to determine time constants for each of the current controller parameter derivations. R_{dTH} and R_{dR} represent the thyristor and MLCR rectifier's internal and switch on state losses respectively.

$$v_{dTH}(t) = R_{dTH}i_{dTH}(t) + L_{dTH}\frac{di_{dTH}(t)}{dt} + L_{dc}\frac{di_{dc}(t)}{dt} + V_o \tag{10.65}$$

$$v_{dR}(t) = R_{dR}i_{dR}(t) + L_{dR}\frac{di_{dR}(t)}{dt} + L_{dc}\frac{di_{dc}(t)}{dt} + V_o \tag{10.66}$$

where V_o is the DC output voltage.

Adding equations 10.65 and 10.66 together, given that $I_{dTH} + I_{dR} = I_{dc}$ and $L_{dR} = L_{dTH}$, the relationships become:

$$\begin{aligned} v_{dR}(t) + v_{dTH}(t) &= R_{dR}I_{dR}(t) + R_{dTH}I_{dTH}(t) + L_{dR}\left(\frac{di_{dR}(t)}{dt} + \frac{di_{dTH}(t)}{dt}\right) + 2L_{dc}\frac{di_{dc}(t)}{dt} + 2V_o \\ &= R_{dR}I_{dR}(t) + R_{dTH}I_{dTH}(t) + (L_{dR} + 2L_{dc})\frac{di_{dc}(t)}{dt} + 2V_o \end{aligned} \tag{10.67}$$

and rearranged to make $\frac{di_{dc}(t)}{dt}$ the subject:

$$\frac{di_{dc}(t)}{dt} = \frac{v_{dR}(t) + v_{dTH}(t) - R_{dR}I_{dR}(t) - R_{dTH}I_{dTH}(t) - 2V_o}{L_{dR} + 2L_{dc}} \tag{10.68}$$

which gives the relationship between the DC side parameters and the rate of current change.

The complete smelter model is thus shown in Figure 10.36.

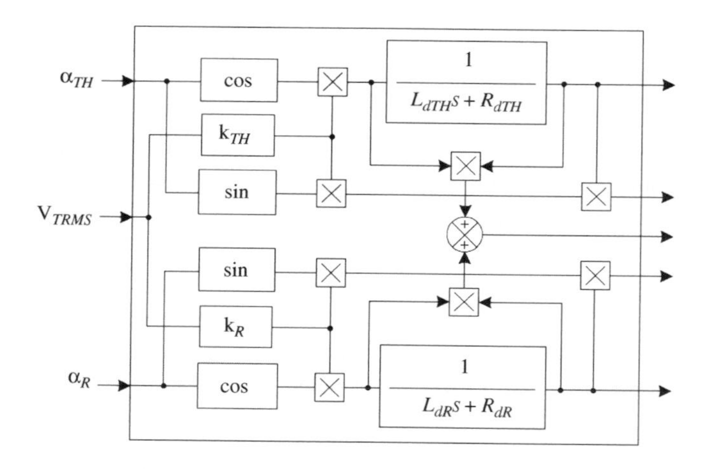

Figure 10.36 Smelter block diagram for control of real and reactive power.

10.6.2 Control system

A diagram of the controller for both the thyristor and MLCR rectifiers is given in Figure 10.37 (a) and (b) respectively. The thyristor rectifiers form the main current path in this configuration. Owing to the passive nature of the DC load, a single-quadrant DC current control process provides sufficient controllability despite small fluctuations in load conditions. The DC current

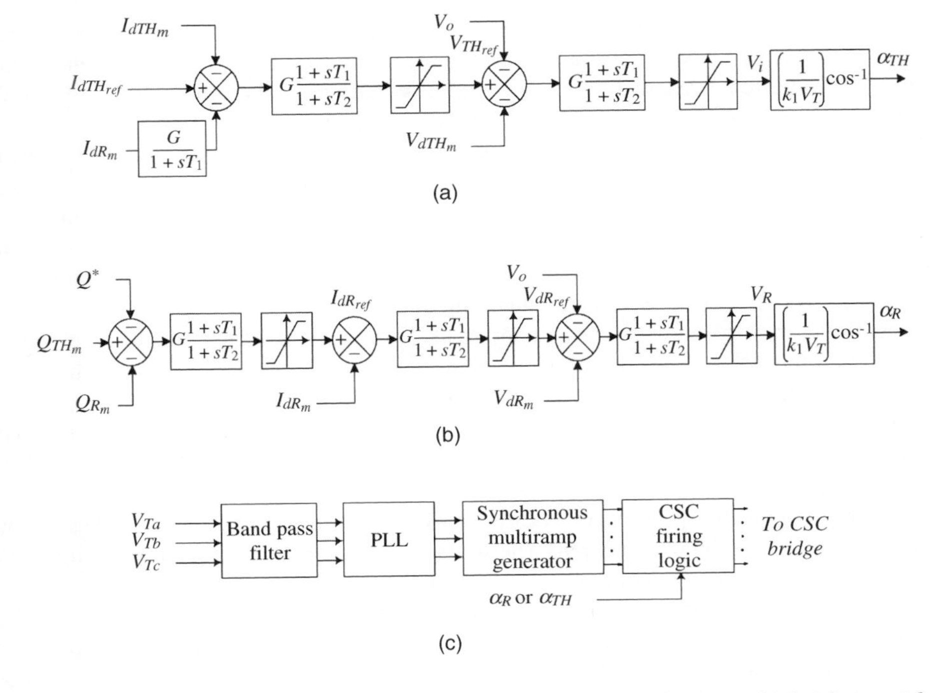

Figure 10.37 Controller layout for both the thyristor (a) and MLCR (b) bridges, with the common firing angle control in (c).

control loop in Figure 10.37(a) is common to all thyristor rectifiers, with a single firing angle command (α_{th}) used to preserve ideal phase shift.

The DC current order ($I_{d_{thRef}}$) is compared to the measured thyristor DC current ($I_{d_{thm}}$), less the contribution of the reactive power compensation (given by $I_{d_{Rm}}$). The error signal is passed to a lead–lag compensator and the resulting signal is hard limited to prevent saturation, the output setting the DC voltage control order $V_{d_{thref}}$. The closed-loop bandwidth of the outer most DC current loop is a function of the DC side time constant, nominally 20–50 Hz.

The thyristor inner control loop error signal is derived from the difference between DC voltage order ($V_{d_{thref}}$), measured DC voltage ($V_{d_{thm}}$) and DC back emf (represented by V_o), and is fed to a second lead–lag compensator, the output of which is hard limited and scaled by the AC (*rms*) terminal voltage (V_T). This incremental voltage is inverse cosined, resulting in firing angle order α_{th}, which is sent to the thyristor firing logic, as given in Figure 10.37(c). Here, phase-locked loops (PLLs) are synchronized to each of the sinusoidal supply voltages ($V_{T_a} - V_{T_c}$) which generate the firing angle references for the six thyristors in each of the 12-pulse bridge rectifiers. The controller bandwidths for each of the 12-pulse DC voltage and firing angle loops are nominally 100–200 Hz and 1 kHz respectively. Correction for the thyristor commutation angle is not made directly in the DC voltage loop; as the terminal voltage *rms* value is reduced by the commutation, the inclusion of V_T in the inner control loop helps to minimize the non-linear effects.

The reactive power controller for the 48-pulse MLCR rectifier in Figure 10.37(b) has a similar configuration to that of the thyristor rectifier, with the addition of a third outer control loop for reactive power compensation. The error signal is derived from the difference between the measured reactive powers of the thyristor and MLCR rectifiers, with an optional boost order (Q^*) if additional reactive power generation is required. The error is then fed into a lead–lag compensator, its closed loop bandwidth being approximately 10 Hz.

The MLCR rectifier firing angle reference is developed using the same PLL as in Figure 10.37(c), but the PLL is synchronized with the three-phase AC terminal voltage taken from the LV side of the converter transformer. The MLCR supply current is quasi-sinusoidal and phase-shifts the LV terminal reference, and thus the true bridge supply voltage, depending on the magnitude of the AC current; with MLCR operation so close to $|90^\circ|$, establishing a true 90° reference relative to V_T is vital for stable operation.

As with the thyristor controller, the bandwidth of the MLCR PLL is 1 kHz. The voltage control loop too has the same control parameters as its thyristor equivalent, but with its gain reduced proportional to the altered MLCR supply transformer ratio. The reinjection current control loop has the same format as its thyristor counterpart, but with a reduction in the DC inductance parameters due to the sinusoidal supply current and associated transformer phase shift. This is shown in Figure 10.36 and equation 10.68.

The control of reactive power is inherently non-linear when related to firing angle (α_R), but as the firing angle is maintained near 90° (between $-90 \leq \alpha_R \leq -80^\circ$ for a transformer ratio reduction of 10), over this small range the control behaviour is almost linear.

The reactive power control works as follows: to increase reactive power generation (Q_R), I_{dR} needs to be increased. I_{dc} is maintained constant by reducing I_{dTH}, which requires an increase in α_{TH} resulting in an increase in Q_{TH}. This process brings the smelter converter into a new operating state which is stable so long as controller interactions are damped. The main interaction path between the MLCR and thyristor controllers is via the measured MLCR DC current (I_{dR_m}) fed into the thyristor current controller. The effect is minimized by low-pass filtering the feedback (I_{dR_m}) when used in the thyristor control. A large time constant is used (in most cases 100–200 mS is sufficient) to decouple the responses.

10.6.3 Dynamic simulation and verification

The simulated test system is shown in Figure 10.38. Four 12-pulse phase-shifted thyristor rectifiers make up the 48-pulse thyristor group, with each thyristor rectifier having its own zig-zag transformer, and phase shifts of 0°, 7.5°, 15°, 22.5° as indicated. The 48-pulse parallel MLCR rectifier is supplied via a three-winding transformer with two secondaries configured in star and delta.

The smelter circuit is simulated using the PSCAD/EMTDC package and its response to a series of step changes recorded. The step change magnitude and order of steps are designed to replicate a typical operating cycle in a smelting process, which may include commissioning, the cell warm-up process, normal production including cell removal for maintenance and smelter shut down.

The simulated time to full production has been compressed to make the dynamic simulation achievable. In practice the commissioning period is specified by how many cells are to be commissioned and how long each takes to reach operating temperature. A smelter consisting of 250 cells may take up to 6 months to commission when one to two cells are commissioned per day. In contrast, a restart may take half an hour if cell temperatures are near operating temperature.

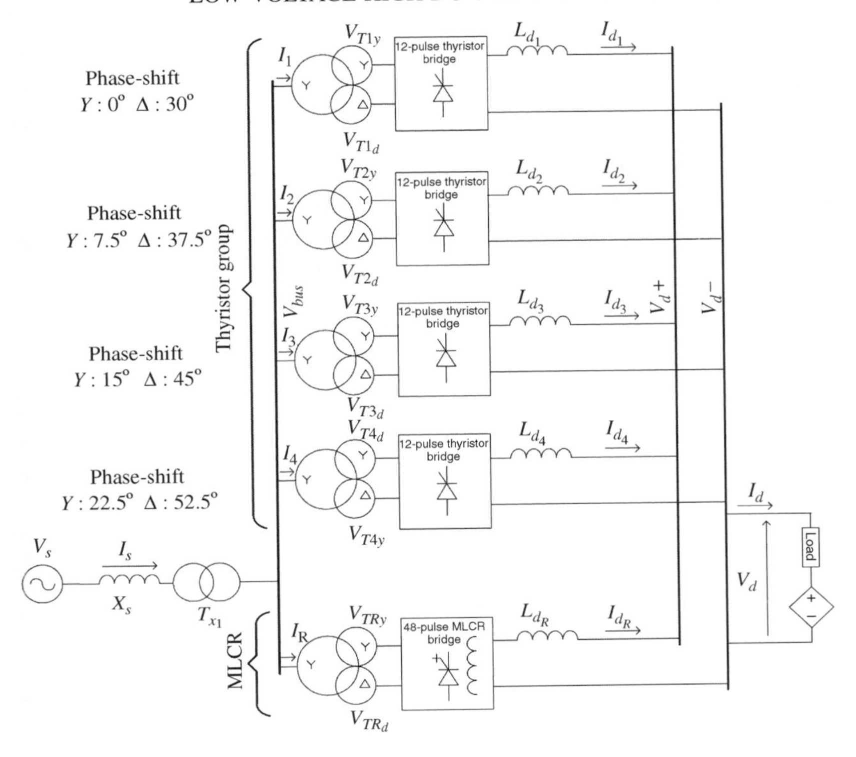

Figure 10.38 Paralleled test system comprised of four 12-pulse thyristor bridges and one 48-pulse MLCR bridge.

It is assumed that during normal operation the smelter requires full DC current regardless of what proportion of the line is in service, and that during commissioning the voltage is increased in stages until the whole smelter DC load is in service. Also, periodically, some cells may be taken out of service for maintenance, with a maximum of 10–15% removed.

10.6.3.1 Test system

The smelter is supplied at 220 kV by a Thevenin source (V_s) through a system impedance (X_s), providing a short-circuit ratio of 2.5. The main supply transformer (Tx_1) is rated at 200 MVA, 220 : 33 kV and feeds each of the thyristor phase-shifted transformers and the MLCR three-winding star–delta transformer, their secondary voltages being 0.8 kV and 7.6 kV respectively. The MLCR transformer ratio is calculated using equation 10.50, and all transformer leakage reactances are specified as 5%.

The DC load is rated at 1000 V and 160 kA with each 12-pulse thyristor bridge capable of delivering 40 kA at rated voltage. The smelter load is modelled as 250 series connected cells, with a total resistance of 4 mΩ, inductance of 5.5 mH and back emf (V_o) of 320 V when all cells are in service.

10.6.3.2 Dynamic response

The results of the simulation are given in Figure 10.39. The simulated smelter is initialized at 0.1 s and the main thyristor groups are enabled with 20% of the installed DC load. The reactive power compensation is enabled at 1 s and maintains unity power-factor until a power rating of 80 MVA is reached at 2 s. The reactive power drawn from the system increases as the DC current rises (in Figure 10.39(d)) until 95% of rated current is reached at t = 4.95 s where a maximum of 46.5 MVAr and 40.5 MW (corresponding to a minimum power-factor of 0.65) is observed due to the large firing angle (Figure 10.39(g)).

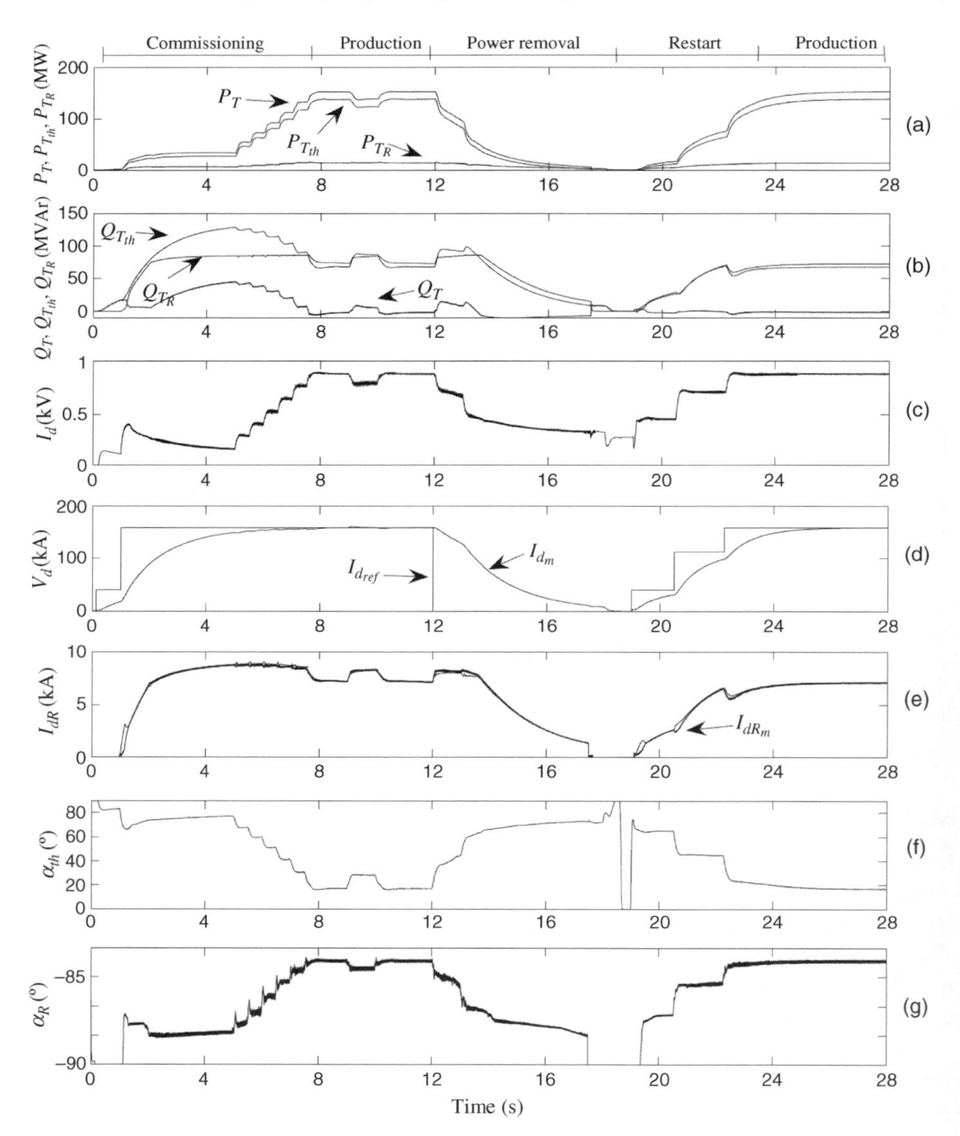

Figure 10.39 Operating response of a 160 MW smelter converter over 28 s period.

At t = 5 s additional cells are added to the smelter load in stages (15% at 0.5 s intervals) until all cells are installed at 8.1 s and full production commences. Figure 10.39(b) shows that the reactive power demands steadily decrease from 5 to 8 s; then 85% of cells are installed and full compensation is possible. The MLCR rectifier MVA output is sized for normal duty (85–100% installed cells at full load current) and thus during the commissioning period (1–8 s) full compensation is not possible. The DC current contribution of the MLCR compensation, shown in Figure 10.39(e), is 8.5 kA, with maximum real power of 14.5 MW, the firing angle to achieve this is maintained between −90° and −83° as shown in Figure 10.39(f).

At t = 9 s, 15% of the smelter load is short circuited (to simulate the maximum removed for cell maintenance) and a corresponding reduction in DC voltage observed in Figure 10.39(c). The DC current on the other hand remains constant in Figure 10.39(d). A small increase in Q_T is noticed, while the reactive power compensation once again is limited to its rated output; high power-factor is maintained despite this (with a minimum terminal power-factor of 0.998). At t = 10 s all cells are reinstated.

At t = 12 s the controlled removal of power begins. The DC current is reduced slowly, so that the absorption of excess reactive power is minimized. The reactive power absorbed by the thyristor bridges is initially larger than the rated MVA of the MLCR transformer and so the power-factor drops momentarily to 0.985 lagging until compensation is made. The DC current order is reduced further with the MLCR rectifier providing the necessary reactive power generation, a slight lag in the MLCR response producing a net reactive power surplus of −7 MVAr as seen from the converter terminal (Figure 10.39(b) from 13.5–17.5 s). The measured DC current falls to zero at 18.1 s, where both the thyristor and MLCR firing circuits are disabled, emulating a full smelter isolation from the AC power system.

The thyristor and MLCR rectifiers are re-enabled at t = 19 s with the full complement of smelter cells installed, a 20% DC current order is given, followed 2.5 s later by a 70% order, and finally 100% at 22.5 s. Full production resumes at 26 s.

10.6.3.3 Reactive power improvement during commissioning

During normal operation the hybrid converter has near unity power-factor, and compensation takes place when a reduction in the number of cells occurs. High pulse and therefore AC current waveform quality is maintained across the entire operating range and the use of thyristors in the main DC current path ensures efficient performance.

The smelter commissioning phase, however, may last for several months and a power-factor of 0.65 is likely to be below the minimum permitted by the supply authority. Simply increasing the MLCR rectifier supply transformer rating and the ratio k_R would provide more compensation, but it may be more cost effective to install an additional 48-pulse MLCR rectifier with the same rating as the first. This would allow correction of twice the reactive power, which would be sufficient in this test case. Two MLCR rectifiers would only be required during commissioning, but could also be used and provide common spare parts and full redundancy.

10.6.3.4 Harmonic performance

The harmonic performance of the smelter converter is taken during the normal operating portion of the simulation shown in Figure 10.39, between 8 and 9 s. The harmonic performance of the thyristor bridge is shown in Figure 10.40. One cycle of the current waveform is shown in Figure 10.40(a) and the harmonic spectra for the first 200 order harmonics are shown in Figure 10.40(b). Very little distortion is seen to be present, with some low-order harmonics being under half a percent, and 48-pulse related ones (47th, 49th) 0.2%. Overall a THD of 1.04% results when the first 1000 harmonic orders are considered.

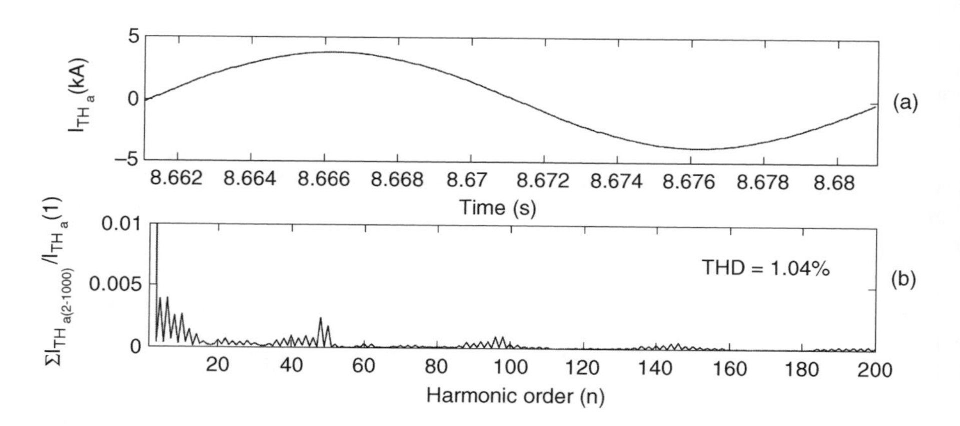

Figure 10.40 AC supply current waveform to the thyristor branch and associated FFT.

In Figure 10.41, the MLCR bridge AC current has a maximum characteristic 48 harmonic magnitude of 1.5%, with lesser magnitudes of 0.8% for the 12 and 96 orders. The overall THD for the first 1000 harmonic orders is 2.9%. No effort has been made to optimize the snubbers on the reinjection circuit which may improve harmonic performance.

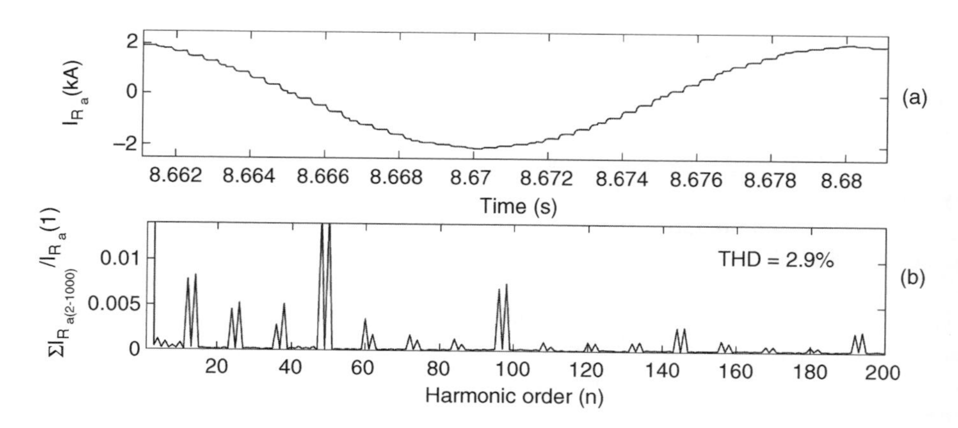

Figure 10.41 AC supply current waveform to the MLCR branch and associated FFT.

Overall the combination of the two converters with characteristic 48-pulse operation is given in Figure 10.42. Total harmonic distortion when considering the first 1000 harmonic

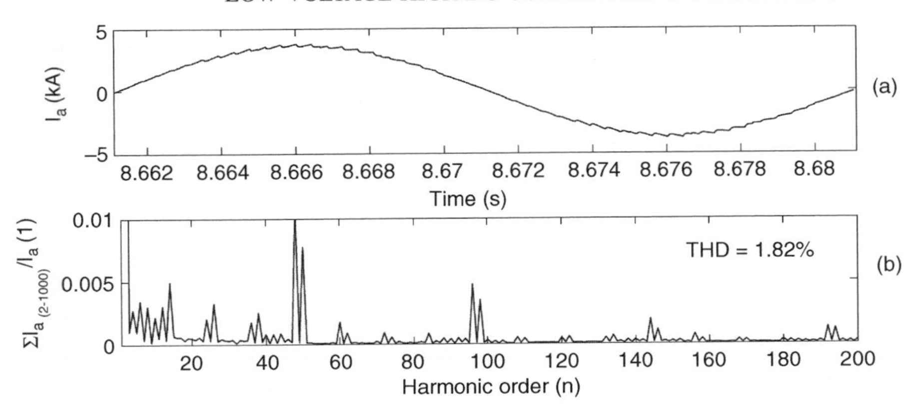

Figure 10.42 Combined AC supply current waveform to the smelter and associated FFT.

orders is calculated at 1.82%. The waveform steps are less distinctive owing to the commutation overlap of the thyristor switches.

10.6.4 Efficiency

The average efficiency of the thyristor rectifiers during normal operation is 96.1% from Figure 10.43(a), while the MLCR rectifier has a much lower average efficiency of 75% (in Figure 10.43(b)), owing to the large transformer size (80 MVA) and relatively low active power output (14.5 MW). The efficiency is calculated as the ratio of output power to input power.

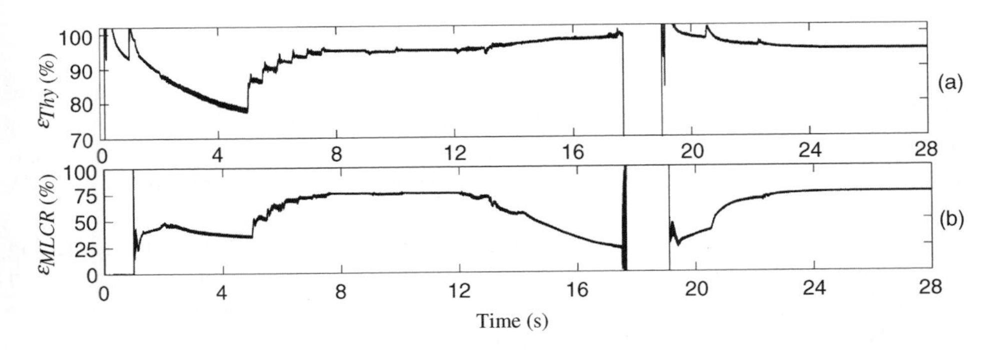

Figure 10.43 Efficiency comparison between the thyristor and MLCR rectifiers.

The combined efficiency is thus 93%, during normal operation, in Figure 10.44. Unlike the thyristor rectifier, the MLCR rectifier's prime purpose is to provide reactive (rather than active) power compensation and thus the conventional efficiency calculation is somewhat misleading. Published STATCOM-related papers [10, 11] state the real power losses in relation to total reactive power compensation capability and a summary of four of these configurations is given in Table 10.3.

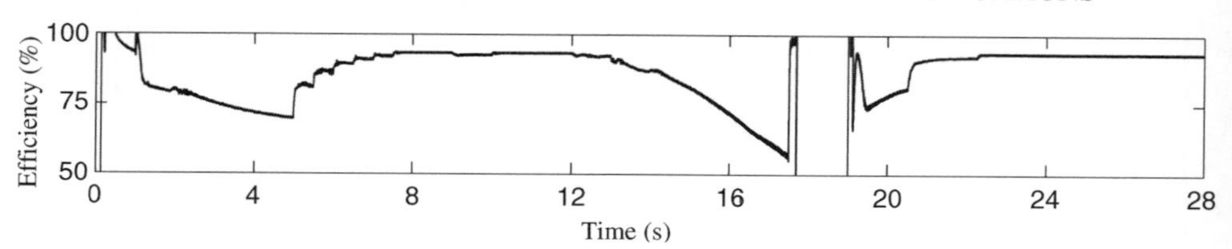

Figure 10.44 Combined efficiency of the hybrid rectifier.

Table 10.3 Losses of Four STATCOM configurations

Converter topology	True 48-pulse	Quasi 48-pulse	Cascade inverter	Binary inverter
Transformer losses	1.00%	1.42%	0.04%	0.04%
Converter losses	0.52%	0.59%	0.71%	0.79%
Total losses	1.52%	2.01%	0.75%	0.83%

The total losses of each topology consist of the STATCOM switch on-state losses (GTOs), transformer resistive losses and simple snubber losses. Using this same method, the MLCR rectifier losses are summarized in Table 10.4.

Table 10.4 Calculated losses in the MLCR rectifier

Transformer losses	1.00%
Converter losses	0.32%

The total calculated losses of the MLCR rectifier (1.32%) are marginally less than those of the true 48-pulse STATCOM example, but much higher than either of the transformerless examples (cascade and binary inverters). The three-winding transformer of the MLCR appears to contribute the main inefficiency, the copper losses scaling with transformer MVA rating.

While Tables 10.3 and 10.4 provide an estimate of the real power losses, the switching frequency and strategy will contribute additional losses, which neither table captures, as will the method of voltage balancing across the switches when many are series connected. This is particularly pertinent to the transformerless topologies (cascade inverter and binary inverter in Table 10.3) as they are fed directly from the medium voltage bus and therefore require more series-connected switches per phase.

The dynamic simulation provides an accurate summation of all MLCR losses and is a much more realistic efficiency benchmark as it includes the switch dynamics, snubbers, non-linear transformer characteristics and control response. The MLCR losses may thus be compared for the reactive power rating of 80 MVAr using:

$$Efficiency(\%) = \left(1 - \frac{P_{T_R} - V_d I_{d_R}}{Q_{R_{rated}}}\right) \times 100\% \qquad (10.69)$$

The revised efficiency, given in Figure 10.45, shows that during normal operation the losses are 3.8% (almost three times those of Table 10.4), and the equivalent efficiency is 96.2%.

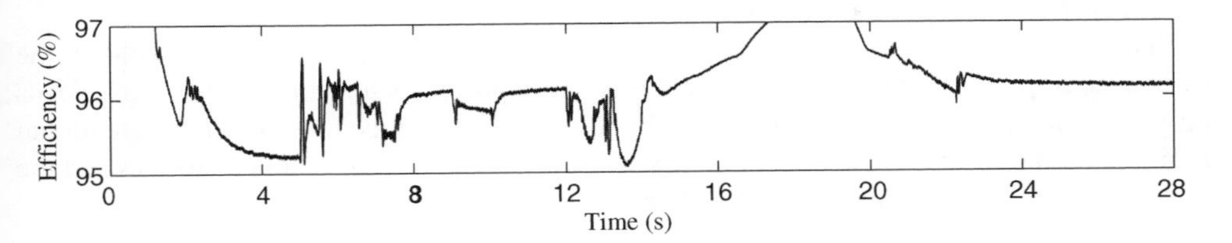

Figure 10.45 MLCR losses relative to reactive power rating.

10.7 Multicell rectification with PWM control

Figure 10.46 shows an IGCT-based rectifier configuration proposed for the copper electro-winning and electro-refining metallurgical processes with current ratings of up to 40 kA [12]. It consists of 24 parallel cells with independent PWM-controlled current-source rectifier modules. The AC side consists of two interfacing transformers converting the 15 kV supply down to 275 V; their primary windings are connected in delta and they have four zig-zag

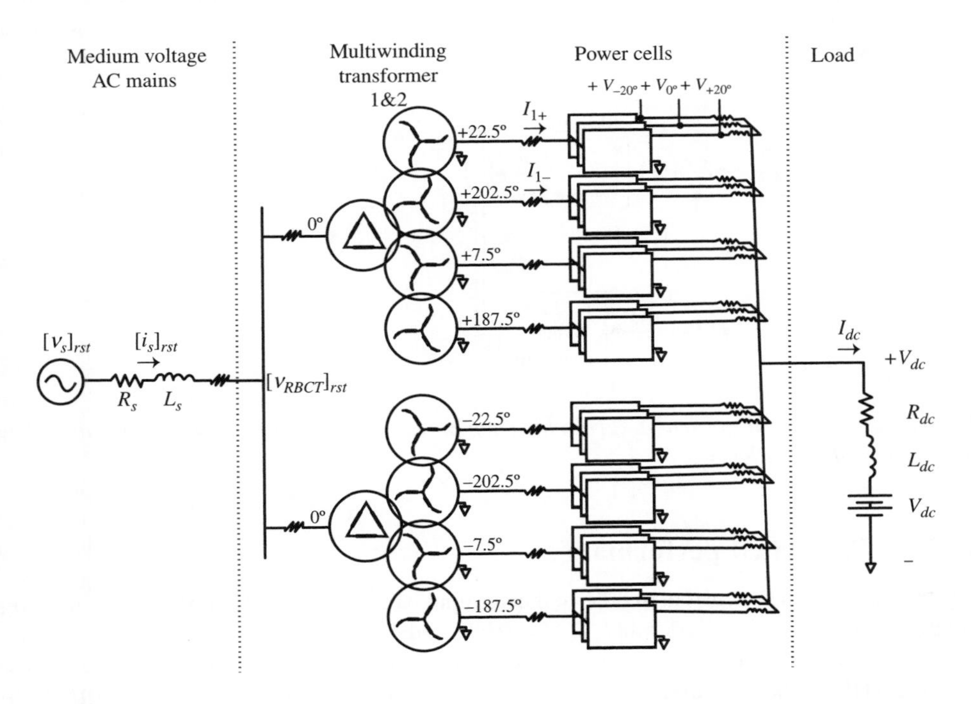

Figure 10.46 A multicell high-current rectifier for the electro-winning industry. (Reproduced by permission of the IEEE.)

secondary windings producing a 15° phase shift. Each secondary winding provides power to three parallel PWM power cells. Their respective PWM carriers are phase-shifted by $+20°$, 0° and $-20°$, producing an 18-pulse current for the transformer. The cells are connected in parallel via interface reactors. Current balance is attained by means of a controller that generates the current references for all the cells.

This configuration draws low distortion currents from the AC system without the assistance of passive filters and operates at near unity power-factor. Each rectifier cell needs only one conducting IGCT (compared with the two SCRs that conduct in the conventional bridge), but the overall switching efficiency is lower due to the higher loss of IGCTs and the higher switching frequency of the PWM process.

10.7.1 Control structure

A block diagram of the control scheme for the power cells is shown in Figure 10.47. It consists of a linear closed loop, which completely eliminates DC current error in the steady state as well as a unity power-factor throughout the operating range. A PI regulator generates the power cell modulating index m for the cell switches. The three-phase synchronizing signals are generated using a $d-q$ frame PLL. These signals are then converted into the $\alpha-\beta$ plane generating the reference current vector used as the input to a synchronous space vector modulator. Each power cell uses independent space vector modulation (SVM) operating at 300 Hz and, thus, the characteristic harmonics are those of a six-pulse converter. However, due to the phase shift between the respective feeding voltages, the six-pulse related harmonics are cancelled through the feeding transformer, which generates, instead, a 24-pulse waveform.

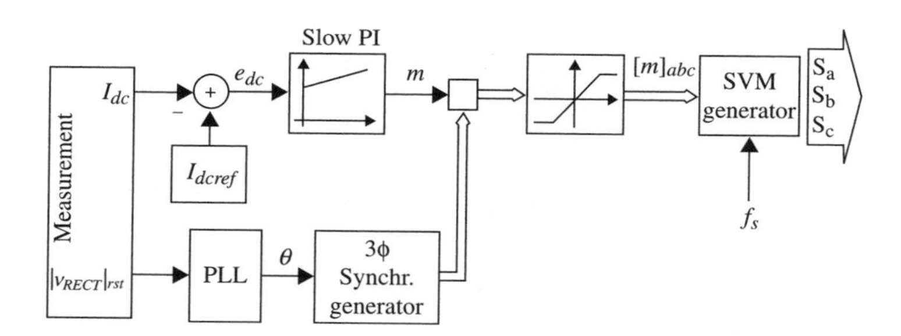

Figure 10.47 Linear control diagram for each power cell in Figure 10.46. (Reproduced by permission of the IEEE.)

10.7.2 Simulated performance

The test case uses a DC current of 80% of the nominal value of a typical copper electro-refining plant, which requires $R_{dc} = 2.5\,\mathrm{m}\Omega$, $L_{dc} = 10\,\mathrm{mH}$ and $V_{ce} = 200\,\mathrm{V}$.

Figure 10.48(a) shows the voltage and current at the input of the converter. The corresponding THD of the AC current is only 2% and the input power-factor 0.99. Figure 10.48(b) shows the input current of a power module, which is the current contributed by a pair of anti-phase secondary windings and their respective six-power cells.

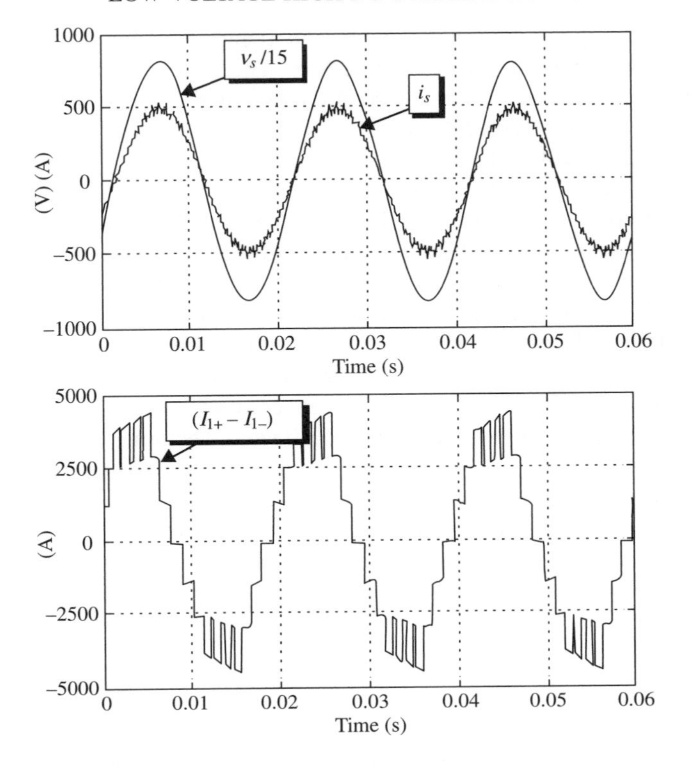

Figure 10.48 Multicell input phase voltage and current waveforms with a 40 kA load. (Reproduced by permission of the IEEE.)

10.8 Summary

With the increasing current rates of the thyristor-based self-commutating switches the thyristor/IGCT double-group 48-pulse rectifier should be a straightforward competitor to the present technology used by the aluminium smelting industry. This configuration provides high power-factor throughout the operating cycle without the assistance of reactive power compensation and transformer OLTCs, while maintaining the traditional high-pulse AC–DC conversion. The simulation results demonstrate the provision of flexible full range control and fast dynamic response during commissioning, normal operation, shut-down and smelter restart.

The multilevel current reinjection (MLCR) configuration based on two five-level MLCR rectifiers (48-pulse equivalent) with firing angles of opposing polarities forms a high current, high power-factor group. Since the groups are modular, several are parallel connected to achieve the high DC current required by the smelter load. In contrast to the 12-pulse rectifiers used by the thyristor/IGCT configuration, each MLCR rectifier has high pulse operation without costly phase-shifting transformers, and can therefore directly control DC current without affecting AC current waveform quality. By increasing DC current in either the leading or lagging rectifiers, additional reactive power control is possible, with reactive power generation used to support terminal voltage and reduce transmission losses. Moreover, multilevel switching reduces the switching stresses. The MLCR rectifier's main bridge zero current switching also removes thyristor commutation angle, although it does require self-commutating switches of full rectifier current rating in the reinjection circuit. Both the MLCR and 12-pulse (phase-shifted) self-commutating switches must be rated at full rectifier current. Self-commutated thyristors have about double the on-state voltage drop of their line-commutated counterparts, so the on-state losses of the MLCR (two thyristor and one IGCT) are about the same as a 12-pulse self-commutated bridge rectifier (two IGCTs). For the added control flexibility of the MLCR, the extra self-commutating switch is easily justified.

The parallel thyristor/MLCR configuration provides full-rated reactive power compensation that is virtually independent of smelter DC load conditions. The use of the multilevel reinjection current source rectifier ensures a high-quality current waveform, and the coupling with the DC load reduces the losses associated with conventional DC side CSC STATCOMs. However, the added complexity with a smelter DC interconnection is expected to increase the installation costs.

The IGCT-based multicell rectification under PWM control option fits somewhere between the multilevel and PWM technologies and is, therefore, more difficult to place in high power conversion. Its technical suitability for the copper electro-winning and electro-refining metallurgical processes (with currents up to 40 kA) has been demonstrated by computer simulation.

References

1. Pollock, P. and Duffey, C. (1998) Power quality and corrective action in an aluminium smelter. *Petroleum and Chemical Industry Conference, Industry Applications Society 45th Annual Conference*, 1998, pp. 181–9.
2. Williamson, A.R.F., Baker, J. and Marshall, D.A. (1996) Voltage quality improvements through capacitor bank tuning in the Eskom transmission network. *Proceedings of IEEE AFRICON*, 1996, vol. 2, pp. 822–6.
3. Rim, C.T., Choi, N.S., Cho, G.C. and Cho, G.H. (1994) A complete DC and AC analysis of three-phase controlled-current PWM rectifier using circuit D-Q transformations. *IEEE Transactions on Power Electronics*, **9** (4), 390–6.
4. Blauh, Y.B. and Barbi, I. (1998) A phase-comtrolled 12-pulse rectifier with unity displacement factor without phase-shifting transformer. *Proceedings of 13th Annual APEC Conference 1998*, vol. 2, pp. 970–6.
5. Kazmierkowski, M.P and Malesani, L. (1998) Current control techniques for three-phase voltage source PWM converters: a survey. *IEEE Transactions on Industrial Electronics*, **45** (5), 691–703.
6. Arrillaga, J. and Smith, B. (1998) *AC-DC Power System Analysis*. IEE Power and Energy Series, London.
7. Paice, D. (1996) *Power Electronic Converter Harmonics: Multipulse Methods for Clean Power*. IEEE Press, Piscataway, N.Y.
8. Bohnert, K., Brandle, H., Brunzel, M.G., Gabus, P. and Guggenbach, P. (2007) Highly accurate fibre-optic DC current sensor for the electrowinning industry. *IEEE Transactions on Industry Applications*, **43** (1), 180–7.
9. Perera, L.B., Liu, Y.H., Watson, N.R. and Arrillaga, J. (2005) Multi-level current reinjection in double bridge self-commutated current source conversion. *IEEE Transactions on Power Delivery*, **20** (2), 984–991.

10. Sato, T., Matsushita, Y., Temma, K., Morishima, N. and Iyoda, I. (2002) Protoytpe test of STATCOM and BTB based on voltage source converter using GCT thyristor. *Proceedings of Transmission and Distribution Conference and Exhibition, 2002: Asia Pacific IEEE/PES*, pp. 2037–42.
11. Lee, C.K., Leung, J.S.K., Hui, Y.R. and Chung, S.H. (2003) Circuit-level comparison of STATCOM technologies. *IEEE Transactions on Power Electronics*, **18** (4), 1084–92.
12. Wiechmann, E.P., Aqueveque, P., Morales, A.S., Acuna, P.F. and Burgos, R. (2008) Multicell high-current rectifier. *IEEE Transactions on Industry Applications*, **44** (1), 238–46.

11

Power Conversion for High Energy Storage

11.1 Introduction

Very high current conversion is not restricted to industrial applications. Complications arising from greater penetration of photo-voltaic and wind generation, and increased diurnal load variability have seen renewed interest in large-scale electrical load-levelling technology, to remove the dependence on gas-fired peaking plant during high demand periods and to better utilize cheaper base load generation when demand is low.

Load-levelling schemes utilizing pumped storage have been successful in over 150 installations worldwide, but with efficiencies as low as 60–70%, owing to the conversion to an intermediate medium for storage. Better efficiencies are reached with battery, flywheel and compressed air energy storage, though they currently lack the energy density necessary for levelling on a large scale.

Superconducting magnetic energy storage (SMES), a technology first proposed during the 1970s oil crisis to reduce the use of expensive oil peaking plant, uses a high-power AC/DC converter to store energy directly from the AC system in the magnetic field of a supercooled inductor. High energy density is made possible with large inductances, and extremely high DC currents (in excess of 100 kA); projected storage levels of 5000 MWh were reported at the time.

Because the energy storage requires no change from one medium to another (other than AC/DC conversion), the SMES exhibits very low losses, with round trip efficiencies of 90–95%. The fast bidirectional operation of the converter permits power-flow direction change in a few cycles, which makes SMES technology suitable for improving voltage stability, providing spinning reserve and in short-duration very high-current applications, such as pulse lasers and fusion reactors.

Early SMES designs had problems with limited converter controllability and involved coils with considerable civil complexity. Although some prototypes were designed [1] and

Self-Commutating Converters for High Power Applications J. Arrillaga, Y. H. Liu, N. R. Watson and N. J. Murray

even commissioned [2], the technology was never fully commercialized, instead being shelved with the end of the oil crisis in the mid 1980s.

Recent improvements in converter control flexibility, ratings and harmonic performance, thanks largely to advances made in HVDC technology and in the motor drive industry, have made four-quadrant operation (a prerequisite for an SMES converter) possible at very high power. Superconducting technology has also evolved, with stronger conductors, greater current densities and increased superconducting temperatures. Thus, many of the major limitations of early SMES designs have been mitigated, and SMES systems are once again feasible.

While the purpose of this chapter is to describe advances in high-power high-current converters with four-quadrant capability suitable for SMES schemes, some attention is given to the SMES coil, as this determines the voltage and current characteristics of any connected power converter, and constitutes a massive mechanical and civil undertaking; one that may account for up to half the total cost of an SMES system.

11.2 SMES technology

Superconduction occurs when a suitable conductor is supercooled in a bath of liquid nitrogen or superfluid helium. At temperatures of around 1.8 K a coil current flows free of any resistance, and when the coil circuit is closed, energy can be stored almost indefinitely. The energy stored is proportional to the coil inductance and to the square of the DC current, and so the higher the current density, the greater the storage capability.

To obtain the energy storage capacities required of a utility-scale load-levelling SMES, the coil supercooled conductors must therefore carry extremely high DC currents, at hundreds of kilo-amperes for the largest systems. The mechanical and civil works required for such a storage coil is understandably extensive. For instance, the coil required for a 1000 MW/5000 MWh utility-scale SMES with an inductance of 67.6 H, and flux density of 7 Tesla, must be 15 m tall, and almost 1.6 km in diameter [3]. Figure 11.1

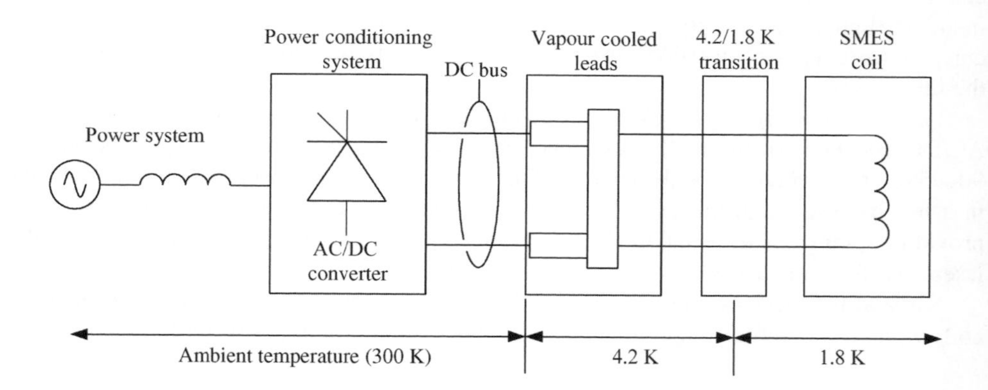

Figure 11.1 Generic SMES layout. (Reproduced by permission of the IEEE.)

illustrates the basic structure of an SMES system. The converter (or power conditioning system (PCS)) interconnects the AC system to the DC coil, with a controlled temperature gradient maintained between the PCS and coil through the use of vapour cooled leads (VCLs).

The SMES power to energy ratio varies greatly with application, and determines the structure (and rating) of both the power converter and SMES coil. For instance, an SMES coil and power converter developed to improve the power system stability, or to provide localized storage for industry [4] would be significantly different to one used for load levelling. Figure 11.2 [5] summarizes the applications for SMES based on storage capability. The

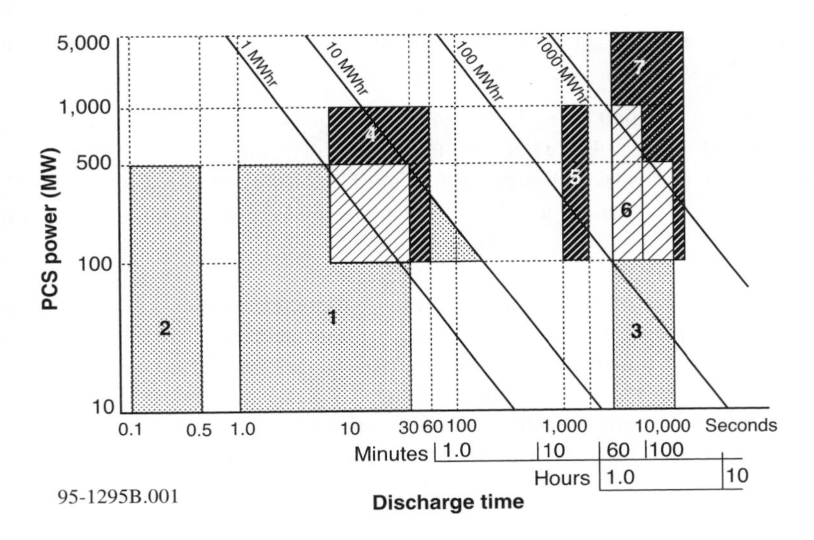

Figure 11.2 Transmission substation applications. (Reproduced by permission of the IEEE.) (1 = transmission stability; 2 = voltage/VAR support; 3 = load levelling; generation system applications; 4 = frequency control; 5 = spinning reserve; 6 = dynamic response; 7 = load levelling.)

boxes in Figure 11.2 signify different SMES applications, with boxes 3, 6, and 7 indicating utility-scale load levelling (the original application of SMES). The lines of constant gradient in the figure specify the amount of energy storage required in the SMES coil. More recent designs for SMES systems have featured smaller coils, designed for high-power low-duration output, with fast-acting (sub-second) response time to improve transient stability (boxes 1, 2 and 4 in Figure 11.2).

11.3 Power conditioning

The SMES PCS provides the power control interface between the DC SMES coil and the high-voltage AV supply network. The PCS may be thought of as a FACTS controller, and shares many similarities with the basic STATCOM structure. However, while a STATCOM provides only reactive power controllability, the PCS must have continuous and independent bidirectional control of both active and reactive power, and therefore possess full four-quadrant control capability.

To manage the energy exchange, the PCS uses three operating states on the DC side. They are:

- Charging – a controlled positive DC voltage is applied to the SMES coil and the DC current increases, thus absorbing energy from the AC network. This may be controlled at a comparatively low power rating, to gradually syphon energy from the AC power system whilst minimizing disturbance.
- Discharging – a controlled negative DC voltage is applied to the SMES coil, the DC current decreases and active power is then supplied to the AC system. The power injected into the AC system is generally the rated SMES power output.
- Freewheeling – when no active power exchange is required with the AC system the coil current uses a freewheeling path, either through the converter or through dedicated switches (which may or may not be at ambient temperature). In this state, the PCS may continue to control reactive power, essentially functioning as a STATCOM.

As the coil charges (or discharges) the DC current will increase (or decrease) and this must be coupled with a corresponding controlled DC voltage decrease (or increase) to maintain constant active power transfer. Figure 11.3 illustrates this for different voltage and current combinations [6].

The active power transfer rate depends on the state of SMES coil charge, the DC voltage rating of the coil and, for charging, the maximum additional load permitted

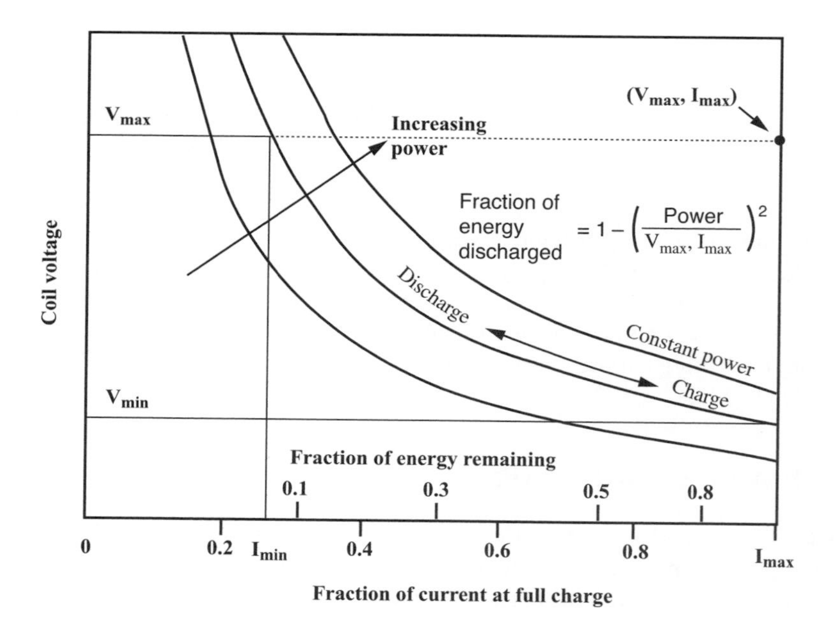

Figure 11.3 SMES coil voltage and current relationship for varied active power.

on the network. Limits on how much or how little energy is in the coil determine the SMES system's 'empty' or 'full' state. Near these limits the PCS may operate at reduced power output. If rated active power output is required across a wide range of charge, the PCS must be rated for both maximum current (corresponding to a near full state of charge) and maximum voltage, required for high-power output at low (near 'empty') currents.

11.3.1 Voltage versus current source conversion

The PCS may either be voltage source or current source converter based. The requirement for four-quadrant converter operation has traditionally favoured voltage source converters over current source converters in SMES conceptual designs, owing to the VSC's direct and independent control of the AC voltage magnitude and phase. The basic voltage source converter is, however, unable to control the DC voltage directly, and so for an SMES system, a DC chopper is needed to modulate the average DC voltage, and thus control the coil charge.

VSC-based SMES designs have tended to dominate the literature, even though current source converters seem the logical choice to interface with the coil's stiff current source. A comparison of VSC and hybrid-CSC PCSs of equal storage capacity, performed by Lasseter *et al.* [7], indicated that fewer switches were needed for the VSC configuration, despite the additional components required for DC voltage control.

The traditional current source converter's main shortcoming had been its reliance on line-commutated switching, which lacked reactive power controllability. Few CSC-based SMES have been proposed, and even fewer have been fully developed. Even with the availability of self-commutating switches, the few CSC SMES examples use PWM to separate active and reactive power control. Although this achieves satisfactory four-quadrant controllability, the high-frequency switching of PWM is inefficient, and the resulting high di/dt makes this solution unattractive for very high powers.

With recent increases in the ratings of self-commutated switches, the traditional restriction of high-power CSC-based switching is less critical. Greater control flexibility is now possible with CSC, at very high power ratings without resorting to PWM. Of particular interest are the multilevel CSC configurations, with desirable attributes such as low-frequency switching, high-quality current waveform, low losses and high current capability.

Within the cadre of proposed multilevel configurations, the multilevel current reinjection (MLCR) current source converter, described in Chapter 4 (section 4.6.1) and the use of multigroup firing-shift control, described in Chapter 9 (section 9.2), can provide four-quadrant controllability as well as switching at zero current. This property permits retaining the use of basic thyristors in the main converter bridges [8, 9]. By connecting multiple MLCR current source converters to a common AC supply, the individual converter firing angles may be shifted to control their summed AC current magnitude and phase, and thus achieve active and reactive power control independence.

The MLCR current source converter topology under firing-shift control presents an excellent base for a high-power SMES PCS, one that provides PWM-like control flexibility but with increased efficiency and rating.

Figure 11.4 illustrates how two converter firing angles (and current phasors) may be coordinated to provide bidirectional active power transfer for a constant reactive power setting.

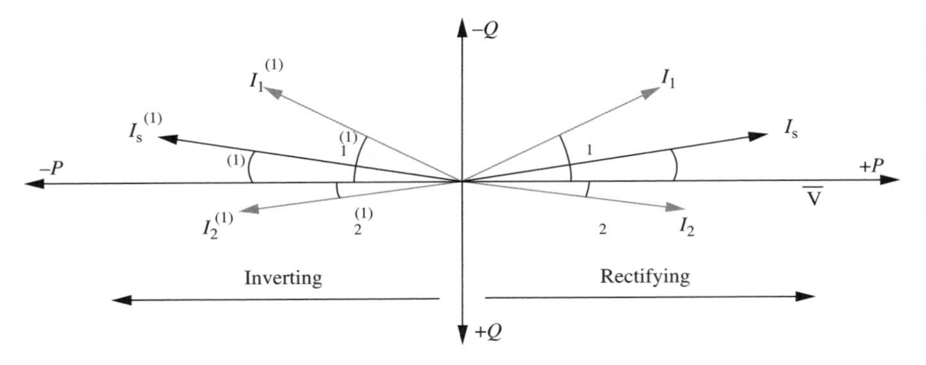

Figure 11.4 Independent current amplitude and phase using firing-shift control to provide bidirectional active power transfer independent of reactive power order.

The left-hand side depicts PCS inverter action during a coil discharge, and the right-hand side, the rectifier action to charge the SMES coil.

11.3.1.1 Power rating considerations

It is inappropriate to compare the rating requirements of the VSC and CSC components required by the PCS based purely on the average power balance. While the average current rating determines the heating effect of the device, the average assumes that all connected components have a large thermal time constant. Though this may be true for the converter transformers and SMES coil, the silicon wafer structures of the power semiconductor switches have a relatively small thermal mass, and their rating must be derived considering both the average and instantaneous currents. Similarly, the switch voltage rating is related to the maximum instantaneous voltage, not the average.

In the case of the MLCR current source converter configuration the converter and transformer ratings depend on the highest DC voltage (also the coil voltage rating) and maximum DC current. For an MLCR of sufficiently high level (say five-level or its 48-pulse equivalent) the voltage and current waveforms are very close to a sinusoide (the current THD being about 4%), and thus the maximum average power is very close to the instantaneous power peak.

However, for the VSC-chopper configuration, the converter and transformer ratings are very difficult to define, as their instantaneous power peak is far from the converter average power. While the voltage source converter DC voltage (which must be higher than the coil average) is not difficult to define, the *rms* currents for different parts of the circuit differ greatly, and are strongly related to the system device parameters and the SMES operating conditions. The maximum coil DC current is chopped in pulses, with the pulsed current distributed between the converter and the DC capacitor. The short pulse duration and large magnitude has a much higher *rms* value than its average and thus to suppress the propagation of these current pulses to the converter and transformer, a very large DC capacitor is needed. A high DC capacitance will then draw its own pulsating current from the AC system, requiring smoothing via a large AC inductance.

If capacitors with high instantaneous current rating were available and the cost of the high capacitance was reasonable, the use of a high voltage rating and low current rating transformer and a voltage source converter with a high voltage and current rating chopper could be viable based on the average power balance. However the following three objections prevent such solution:

1. If the coil voltage rating is fixed, so is the DC capacitor voltage. The widths of the charging and freewheel periods are related to the average voltage of the coil, which is always less than its actual voltage rating and, therefore, never reaches its rated power.
2. The fact that the average power is less than the instantaneous power means that there is some energy buffer in the system between the coil and the converter; this extra power buffering has no energy storage function.
3. The DC capacitor plays a role in the energy buffer and could in theory be made to store energy directly. In practice, however, the VSC capacitor storage system has low utilization of the capacitor energy storage capability. Of the $E = \frac{1}{2}CV^2$, only a small proportion can be used for power exchange with the AC system; the energy density of a capacitor is much lower than that of the SMES coil.

Thus the component ratings of the voltage source converter cannot be derived from the average power balance, instead requiring much larger current ratings due to high instantaneous current peaks; moreover the component ratings are also related to the component parameters. Conversely, for the same average power, the MLCR current source converter with power ratings close to the average power balance would constitute a better option, one that would scale to virtually any size.

11.4 The SMES coil

Coil size, Lorentz forces and power delivery requirements present design challenges not faced in smaller superconducting applications. The coil must be strong enough to withstand the large Lorentz forces occurring during energization. To protect the coil against failures (such as the breakdown of the vacuum or helium vessel) the stored energy needs to be dissipated in a controlled manner to avoid insulation breakdowns.

The energy storage and the rate of energy exchange with the power system largely determines the coil construction, though the cost of the SMES coil is primarily determined by the amount of energy to be stored. Early coil designs considered for load levelling called for large earth-supported monolithic type conductors to be partially buried and constructed *in situ*. With the technology in its infancy, each design was a one off, and shared few similarities with other SMES concepts of the time.

The conductors were built of a niobium–titanium alloy and suspended in a bath of liquid helium, with high-purity aluminium supporting beams linking the cold conductor to the warm supporting structure. The size of the largest SMES, as mentioned in the introduction, used designs based on massive solenoidal (as opposed to toroidal) coils due to the relative simplicity of construction, better suitability to very high currents and good utilization of the land and materials. They also required less supporting structure and generally lower cost for a specified energy storage.

A solenoid-based SMES does however have a significant external magnetic field and must therefore be constructed away from populated areas. With potential health risks associated with long-term exposure, the power converter and all technician-serviceable equipment also needs to be sufficiently distant from the high magnetic fields.

Monolithic conductors, the basis of most early coil designs, were superseded by the stronger, and self-supported, cable-in-conduit conductor (CICC), which allowed SMES coils to be built above ground, thus eliminating much of the heat loss through the cold to warm support structure. By eliminating the supports and trench, the CICC design also possessed an immediate cost advantage.

The resilience to stresses placed on a conductor during the initial cooling process also favoured the CICC concept, as the strain was automatically spread evenly between each of the coil sections. The CICC also had a cost advantage over an equivalent monolithic conductor up to storage levels of at least 1000 MWh, though this depended largely on design [5].

The conductors themselves are made from filaments which superconduct at around 1.8 K for low temperature superconductors (LTS) or about 20 K for high temperature superconductors (HTS). The conductors are typically made of Nb–Ti (niobium–titanium) for LTS and copper oxide (Cuprates) or iron-based superconductors for HTS.

The perceived advantage of HTS wound SMES coils is that the overall costs for a given storage will be less, owing to the significant reduction in refrigeration. However, HTS wire is more expensive than LTS, and is limited by a lower critical current density. HTS wire is also more brittle, with a lower strain tolerance, and is thus more susceptible to thermal stresses. The preference for HTS in an SMES design therefore depends on whether the reduction in refrigeration costs outweighs the increases in HTS wire and bracing material. In this respect HTS-based SMES may be cost effective for storage levels under 10 MWh.

Though coil resistance is negligible, eddy currents are produced during each charge/discharge cycle, which result in losses that are a function of the cyclic frequency and magnitude [10]. Additional losses are present in small SMES coils, as a discharge resistor is often parallel connected to the coil to protect the coil from quench.

Quench is a phenomenon that occurs when a conductor suddenly moves from superconducting to resistive state, and results in sudden over-voltages that may damage the converter and coil. In large coils a discharge resistor is impractical, so the coils are designed with quench self-support [11]. The converter voltage rating must be increased to account for this. Though power converters are capable of DC output voltages of several hundred kilo-volts in HVDC applications, the insulation of SMES coils restricts the DC voltage to, generally, under 30 kV.

11.5 MLCR current source converter based SMES power conditioning system

Figure 11.5 shows the prospective MLCR current source converter SMES scheme. Each MLCR current source converter group is a five-level parallel configuration, providing 48-pulse equivalent operation; the two groups are series-connected and independently controlled to

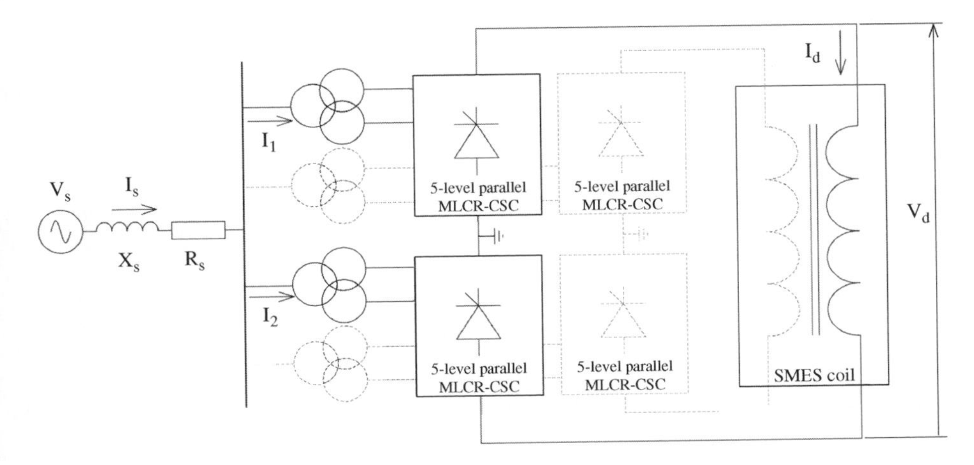

Figure 11.5 SMES system with MLCR current source converter based PCS. Possible second MLCR current source converter PCS and closely coupled (but isolated) coil in dashed line.

provide a firing shift between them. Several such groups would be connected in parallel to a common DC bus to increase the current capacity (and therefore the storage capability) of the system and would either share a common coil or closely coupled, but isolated, coils as shown by dashed trace in the figure.

11.5.1 Control system design

The relationship between active power and firing angle is strongly non-linear, and is further complicated by two firing angles and concurrent reactive power control. A non-linear controller could be developed, but a simpler method is to linearize the converter firing logic inputs or feedback, which then allows linear controllers to be used for the active and reactive power control channels. This forms an integral part of the control system design.

The two firing angles are related to the AC terminal active power (P_T) by

$$P_T = 3k_1 V_T I_d(\cos\alpha_1 + \cos\alpha_2) \tag{11.1}$$

and to the reactive power by

$$Q_T = 3k_1 V_T I_d(\sin\alpha_1 + \sin\alpha_2) \tag{11.2}$$

where k_1 is a constant defined by converter topology and relates AC terminal voltage (V_T) to maximum DC voltage (V_d).

The equations are very similar to those of an HVDC terminal but as the converter and DC side losses are practically negligible, and the power required by the coil refrigeration is assessed separately, the AC terminal power (P_T) may approximate the DC power (P_d).

The DC voltage is thus given by

$$V_d = k_1 V_T(\cos\alpha_1 + \cos\alpha_2) \tag{11.3}$$

By making

$$U_x = (\cos(\alpha_1) + \cos(\alpha_2)) \tag{11.4}$$

and

$$U_y = (\sin(\alpha_1) + \sin(\alpha_2)) \tag{11.5}$$

and replacing Q_T and V_d in equations 11.2 and 11.3, with Q_y and V_x, the two input (V_x,Q_y), two output (α_1, α_2) translation becomes

$$U_x = \frac{V_x}{k_1 V_T} = \cos(\alpha_1) + \cos(\alpha_2) \tag{11.6}$$

$$U_y = \frac{Q_y}{k_1 V_T I_d} = \sin(\alpha_1) + \sin(\alpha_2) \tag{11.7}$$

To isolate α_1 and α_2

$$\frac{U_y}{U_x} = \frac{\sin(\alpha_1) + \sin(\alpha_2)}{\cos(\alpha_1) + \cos(\alpha_2)} = \tan\left(\frac{\alpha_1 + \alpha 2}{2}\right) = \frac{Q_y}{V_T I_d} \tag{11.8}$$

and

$$U_x^2 + U_y^2 = (\sin(\alpha_1) + \sin(\alpha_2))^2 + (\cos(\alpha_1) + \cos(\alpha_2))^2 = \left(\frac{V_x}{k_1 V_T}\right)^2 + \left(\frac{Q_y}{k_1 V_T I_d}\right)^2$$
$$= 4\cos^2\left(\frac{\alpha_1 - \alpha_2}{2}\right) = \frac{V_x^2 I_d^2 + Q_y^2}{9k_1^2 V_T^2 I_d^2} \tag{11.9}$$

Using two further intermediate variables (A_1 and A_2), the firing angles are obtained as follows

$$A_1 = \tan^{-1}\left(\frac{Q_y}{V_x I_d}\right) = \frac{\alpha_1 + \alpha_2}{2} \tag{11.10}$$

$$A_2 = \cos^{-1}\left(\sqrt{\frac{V_x^2 I_d^2 + Q_y^2}{36k_1^2 V_T^2 I_d^2}}\right) = \frac{\alpha_1 - \alpha_2}{2} \tag{11.11}$$

$$\alpha_1 = A_1 + A_2; \quad \alpha_2 = A_1 - A_2. \tag{11.12}$$

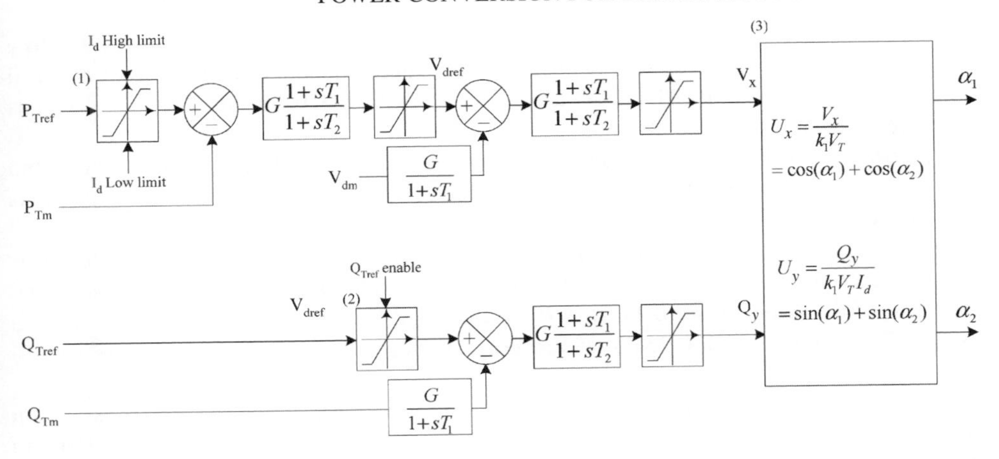

Figure 11.6 Control system for MLCR current source converter SMES independent active and reactive power.

The MLCR-based SMES power conditioning system is thus described in terms of active power, DC voltage and reactive power in Figure 11.6.

The coil DC current cannot be directly controlled (due to the inductor's very long time constant), though it is influenced by DC voltage polarity and magnitude. The DC voltage is thus selected as the major controlled variable, and is cascaded with an active power outer control loop. The active power control order has upper and lower current limits imposed (indicated by (1) in the diagram), to prevent rectification when 100% coil capacity is reached, and prevent further inversion when coil charge is at minimum level.

The reactive power channel consists of a single control loop with hard limited output, and forms the lower half of Figure 11.6. During active power direction changes, the reactive power control order is forced to zero (indicated by block (2) in the diagram) to ensure that firing angles α_1 and α_2 shift quadrants ($2 \rightarrow 1$ and $3 \rightarrow 4$ respectively during inversion to rectification transition) in unison.

This is one limitation of the firing-shift four-quadrant control method, as zero active power and reactive power is achieved whenever α_1 and α_2 are 180° out of phase, and thus an infinite number of combinations arise that may constitute a freewheeling ($P_T = Q_T = 0$) state.

Within the linearization block, care must be taken to limit V_x to non-zero values in equation 11.10, and an additional -1 multiplier must be included with the square root operator in equation 11.11 for negative values of V_x to preserve correct bidirectional control action.

11.6 Simulation verification

A 1000 MW/100 MWH MLCR current source converter based SMES is used in PSCAD/EMTDC to verify the scheme's dynamic performance in response to a series of active and reactive power step changes. The SMES system (as given in Figure 11.5) is connected to a 220 kV transmission network, represented by a Thevenin equivalent consisting of an ideal source in series with an 0.015 H inductance and 0.4 Ω resistance. Each MLCR transformer has a leakage reactance specified at 5%.

The SMES coil is represented by a 70 H inductance and 0.001 Ω resistance, with rated DC current and voltage of 100 kA and 20 kV respectively. The power base for the simulation is selected as 1000 MVA.

Figure 11.7 summarizes the four-quadrant control capability of the MLCR current source converter SMES over a half second period, for varied storage levels and constant power output. Figure 11.7(a)–(c) show the dynamic performance of the SMES for a DC current (I_d) of 0.25 pu which corresponds to a coil energy storage of 5% remaining.

The controlled variables P_{Tref} and Q_{Tref} (Figure 11.7(a)) are specified at t = 0.0 s as 0.25 pu and −0.15 pu respectively. The reactive power order is changed to 0.15 pu at t = 0.01 s to illustrate active and reactive power control independence. The firing angles (α_1 and α_2) in Figure 11.7(c) change from −3° and −52° to 47° and −4° respectively, with negligible change to the DC voltage (V_d) in Figure 11.7(b). The sinusoidal AC waveform (I_{Sa}) in Figure 11.7(b) changes phase with reactive power change but the magnitude remains virtually constant as expected.

At t = 0.15 s the power order (P_{Tref}) is changed to −0.25 pu to simulate the transition from rectification to inversion, the process taking approximately 0.15 s. At the same time the reactive power order is changed to 0.1 pu to illustrate the complete dynamic control independence. The AC waveform changes polarity at t = 0.19 s and retains its sinusoidal wave shape.

The firing angles at t = 0.3 s settle to 127° and −155° for α_1 and α_2 respectively, before both the active and reactive power orders are set to zero. At t = 0.4 s the active and reactive power orders return to 0.25 pu and −0.15 pu respectively, simulating recharging of the coil at constant power from the AC network.

Figure 11.7(d)–(f) illustrate the same control capability but with coil charge near 100% capacity. At t = 0.0 s the active power is changed from 1 pu rectifying to inverting, to simulate the type of active power burst that may be required for maintaining voltage stability. The reactive power order is set to zero during the transition at 0.01 s to ensure that firing angles pass from 1 to 2 and 4 to 3 in unison.

At 0.15 s the active and reactive power orders are set to zero to demonstrate the MLCR current source converter's freewheeling capability. Some switching spikes are evident owing to the large circulating currents between the two series-connected converters and their transformers.

At 0.22 s the active and reactive power orders are set to 0.2 pu and −0.15 pu respectively, to simulate the same power output as in Figure 11.7(a) but with far larger DC current. At 0.325 s the reactive power order is switched to 0.15 pu to again highlight the control independence at different power settings. Interestingly, given the complete four-quadrant capability of the MLCR controller, the firing angles from 0.33 s to 0.5 s are in quadrants 4 and 2 for reactive power absorbtion, and 1 and 3 for reactive power generation at this low active power setting.

The simulation thus confirms the four-quadrant controllability and the MLCR current source converter's suitability as a power conditioning system for SMES.

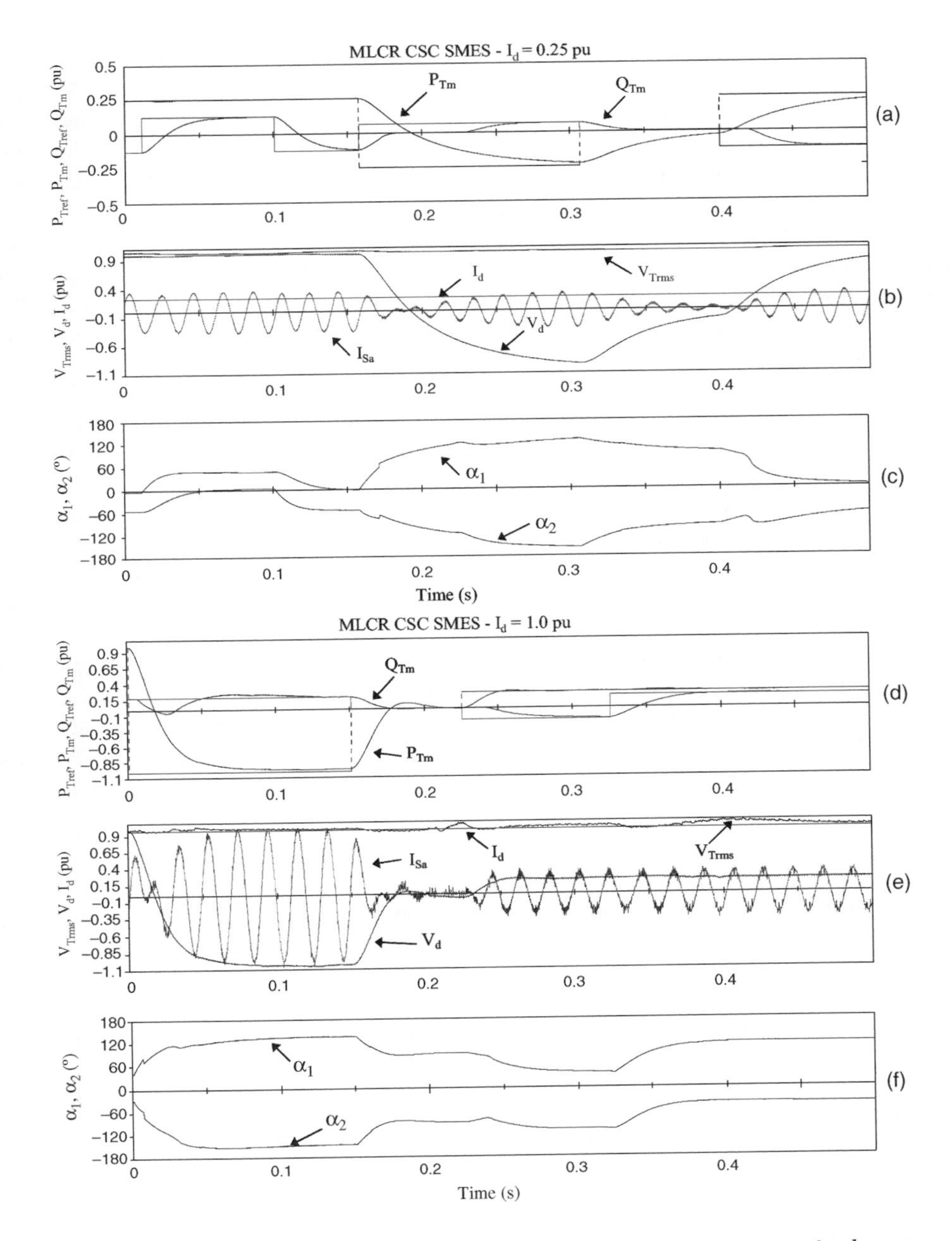

Figure 11.7 MLCR current source converter SMES dynamic performance for low storage level ((a)–(c), $I_d = 0.25\,pu$) and high storage level ((d)–(f), $I_d = 1.0\,pu$).

11.7 Discussion – the future of SMES

There is no denying that the control flexibility and efficiency of SMES schemes is far superior to that of pumped storage. The near instantaneous power flow control gives SMES an unrivalled advantage for fast turn-around energy exchange. By harnessing the averaging capability of an SMES at the point of generation, large remote renewable photo-voltaic and wind power installations could be turned into pseudo-baseload generation, at the same time allowing rationalization of their transmission rating.

The PCS technology is basically mature, thanks to advances in semiconductor ratings and lessons learnt in HVDC and motor drive systems. Converter topologies with four-quadrant thyristor-based control capability, as demonstrated with the MLCR current source converter under-firing shift control, present an efficient high-power alternative to VSC-chopper configurations.

The major limitation with large load-levelling schemes is in the coil design. The coil technology has not yet seen commercial success on a utility scale, and in a deregulated network, there is little incentive for investors, given the one-off bleeding-edge nature of coil storage design.

The future from an engineering perspective seems to lie in SMES with high power to energy storage ratios, with converters of several hundred mega-watts, but with only very short-duration storage capacity. The lessons learnt from small high-power SMES coils will undoubtedly help future development of much bigger longer-term storage coils.

From an investor's perspective, an SMES system would seem a risky venture, as it is difficult to forecast a return on investment. With this in mind, it is likely to be large-power quality-sensitive industries that consider installing SMES, where the cost savings through insulating their business from voltage disturbances are more easily calculated.

References

1. Loyd, R.J., Walsh, T.E., Kimmy, E.R. and Dick, B.E. (1989) An overview of the SMES ETM program: the Bechtel team's perspective. *IEEE Transactions on Magnetics*, **25** (2), 1569–75.
2. Rogers, J.D., Boenig, H.J., Schemer, R.I. and Hauer, J.F. (1985) Operation of the 30 MJ SMES system in the Bonneville Power Administration electrical grid. *IEEE Transactions on Magnetics*, **21**, 752–5.
3. Masuda, M. and Shintomi, T. (1987) The conceptual design of utility-scale SMES. *IEEE Transactions on Magnetics*, **23** (2), 549–52.
4. Karasik, V., Dixon, K., Weber, C., Batchelder, B., Campbell, G. and Ribeiro, P. (1999) SMES for power utility applications: a review of technical and cost considerations. *IEEE Transactions on Applied Superconductivity*, **9** (2), 541–6.
5. Luongo, C.A. (1996) Superconducting storage systems: an overview. *IEEE Transactions on Magnetics*, **32** (4), 2214–23.
6. Loyd, R.J., Walsh, T.E. and Kimmy, E.R. (1991) Key design selections for the 20.4 MWh SMES/ETM. *IEEE Transactions on Magnetics*, **27** (2), 1712–15.
7. Lasseter, R.H. (1991) Power conditioning systems for superconductive magnetic energy storage. *IEEE Transactions on Energy Conversion*, **6** (3), 381–7.

8. Arrillaga, J., Liu, Y.H., Perera, L.B. and Watson, N.R. (2006) A current reinjection scheme that adds self-commutation and pulse multiplication to the thyristor converter. *IEEE Transactions on Power Delivery*, **21** (3), 1593–9.
9. Arrillaga, J., Watson, N.R. and Liu, Y.H. (2007) *Flexible Power Transmission – The HVDC Options*. John Wiley & Sons, London.
10. Gurol, H., Motowidlo, L. and Luongo, C. (1989) AC losses in the SMES conductor and coil structure. *IEEE Transactions on Magnetics*, **25** (2), 1582–5.
11. Abdelsalam, M.K. and Eyssa, Y.M. (1991) Quench protection for the 21 MWh ETM coil. *IEEE Transactions on Magnetics*, **27** (2), 2316–19.

Index

Self-Commutating Converters for High Power Applications J. Arrillaga, Y. H. Liu, N. R. Watson and N. J. Murray